Micellar Solutions and Microemulsions

S.-H. Chen   R. Rajagopalan
Editors

Micellar Solutions and Microemulsions

Structure, Dynamics, and
Statistical Thermodynamics

With 111 Illustrations

Springer-Verlag
New York Berlin Heidelberg
London Paris Tokyo Hong Kong

S.-H. Chen
Department of Nuclear Engineering
Massachusetts Institute of Technology
Cambridge, MA 02139
USA

R. Rajagopalan
Department of Chemical Engineering
University of Houston
Houston, TX 77204-4792
USA

Library of Congress Cataloging-in-Publication Data
Micellar solutions and microemulsions: structure, dynamics, and
    statistical thermodynamics/S.-H. Chen, R. Rajagopalan, editors.
        p.   cm.
    ISBN 0-387-97106-8 (alk. paper)
    1. Micelles. 2. Solubilization. 3. Emulsions. I. Chen, Sow
-Hsin II. Rajagopalan, Raj.
    QD549.M64   1990
    541.3'45--dc20   89-21842

Printed on acid-free paper.

Typeset by APS Ltd., Salisbury, Wiltshire.
Printed and bound by Edwards Brothers, Inc., Ann Arbor, Michigan.
Printed in the United States of America.

9 8 7 6 5 4 3 2 1

ISBN 0-387-97106-8 Springer-Verlag New York Berlin Heidelberg
ISBN 3-540-97106-8 Springer-Verlag Berlin Heidelberg New York

Preface

During the last decade there has been a renewed interest in research on supramolecular assemblies in solutions, such as micelles and microemulsions, not only because of their extensive applications in industries dealing with catalysts, detergency, biotechnology, and enhanced oil recovery, but also due to the development of new and more powerful experimental and theoretical tools for probing the microscopic behavior of these systems. Prominent among the array of the newly available experimental techniques are photon correlation spectroscopy, small-angle neutron and X-ray scattering, and neutron spin-echo and nuclear magnetic resonance spectroscopies. On the theoretical side, the traditionally emphasized thermodynamic approach to the study of the phase behavior of self-assembled systems in solutions is gradually being replaced by statistical mechanical studies of semi-microscopic and microscopic models of the assemblies. Since the statistical mechanical approach demands as its starting point the microscopic structural information of the self-assembled system, the experimental determination of the structures of micelles and microemulsions becomes of paramount interest. In this regard the scattering techniques mentioned above have played an important role in recent years and will continue to do so in the future. In applying the scattering techniques to the supramolecular species in solution, one cannot often regard the solution to be ideal. This is because the inter-aggregate interaction is often long-ranged since it is coulombic in nature and the interparticle correlations are thus appreciable. Even if the aggregates are charge-neutral, there are other forces such as the van der Waals force and the solvation forces, which are sufficiently important for the thermodynamics of dense colloidal solutions. Thus the fundamental task of the statistical mechanical theories is to elucidate these basic interactions and to use them as a starting point for the development of the statistical mechanical models. This monograph discusses these fundamental issues.

Part One of the book, which contains eight chapters, deals with basic interactions, statistical mechanics, and computer simulation, and Part Two, which contains another eight chapters, gives selected applications of statistical thermodynamics to the phase behavior, critical phenomena, structure, interface tension, and transport properties of micellar solutions and micro-emulsions. In both parts, experimental and theoretical chapters are intermingled with each other. The choice of the subject matter is of course determined by the timeliness of the topics and the taste of the editors.

In Chapter 1, Chen and Sheu discuss the method of analysis of small-angle neutron scattering data when the inter-aggregate interaction is significant, such as in charged micellar solutions or in charged globular protein solutions. For these solutions the dominant interaction is the repulsive double-layer interaction. The interparticle structure factor in this case can be calculated analytically by a generalized one-component macroion (GOCM) theory using the generalized Derjaguin-Landau-Verwey-Overbeek (DLVO) interaction. A detailed derivation of the generalized DLVO interaction using a standard liquid-state theory formalism is given in this chapter. In Chapter 2, Lozada-Cassou and Henderson have developed a liquid-theoretic method for calculating the force between two flat interacting electrical double layers. This new method can be regarded as an extension of the well-known Poisson-Boltzmann (PB) theory and predicts that the electrostatic interaction between colloidal parts can, under certain conditions, be attractive. In Chapter 3, Jönsson and Wennerström use computer simulation techniques to study the electrical double layer with respect to both equilibrium and transport properties. This method can be used to test and complement the results obtained from the conventional use of the Poisson-Boltzmann equation. For a system with high surface charge densities and divalent counterions, the PB approach predicts a repulsive interaction between overlapping double layers, with the Monte Carlo simulations show a net attraction due to the strong ion-ion correlation effects. Rossky, Murthy, and Bacquet describe, in Chapter 4, a series of integral equation calculations and Monte Carlo computer simulations of ion distributions around rod-like polyelectrolytes such as DNA in aqueous solutions. The results of the two methods agree in general for both monovalent and divalent counterions but are sufficiently different from those of the PB equation, especially in the case of divalent counterions. The calculations also show the range of validity of the popular phenomenological charge-condensation model. The present status of the experimental studies of solvation forces, namely, the forces arising from the structural rearrangement of the solvent molecules near surfaces, is reviewed in Chapter 5 by Horn and Ninham. In the classical DLVO theory of colloidal stability, there are only the repulsive double-layer force and the attractive van der Waals interaction. The discovery of solvation forces is more recent and their origin less certain. Horn and Ninham discuss the circumstances under which the solvation forces can be expected to be significant. In recent years, Israelachvili and coworkers have made direct measurements of forces between two mica plates immersed in liquids at very close separations. These solvation forces arise when the plates are separated by a gap of the order of the diameter of the liquid molecules. The origin of such forces is examined in Chapter 6 by Wertheim, Blum, and Bratko using a model system consisting of hard spheres confined between two plates. Because of the simple geometry of their system, Wertheim, Blum, and Bratko have been able to carry out some rigorous statistical mechanical calculations of the density profile between the plates. The results are compared with exact profiles generated by Monte Carlo simulations. For high densities, oscillatory profiles similar to those obtained experimentally are seen. In Chapter 7, Hirtzel and Rajagopalan review the use of some of the standard Monto Carlo and Brownian dynamics simulation techniques, with a focus on their use in determining the interparticle structure and thermodynamic and transport properties of colloidal dispersions. These are powerful techniques for carrying out exact statistical mechanical calculations of complex fluids and are becoming popular in the studies of supramolecular systems. This chapter is written specifically in a tutorial style accessible to graduate students entering this field. Since scattering experiments also measure, besides the particle structure of colloids, the interparticle structure factor, which is intimately related to the thermodynamic properties of the solution, theoretical calculation of the structure factors for given interparticle interaction potentials becomes important. Although the GOCM discussed in Chapter 1 is suitable for calculating the structure factor of monodispersed charge-stabilized systems, there are some difficulties in applying it to very polydispersed systems. Fortunately, it is possible to carry out an analytical calculation of the structure factor for polydispersed hard-sphere systems in the Percus–Yevick approximation. Vrij and de Kruif present a detailed exposition of such a calculation in Chapter 8. Many of the analytical results summarized in this chapter are not available in the published literature. We hope that this formalism will find good use in the future by those doing scattering experiments.

Part Two of the book begins with a review by Safran, in Chapter 9, of a phenomenological theory for the structure and phase behavior of dilute three-component microemulsions. Microemulsions in this region are likely to have globular structures such as spherical, cylindrical, or lamellar. This phenomenological theory starts with an appropriate expression for the free energy of the system in terms of a few characteristic parameters such as the spontaneous curvature and the bending constant of the surfactant terms film and then adds to it the mean-field interaction and the entropy terms. It is rather remarkable that this simple theory is able to predict a fairly realistic structure of the aggregates and the phase transitions from one structure to the other as observed in experiments. Another type of mean-field theory is formulated in Chapter 10, by Blankschtein, Thurston, Fisch, and Benedek, for two-component surfactant-water systems by taking into account the monomer–micellar association chemical equilibrium. This theory is successful in accounting for the micellar–micellar phase separation phenomena often found in zwitter-ionic and nonionic micellar solutions much like the well-known Flory–Huggins theory for polymer solutions. In Chapter 11, Taupin, Auvray, and di Meglio review an extensive series of theoretical and experimental work by a French group on five-component microemulsions in the region of the phase diagram where the structure is bicontinuous. Small-angle X-ray and neutron scattering experiments are used to show, for the first time, that the surfactant film has a low rigidity constant when a cosurfactant is present and has, on the average, zero mean curvature in the bicontinuous microemulsions. Chapter 12, by Langevin, goes on to discuss the theory and experiments on the characteristic properties of ultralow interfacial tension in five-component microemulsions in the droplet and bicontinuous phases. Understanding of the ultralow surface tensions between a microemulsion and water-rich or oil-rich phases is vital for industrial applications of microemulsions. In Chapter 13, Huang, Kotlarchyk, and Chen describe a microemulsion-to-microemulsion phase separation phenomenon in an AOT/water/alkane three-component microemulsion system. This popular system behaves like a typical binary fluid mixture showing a lower consolute point, when the molar ratio of water-to-AOT is kept constant. Thus, one can apply the well-developed theory of static and dynamic phenomena quite successfully. In Chapter 14, Roux and Bellocq continue the exploration of critical phenomena in another ternary microemulsion system composed of SDS/water/pentanol from the point of view of identifying a new field variable. For both the AOT-based and SDS-based ternary microemulsion systems the critical phenomena have been demonstrated to belong to an Ising-like universality class. Light scattering and neutron scattering investigations of a five-component microemulsion follow in Chapter 15, by Chang, Billman, Licklider, and Kaler. The microstructure of these microemulsions is identified to be droplets at low volume fractions of water or oil, while microemulsions with comparable water and oil volume fractions are shown to have bicontinuous structures. In the last chapter, Kim and Dozier discuss the measurements of transport properties, such as the self-diffusion coefficient, mutual diffusion coefficient, and electrical conductivity of the AOT-based ternary microemulsions. In particular, a striking electrical percolation threshold, both in temperature and in volume fraction, is found. The conductivity is shown to have a power-law behavior below and above the percolation threshold.

In preparing this volume our primary intention was to assemble a set of coherent, up-to-date articles written by some of the most active researchers to cover the present status of progress in statistical mechanical theories of colloidal, micellar, and microemulsion systems. This monograph has turned out to be an exciting collection of original articles, accurately reflecting the direction of current developments in the field. This would not have been possible without the cooperation of all the authors, and we would like to express our sincere gratitude for their efforts and patience.

This monograph is intended for graduate students and professional scientists and engineers who wish to have an overview of the present status of the fundamental issues in the field of micellar solutions and microemulsions.

To facilitate the readers' understanding of the material included in this book, the following three references are recommended:

- J.N. Israelachvili, *Intermolecular and Surface Forces*, Academic Press, London, 1985.
- C. Tanford, *The Hydrophobic Effect: Formation of Micelles and Biological Membranes*, 2nd Ed., Wiley, New York, 1980.
- J.-P. Hansen and I.R. McDonald, *Theory of Simple Liquids*, 2nd Ed., Academic Press, New York, 1986.

Israelachvili's book presents a good introduction to the various forces relevant to colloidal solutions. Tanford's book explores the thermodynamic principles of self-assembly of amphiphilic molecules in aqueous media. The book by Hansen and McDonald presents a detailed treatment of integral equation theories of liquids.

We are grateful to the Petroleum Research Funds administered by the American Chemical Society for financial support during the initial planning stage of this monograph. We would also like to thank the editors and staff at Springer-Verlag, New York, who have been most helpful and patient during the compilation and production of this volume.

Cambridge, MA   Sow-Hsin Chen
Houston, TX   Raj Rajagopalan
February 1990

Contents

Contributors

L. AUVRAY   Laboratoire de Physique de la Matière Condensée, Collège de France, 75231 Paris Cedex 05, France.

R. BACQUET   Department of Chemistry, University of Texas at Austin, Austin, TX 78712, USA.

A.M. BELLOCQ   Centre de Recherche Paul Pascal and GRECO "Micro-emulsions" CNRS, Domaine Universitaire, 33405 Talence Cedex, France.

G.B. BENEDEK   Department of Physics, Massachusetts Institute of Technology, Cambridge, MA 02139, USA.

J.F. BILLMAN   Department of Chemical Engineering, University of Washington, Seattle, WA 98195, USA.

D. BLANKSCHTEIN   Department of Chemical Engineering, Massachusetts Institute of Technology, Cambridge, MA 02139, USA.

L. BLUM   Physics Department, University of Puerto Rico, Rio Piedras, Puerto Rico 00931.

D. BRATKO   Faculty of Natural Sciences and Technology, E. Kardelj University, 61000 Ljubljana, Yugoslavia.

N.J. CHANG   Miami Valley Research Laboratory, Proctor and Gamble, Cincinnati, OH 44717, USA.

S.-H. CHEN   Department of Nuclear Engineering, Massachusetts Institute of Technology, Cambridge, MA 02139, USA.

C.G. DE KRUIF   Van 't Hoff Laboratory, University of Utrecht, 3584 CH Utrecht, The Netherlands.

J.-M. DI MEGLIO   Laboratoire de Physique de la Matière Condensée, Collège de France, 75231 Paris Cedex 05, France.

W.D. DOZIER   TRW, Redondo Beach, CA 90278, USA.

M.R. FISCH   Department of Physics, John Carroll University, Cleveland, OH 44118, USA.

D. HENDERSON   IBM Almaden Research Center, San Jose, CA 95120–6099, USA.

C.S. HIRTZEL   Department of Chemical Engineering and Materials Science, Syracuse University, Syracuse, NY 13244–1190, USA.

R.G. HORN   United States Department of Commerce, National Institute of Standards and Technology, Gaithersburg, MD 20899, USA.

J.S. HUANG   Exxon Research and Engineering Co., Annandale, NJ 08801, USA.

B. JÖNSSON   Department of Physical Chemistry 2, Chemical Center, S–22100 Lund, Sweden.

E.W. KALER   Department of Chemical Engineering, University of Delaware, Newark, DE 19716, USA.

M.W. KIM   Exxon Research and Engineering Co., Annandale, NJ 08801, USA.

M. KOTLARCHYK   Department of Physics, Rochester Institute of Technology, Rochester, NY 14623, USA.

D. LANGEVIN   Laboratoire de Spectroscopic Hertzienne de l'E.N.S., 75231 Paris Cedex 05, France.

R.A. LICKLIDER   Naval Weapons Station, China Lake, CA 93550, USA.

M. LOZADA-CASSOU   Departamento de Fisica, Universidad Autonoma Metropolitana-Iztapalapa, 09340 Mexico DF, Mexico.

C.S. MURTHY   Department of Chemistry, University of Texas at Austin, Austin, TX 78712, USA.

B.W. NINHAM   Department of Applied Mathematics, Research School of Physical Sciences, The Australian National University, Canberra, 2601, Australia.

R. RAJAGOPALAN   Department of Chemical Engineering, University of Houston, Houston, TX 77204–4792, USA.

P.J. ROSSKY   Department of Chemistry, University of Texas at Austin, Austin, TX 78712, USA.

D. ROUX   Centre de Recherche Paul Pascal and GRECO "Microemulsions" CNRS, Domaine Universitaire, 33405 Talence Cedex, France.

S.A. SAFRAN   Exxon Research and Engineering Co., Annandale, NJ 08801, USA.

E.Y. SHEU   Texaco Research Laboratory, Beacon, NY 12508, USA.

C. TAUPIN   Laboratoire de Physique de la Matière Condensée, Collège de France, 75231 Paris Cedex 05, France.

G.M. THURSTON   Department of Physics, Massachusetts Institute of Technology, Cambridge, MA 02139, USA.

A. VRIJ   Van 't Hoff Laboratory, University of Utrecht, 3584 CH Utrecht, The Netherlands.

H. WENNERSTRÖM   Department of Physical Chemistry 1, Chemical Center, S–22100 Lund, Sweden.

M.S. WERTHEIM   Mathematics Department, Rutgers University, New Brunswick, NJ 08903, USA.

Part One Basic Interactions, Statistical Mechanics, and Computer Simulations

1
Interparticle Correlations in Concentrated Charged Colloidal Solutions—Theory and Experiment

S.-H. CHEN and E.Y. SHEU

We present a liquid theoretic approach to analyses of small-angle scattering data when the interparticle correlations effect is important. In particular, the one-component macroion (OCM) theory for calculating the interparticle structure factor is shown to be an appropriate theory for analyses of small-angle neutron scattering (SANS) and small-angle X-ray scattering (SAXS) data on dilute-charged colloidal solutions. A detailed derivation of a generalized one-component macroion (GOCM) theory is given. The GOCM theory extends the range of validity of OCM theory to higher concentrations. The result of GOCM theory indicates that the classical Derjaguin–Landau–Verwey–Overbeek (DLVO) double-layer interaction should be interpreted as the potential of mean force in the dilute concentration limit. Two examples, a series of moderately concentrated cytochrome-C protein solutions and sodium dodecyl sulfate micellar solutions, are used to illustrate the applicability of GOCM theory.

I. Introduction

During the last decade, various experiments using light scattering, both static and dynamic [1], small-angle neutron scattering (SANS) as well as small-angle X-ray scattering (SAXS) on polystyrene latex spheres [2,3], ionic micelles [4,5], and hydrophilic proteins [5] in aqueous solutions, have identified a pronounced correlation peak in the scattering intensity distributions. The position of the correlation peak is generally located at the $|\mathbf{Q}|$ value ($\mathbf{Q}$ is the scattering vector) approximately equal to 2π times the reciprocal of the mean interparticle separation distance. This experimental fact implies that there exists significant local ordering around a given macroion in the electrostatically stabilized colloidal suspensions due to their strong mutual electrostatic interactions, even at very low particle volume fraction. Since this electrostatic interaction is screened by the available counterions, which keep the solution neutral as a whole, the macroion-macroion interaction in this case is dominated by the double-layer repulsive interaction. This interaction can be long-ranged and can be many $k_B T$ at contact depending on the ionic strength of the solution. Therefore, elucidation of the functional form of this double-layer interaction and its strength as a function of the macroion charge, volume fraction, and the added salt are of primary importance to problems such as the stability of the colloid [6], calculation of thermodynamic and transport properties of the solution, and determination of the aggregation behavior of supramolecular systems such as the ionic micellar solution [7].

From a liquid state theory point of view, the ionic correlations in colloidal solutions are analogous to the problem in binary electrolytes [8] or in molten salt [9]. However, there is a fundamental difference between them: The size and charge of macroions in colloids are usually much larger than those of the counterions and therefore colloidal solutions belong to a class of highly asymmetric electrolytes in which the usual theories for simple electrolytes may not be applicable. For such a highly asymmetric electrolyte, it is theoretically very difficult to treat the counterion–macroion correlation accurately. This is because the macroion surface charge density is so high in this case that there is a pileup of counterions along the surface of the macroion, a phenomenon that requires a more sophisticated treatment of the macroion–counterion correlation at short range than the conventional liquid theory approximation. Because the macroion–counterion correlation is coupled to the macroion–macroion correlation via the Ornstein–Zernike (OZ) equation, a poor approximation in evaluation of the macroion–counterion correlation would severely affect the accuracy in the calculation of the macroion–macroion pair correlation function and thus the effective macroion–macroion interaction, which is related to the potential of mean force obtainable through the macroion–macroion pair correlation. In this article, we shall discuss some of our recent work in trying to improve our theoretical ability to quantitatively determine the correlations in colloidal solutions and how these efforts impact on our capability in analyzing SANS data from these highly asymmetric electrolytes. We shall illustrate the interrelation between the theory and experiment with two concrete systems: ionic micelles formed by sodium dodecyl sulfate (SDS) and a globular protein, cytochrome-C, in aqueous solution.

II. Phenomenological Description of the Correlation Peak

Before describing the details of statistical mechanical theories for the correlation peak, we shall start by giving a simplistic interpretation of the observed correlation peak in SANS experiments. In this simple zeroth-order approximation, we shall concentrate only on the interpretation of the peak position and disregard the width and height of the peak. Although this is a phenomenological interpretation, it nevertheless provides useful information, particularly when there are no statistical mechanical theories available, such as in the

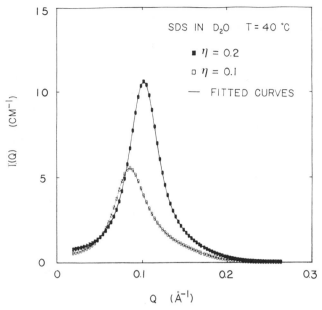

FIGURE 1.1. SANS intensity distributions from SDS micellar solutions of volume fractions 0.1 and 0.2 at $T = 40°C$. Both peak height and peak position increase as the micellar volume fraction increases. The peak position, Q_{max}, is a unique function of the mean intermicellar distance d according to Eq. (2).

case of cylindrical micellar solutions [10]. In Fig. 1.1, we show two intensity distributions of neutrons scattered by SDS micellar solutions at $T = 40°C$. The ordinate $I(Q)$ is expressed in scattering cross section per unit volume of the sample. The coordinate is the magnitude of the scattering vector

$$|\mathbf{Q}| = \frac{4\pi}{\lambda} \sin \frac{\theta}{2} (\text{Å}^{-1}), \qquad (1)$$

where λ is the wave length of neutrons in angstrom ($\lambda \sim 5$ Å), and θ is the scattering angle. The open squares refer to an 8% SDS solution in D_2O with an effective volume fraction of the micelle equal to 0.1. The closed squares refer to a solution with twice the concentration. We see that as the particle volume fraction doubles the peak intensity also doubles but the peak position Q_{max} increases only slightly. To relate the mean interparticle distance d to the peak position Q_{max}, we assume that the micelles are quasispherical with an axial ratio less than two and the local ordering around a micelle follows a face-centered-cubic-like, close-packed structure. Then, d can be expressed in terms of Q_{max} as [11]

$$d = \sqrt{6}\pi/Q_{max} = \frac{1}{\sqrt{2}} \left[\frac{4000\bar{n}}{N_A [C]} \right]^{1/3} \times 10^8 \text{ Å}, \qquad (2)$$

where $\bar{n}$ is the mean aggregation number of the SDS micelle, N_A is Avogadro's number, and $[C]$ is the molar concentration of the surfactant molecules that form micelles. This equation is, however, not accurate since the correlation peak is not a Bragg peak, because there is no long-range order existing in an isotropic micelle phase. Prins and Peterson [12] and Bahe [13] argued that this face-centered cubic (FCC)-like structure is only realized locally, and as one moves away from the first nearest neighbor shell to the second and successively further neighbor shells, the disorder gradually sets in and destroys the long-range order. To evaluate such a distorted FCC structure, one can first write down the structure factor $S(Q)$ in terms of the pair correlation function $g(r)$ as

$$S(Q) = 1 + \int_0^\infty dr\, 4\pi r^2 n_o [g(r) - 1] \frac{\sin Qr}{Qr}, \tag{3}$$

where n_o is the number density of micelles in solution. Because $4\pi r^2 n_o g(r)\, dr$ respresents the number of micelles at a distance $(r, r + dr)$ from the one at the origin, $S(Q)$ can be rewritten as a discrete sum over different neighboring shells:

$$S(Q) = 1 + \sum_k n_k \frac{\sin Qr_k}{Qr_k} \exp[-0.0025(Qr_k)^2], \tag{4}$$

where the summation k is over the successive neighbors in a face-centered-cubic structure and n_k and r_k are, respectively, the numbers of and distance to the kth neighbors. We note that in this approximation a Debye–Waller-like factor is multiplied to each term in the summation. This factor effectively attenuates the contribution of the further neighbors to the correlation peak because of the distortion. Equation (4), when computed up to the 13th neighbor, gives the first diffraction peak occurring at

$$Q_{max}d = 7.884. \tag{5}$$

With this equation and Eq. (2), Bahe [13] predicts the Q_{max} of lithium dodecyl sulfate (LDS) micellar solutions and found a striking agreement with the data of Bendedouch, Chen, and Koehler [14]. However, this comparison was not based on a first principle calculation because $\bar{n}$ in Eq. (2) was taken from a separate analysis of the experiment using Hayter and Penfold's one-component macroion (OCM) theory, which gives not only the peak position, but also the peak height and width, as well as the mean aggregation number $\bar{n}$. Recently, we have arrived at a more accurate expression using more than 50 data sets [11]:

$$Q_{max}d = 6.8559 + 0.0094d. \tag{6}$$

From Eqs. (2)–(6), it suffices to say that in the absence of a statistical mechanical theory Eq. (6) can be combined with Eq. (2) to give a quick estimate of $\bar{n}$ once Q_{max} is known experimentally. Also, the phenomenological

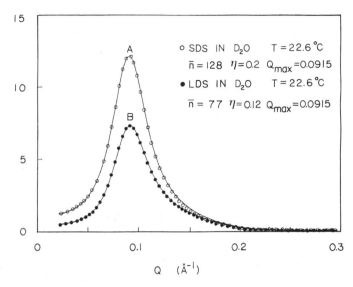

FIGURE 1.2. SANS intensity distributions from SDS (open circle) and LDS (closed circle) micellar solutions at $T = 22.6°C$. The Q_{max} of these two systems are exactly the same, even though the detailed analysis shows that $\bar{n} = 128$ in the case of A, and $\bar{n} = 77$ in the case of B. This is indicative of the fact that Q_{max} is d dependent and that d is a function of the ratio of $\bar{n}$ to η only.

argument presented previously implies that the mean aggregation number is essentially determined by the peak position only. This important fact thus assures that decomposition of the scattering intensity distribution function $I(Q)$ into a product of the particle structure factor $P(Q)$ and interparticle structure factor $S(Q)$ is an appropriate treatment. This is because $P(Q)$ is largely determined by $\bar{n}$, and hence by Q_{max}, while $S(Q)$ is mainly governed by the particle volume fraction η and the charge Z_0 of the macroion. Therefore, one does not need to eliminate $S(Q)$, by adding salt or by decreasing particle volume fraction, to determine $P(Q)$ [15].

Figure 1.2 illustrates this point vividly. Samples A and B give rise to exactly the same Q_{max} even though the detailed analysis shows that $\bar{n} = 128$ in the case of A, and $\bar{n} = 77$ in the case of B. This is because Q_{max} depends on d, which is a function only of the ratio of $\bar{n}$ to the volume fraction η.

III. SANS Cross-Section Formula

The real advantage of the SANS technique lies in the fact that its scattering intensity distribution function $I(Q)$, or the differential cross section, can be unambiguously written down and computed once the structural models of micelle and intermicellar interaction are specified.

For a system of monodispersed, or nearly monodispersed micelles in solution, $I(Q)$ can be written as [4]

$$I(Q) = (C - CMC)\bar{n}(b_m - v_m\rho_s)^2 P(Q)S(Q), \tag{7}$$

where C is the surfactant concentration, CMC is the critical micellar concentration, b_m is the total scattering length of a surfactant monomer, v_m is the dry volume of the surfactant monomer, and ρ_s is the scattering length density of the solvent. Important points to notice from this formula are that the absolute intensity of SANS is proportional to, aside from the known concentration factor and the molecular parameters of the surfactant monomer and the solvent, the weight-averaged aggregation number $\bar{n}$, the particle structure factor $P(Q)$, and the interparticle structure factor $S(Q)$. Furthermore, $P(Q)$, as shown in the previous section, depends only on $\bar{n}$ and the number of solvent molecules associated with the head group of a surfactant molecule for a globular shaped micelle [16]. $S(Q)$, as will be discussed in the following, can be calculated using various statistical mechanical theories by assigning an effective charge Z_o to the micelle. These statistical mechanical approaches are much more satisfactory than the phenomenological approach given in the last section because the former approach gives absolute intensities at all Q, not only Q_{max}. In the case of globular protein, the prefactor in Eq. (7), $A_p = (C - CMC)\bar{n}(b_m - v_m\rho_s)^2$ can be put into a more explicit form by writing it as [17]

$$A_p = 10^{-3}[p]N_A\left[\sum_{i}^{p} b_i + N_{ex}(b_D - b_H)\chi - v_d\frac{[\chi b_{D_2O} + (1 - \chi)b_{H_2O}]}{v_w}\right], \tag{8}$$

where $[p]$ is the protein concentration in mM, N_A is the Avogadro number, $\sum_{i}^{p} b_i$ is the sum of scattering lengths of all atoms in a protein molecule, N_{ex} is the number of labile protons in a protein molecule that have been exchanged, χ is the volume fraction of D_2O in the solvent, v_d the dry volume of the protein molecule, and v_w is the volume of a water molecule. This detailed form of A_p is helpful in a contrast variation experiment, from which one can extract two parameters, N_{ex} and v_d. We shall show this point later in Section V.

IV. One-Component Macroion Theory for $S(Q)$

The intermacroion structure factor $S(Q)$, which depends on the macroion-macroion interaction, the particle diameter σ_o, and the number density n_o of the macroion, can be calculated by methods based on standard liquid theory. The most successful statistical mechanical theory in treating a highly asymmetric electrolyte solution, such as the systems treated here, has been based on solving the OZ integral equation for the total correlation function $h(r)$, which is given by

$$h(r) = c(r) + n_o \int c(r')h(|r - r'|)\, d^3 r'. \tag{9}$$

From the total correlation function $h(r)$, the pair correlation function, $g(r) = h(r) + 1$, and thus $S(Q)$, can be calculated from Eq. (3). To solve the OZ equation, one needs to impose a closure relation for the direct correlation function $c(r)$. For a charged hard sphere system, two closures are often used, namely, the hypernetted chain approximation (HNCA) and the mean sperhical approximation (MSA). They can be expressed respectively, by the following equations. For the HNCA, one has

$$c(r) = -\beta u(r) + h(r) - \ln[h(r) + 1], \tag{10}$$

and for the MSA,

$$c(r) = -\beta u(r), \qquad r > \sigma_o, \tag{11}$$

and

$$h(r) = -1, \qquad r < \sigma_o. \tag{12}$$

In these equations, $\beta = 1/k_B T$ is the Boltzmann factor and $u(r)$ is the interparticle potential. The HNCA has been known to be the most accurate theory for a charged hard sphere system [18]. However, it is a nonlinear theory and has to be solved by numerical methods. On the other hand, MSA is a lincarized version of HNCA and is less accurate, but has a great advantage of having an analytical solution when the potential function is of the Yukawa form or screened Coulomb form [19]. Since the pioneering work of Hayter and Penfold [20], it has been customary in SANS literature to use a so-called Derjaguin–Landau–Verwey–Overbeek (DLVO) double-layer repulsive potential as the intermacroion potential function in the MSA. This potential has the form

$$V_R(r) = \frac{Z_o^2 e^2}{\varepsilon(1 + \kappa\sigma_o/2)^2} \frac{e^{-\kappa(r - \sigma_o)}}{r}, \tag{13}$$

where Z_o is the macroion charge number, e the electronic charge, ε the dielectric constant of the solvent, and κ the Debye screening constant. In the case of ionic micellar solution without added salt, κ is completely determined by the degree of counterion dissociation. Thus, Z_o and κ are related via a charge-neutrality condition. In this case, the contact potential $Z_o^2 e^2/[\varepsilon(1 + \kappa\sigma_o/2)^2]$ is strongly coupled to the decay constant of the exponential function. Using Eq. (13) as the pair potential and the MSA ansatz, Eqs. (11) and (12), an analytical solution for the OZ equation can be obtained. A convenient algebraic form of $S_{OCM}(Q)$ for macroion–macroion interactions has been given by Hayter and Penfold [20]. We shall call this approximation the OCM theory. Although it has been widely used in analysis of SANS data from ionic micellar solutions, it is somewhat of a puzzle why $V_R(r)$ is called the DLVO potential. Moreover, it is surprising that $V_R(r)$, when used in conjunction with MSA, has been so successful in fitting correlation peaks of SANS data from interacting charged colloidal solutions [4]. To understand these two points requires a digression and a brief description of a new extension of OCM theory.

A. Meaning of the DLVO Interaction

In the so-called DLVO theory of colloidal stability [6], these authors considered the main interactions between two colloidal particles in solution to consist of two parts: the first part is the repulsive component arising from the overlapping of the double layers of the two colloidal particles; and the second part is an attractive component arising from van der Waals interactions between constituent molecules of the colloidal particles. The macroscopic manifestation of this latter attractive interaction is usually called Hamaker interaction with a Hamaker constant A, the magnitude of which is of the order of $k_B T$ [6]. In colloidal solutions at low ionic strengths or in ionic micellar solutions without added salt, the magnitude of the double-layer interaction is usually several $k_B T$ at contact. Therefore, at equilibrium at temperature T, colloidal particles seldom approach close enough to sense the deep attractive Hamaker interaction near the surface of the particle. Thus, as far as the calculation of the structure factor $S(Q)$ is concerned, one can neglect the presence of the Hamaker interaction and consider only the effect of the double-layer repulsion. We shall briefly recall the so-called double-layer repulsive interaction as presented in the classic work of Verwey and Overbeek [6].

In a linear approximation the electrostatic potential ψ surrounding a colloidal particle can be obtained by solving the Debye–Hückel equation [6]

$$\nabla^2 \psi = \kappa^2 \psi. \tag{14}$$

In this approximation, the surface potential ψ_o of a charged colloidal particle of radius a surrounded by its counterions immersed in a medium with a dielectric constant ε is related to its charge q by a relation

$$\psi_o = \frac{q}{\varepsilon a} \frac{1}{1 + \kappa a}, \tag{15}$$

where κ is the Debye screening length of counterions, which are by the subscript 1:

$$\kappa^2 = \frac{8\pi e^2 z_1^2 n_1}{\varepsilon k_B T}. \tag{16}$$

We note from Eq. (15) that the surface potential is proportional to the surface charge in this approximation. Verwey and Overbeek then showed that in this linear approximation the free energy (the work done at constant temperaure) of bringing two such colloidal particles from infinity to a finite separation distance r is given by

$$\nabla F = V_R(r) = q[\psi(r) - \psi_o], \qquad (17)$$

where $\psi(r)$ is the surface potential of the sphere at a separation r. This relation is obtained by assuming that the charge of the particle is held constant in the process. In order to obtain an explicit functional dependence of $V_R(r)$ on r, a relation between $\psi(r)$ and q is established by solving Eq. (14) for two spheres separated by a distance r and applying a boundary condition requiring the charge of the spheres to be q. An approximate solution of the problem can be put into the form

$$\psi(r) = q \cdot F(r/a, \kappa a), \qquad (18)$$

where the function $F(r/a, \kappa a)$ is given explicitly in Ref. 6, Eq. (79). Upon substitution of Eqs. (12) and (18) into Eq. (17), one finally obtains the well-known result

$$V_R(r) = \frac{q^2}{\varepsilon(1 + \kappa a)^2} \frac{e^{-\kappa(r - 2a)}}{r} \gamma(r/a, \kappa a). \qquad (19)$$

The factor γ is quoted to be ranging from about 0.78 to 1.0, depending on the magnitudes of r/a and κa [see Ref. 6, Eq. (83)].

In Hayter and Penfold's treatment, they interpreted $V_R(r)$ as an effective colloidal interaction potential and took a simplified version of Eq. (1.19) by assuming the factor $\gamma(r/a, \kappa a)$ to be unity. Introducing a notation $x = r/\sigma_o$, $k = \kappa \sigma_o$, and $\sigma_o = 2a$, they wrote

$$V_R(r) = V_1 \frac{e^{-k(x - 1)}}{x}, \qquad (20)$$

where $V_1 = q^2/[\varepsilon \sigma_o(1 + \kappa \sigma_o/2)^2]$. They then invoked the MSA ansatz given in Eqs. (11) and (12) to obtain the direct correlation function inside the core in a form

$$c(x) = a + bx + \frac{1}{2}\eta a x^3 + v\left(\frac{1 - e^{-kx}}{x}\right) + v^2\left(\frac{\cosh x - 1}{2V_1 k^2 x e^k}\right), \qquad x \leqslant 1, \qquad (21)$$

where $\eta = (\pi/6)\sigma_o^3 n_o$ is the volume fraction of the macroions and the constants a, b, and v are explicitly given in Ref. 16. Since $c(r)$ for both inside and outside of the core σ_o is now analytically known [Eqs. (20) and (21)], one can easily take a Fourier transform to obtain $S(Q)$ via a rigorous relation:

$$S(Q) = /[1 - n_o C(Q)]. \qquad (22)$$

This procedure for computing $S(Q)$ has been shown to produce excellent agreement with SANS data for ionic micellar solutions with $q = Z_0 e$ and σ_o as adjustable parameters [4,21,22]. In the cases of low particle volume fraction and high macroion charge, this procedure will yield a negative $g(r)$ value at contact. To remedy this unphysical result and to obtain the correct interparticle structure factor, Hansen and Hayter [22] proposed a rescaling procedure based on Gillan's condition [23]. From a theoretical point of view, there are two criticisms about the OCM theory of Hayter and Penfold. First, the effective potential [Eq. (20)] is clearly an approximation to the original DLVO form. Second, the MSA ansatz may or may not be a good approximation, because it is often mentioned in the literature that the HNCA closure [Eq. (10)] is more accurate than that of the MSA [24,25]. We shall show in the next section that the thermodynamic function $-\beta V_R(r)$ in Eq. (20) is more appropriately interpreted as a direct correlation function outside the core in the dilute concentration limit. If we accept this interpretation, then the MSA ansatz of Eq. (11) becomes unnecessary. In this regard, to further impose an HNCA ansatz by treating $V_R(r)$ as an intermacroion potential in order to solve the one-component OZ equation would be an inconsistent approximation [25].

B. Generalized One-Component Macroion Theory

The idea of a generalized one-component macroion (GOCM) theory [26,27,28] originated from a paper by Beresford-Smith and Chan [25] in 1982. These authors showed that it is possible to first write down a set of multicomponent-coupled OZ equations involving the correlation functions of the macroions and counterions and then to contract these equations to obtain an effective one-component OZ equation for the macroion alone. In this process, one is naturally led to a definition of an effective direct correlation function, $c_{00}^{eff}(r)$, for the macroions. We shall outline the formulation of the GOCM in the following paragraphs [26,27,28].

For a system of multicomponent coulomb fluid in the primitive model [8], the OZ equations in Q space (the Fourier transform space) and their associated MSA closure relations for $c_{ij}(r)$ are given by

$$\tilde{h}_{ij}(Q) = \tilde{C}_{ij}(Q) + \sum n_k \tilde{C}_{ik}(Q)\tilde{h}_{kj}(Q), \tag{23}$$

and

$$c_{ij}(r) = \frac{-\beta Z_i Z_j e^2}{\varepsilon r_{ij}}, \, r_{ij} > \sigma_{ij} \left(\equiv \frac{\sigma_i + \sigma_j}{2} \right). \tag{24}$$

The Fourier transform is defined as

$$h_{ij}(Q) = \int d^3r \exp(-\mathbf{Q} \cdot \mathbf{r}) h_{ij}(r) = \int_0^\infty dr \, 4\pi r^2 \frac{\sin Qr}{Qr} h_{ij}(r). \tag{25}$$

Here, $h_{ij}(r)$ is the total correlation function between particles i and j, and ε is the dielectric constant of the solvent. Denoting by 0 the macroion and by 1, 2, etc., the small ions, the effective one-component OZ equation can be obtained by contracting Eq. (23) [28], namely,

$$\tilde{h}_{00}(Q) = \tilde{C}_{00}^{\text{eff}}(Q) + n_0 \tilde{C}_{00}^{\text{eff}}(Q)\tilde{h}_{00}(Q), \tag{26}$$

where

$$\tilde{C}_{00}^{\text{eff}}(Q) = \tilde{C}_{00}(Q) + \bar{C}_0^T[\bar{I} - \bar{C}]^{-1}\bar{C}_0 \tag{27}$$

represents the effective one-component direct correlation function in Q space for the macroion. $\bar{C}_0$ in Eq. (27) is a column matrix having elements

$$(\bar{C}_0)_i = \sqrt{n_0}\bar{C}_{i0}(Q), \qquad i = 1,2,3,\ldots, \tag{28}$$

and $\bar{C}$ is a two-dimensional matrix with elements

$$(\bar{C})_{ij} = \sqrt{n_i n_j}\,\tilde{C}_{ij}(Q), \qquad i = 1,2,3\ldots. \tag{29}$$

One sees from Eq. (27) that the Fourier transform of the effective direct correlation function $\tilde{C}_{00}^{\text{eff}}(Q)$ between two macroions contains two terms: a bare term representing the contributions from these two macroions and an indirect term containing the contributions of macroion–small ion and small ion–small ion bare direct correlations. In order to derive the effective direct correlation between macroions from Eq. (27), we first write an appropriate form for $c_{ij}(r)$, then we take the Fourier transform of $c_{ij}(r)$ and substitute the result into Eq. (27). Finally, we perform the inverse Fourier transform of $\tilde{C}_{00}^{\text{eff}}(Q)$ to get $\tilde{C}_{00}^{\text{eff}}(r)$. To write an appropriate form for $c_{ij}(r)$, one has to note that the asymptotic form of the direct correlation functions are the genuine coulomb potential between ion, not the effective potential as used in OCM theory. Thus, we put

$$c_{ij}(r) = -\beta\frac{Z_i Z_j e^2}{\varepsilon r} + c_{ij}^s(r), \qquad i,j = 0,1,2\ldots, \tag{30}$$

where the term $c_{ij}^s(r)$ takes into account the deviation of the direct correlation between particles i and j from the coulomb potential because of a short-range effect. Moreover, because in a highly asymmetric polyelectrolyte, the size of the small ion is generally much smaller than that of the macroion, we can safely make an approximation in the course of deriving $\tilde{C}_{00}^{\text{eff}}(Q)$ by neglecting the small ion size (i.e. taking $\sigma_i = 0$ for $i = 1,2,3\ldots$). Based on this approximation, one obtains, according to Eqs. (24) and (30), these relations:

$$c_{00}^s(r) = 0 \quad \text{for} \quad r > \sigma_0, \tag{31}$$

$$c_{0i}^s(r) = 0 \quad \text{for} \quad r > \sigma_0, \tag{32}$$

and

$$c_{ij}^s(r) = 0 \quad \text{for} \quad r > 0. \tag{33}$$

Taking the Fourier transform (FT) of Eq. (30), employing Eqs. (31)–(33), and then substituting the results into Eq. (27), one gets

$$\tilde{C}_{00}^{\text{eff}}(Q) = \tilde{C}_{00}^{s}(Q) + \sum_{i=1}(\tilde{C}_{0i}^{s})^2 - \frac{\left(\tau_0 + \sum_{i=1}\tau_i\tilde{C}_{0i}^{s}\right)^2}{Q^2 + \kappa^2}, \tag{34}$$

where

$$\tau_i = \left[\frac{4\pi\beta e^2 n_i Z_i^2}{\varepsilon}\right] \qquad i = 1,2,3\ldots, \tag{35}$$

are Q-independent quantities and thus do not cause any difficulty for the subsequent inverse Fourier transform of $\tilde{C}_{00}^{\text{eff}}(Q)$ to determine $c_{00}^{\text{eff}}(r)$ outside the core. Also, the first and second terms on the right-hand side of Eq. (34) give no contribution to $c_{00}^{\text{eff}}(r)$ outside the core since $c_{ij}^{s}(r) = 0$ outside the core for $i,j = 1,2\ldots$ [see Eqs. (31)–(33)]. Thus, $c_{00}^{\text{eff}}(r)$ is equal to the inverse Fourier transform of the last term on the right-hand side of Eq. (34). However, it involves $\tilde{C}_{0i}^{s}(Q)$, which is a Q-dependent quantity and should be in a proper form for an analytic inverse Fourier transform to be possible. The explicit expression for $\tilde{C}_{0i}^{s}(Q)$ can be obtained via a Fourier transform of $C_{0i}^{s}(r)$ given by Hiroike [29]. The result is

$$c_{0i}^{s}(r) = \frac{1}{\eta - 1} - \frac{8\pi\beta e^2 \xi}{\varepsilon\sigma_0} + \frac{Z_i Z_j e^2 \beta}{\varepsilon r} \qquad \text{for} \qquad r < \sigma_0/2, \tag{36}$$

where

$$\xi = (\Gamma\sigma_0 + \mu)/(1 + \Gamma\sigma_0 + \mu), \qquad \mu = \frac{3\eta}{1 - \eta}, \tag{37}$$

and Γ is equivalent to a generalized Debye screening constant that takes into account, besides the contribution of counterions, that of the macroions [30]. It can be obtained by solving the following implicit equations:

$$4\Gamma^2 = \kappa^2 + \frac{\tau_0^2}{(1 + \Gamma\sigma_0 + \mu)^2}, \tag{38}$$

and

$$\tau_0^2 = 24\beta Z_0 e^2 \eta/(\varepsilon\sigma_0). \tag{39}$$

Since the first two terms in Eq. (36) are r-independent, they can be treated as constants in the process of a FT. The resultant $\tilde{C}_{i0}^{s}(Q)$ obtained by the FT of Eq. (36) is

$$\tilde{C}_{0i}^{s}(Q) = C_i \frac{\sin(QR)}{Q^3} + D_i \frac{R}{Q^2} \cos(QR), \tag{40}$$

where constants

$$C_i = -\left[\frac{4\pi}{1 - \eta} + \frac{4\pi Z_0 Z_i e^2}{\varepsilon k_{\text{B}} T R}\right] \tag{41}$$

and

$$D_i = C_i + \frac{4\pi z_o Z_i e^2}{\varepsilon k_B T R} \tag{42}$$

are Q-independent, and R is the radius of the macroion. Upon substitution of $\tilde{C}^s_{0i}(Q)$ into Eq. (34), one sees that $\tilde{C}^s_{0i}(Q)$ is the only term that would contribute to $\tilde{C}^{eff}_{00}(r)$ in performing the inverse Fourier transform. Since $\tilde{C}^s_{0i}(Q)$ is a trigonometric function, its inverse Fourier tranform can be performed analytically. The effective direct correlation function outside the core so obtained is

$$c^{eff}_{00}(r) = -\frac{\beta Z_0^2 e^2}{\varepsilon} X^2 \frac{e^{-\kappa r}}{r}, \qquad r > \sigma_o, \tag{43}$$

where

$$X = \cosh(\kappa R) + U[\kappa R \cosh(\kappa R) - \sinh(\kappa R)], \tag{44}$$

and

$$U = \mu/(\kappa R)^3 - \xi/(\kappa R). \tag{45}$$

From Eqs. (43)–(45), one can show that in the dilute limit, that is, $n_o \to 0$,

$$\lim_{n_o \to 0} X = \frac{e^{\kappa \sigma_o/2}}{1 + \kappa \sigma_0/2}. \tag{46}$$

Therefore, by substituting Eq. (46) into Eq. (43), one obtains

$$\lim_{n_o \to 0} \tilde{C}^{eff}_{00}(r) = -\beta V_R(r). \tag{47}$$

This equation clearly indicates that in the dilute limit the effective direct correlation function between macroions becomes identical to the negative of the DLVO interaction divided by $k_B T$. This point has already been noted by Medina-Noyola and McQuarrie [31] in 1980. They pointed out that the so-called DLVO interaction, $V_R(r)$, is essentially the potential of mean force between macroions in the dilute limit. Equation (43) is an explicit extension of their results to finite concentrations.

From Eq. (43), one also notices that under the assumption of the MSA closure in a multicomponent primitive model of a charged colloidal solution, the effective direct correlation function outside the core has a Yukawa form when the counterions are taken as point charges and, this form is valid at any concentration of macroions. This theorem, when used together with the hard-core condition of Eq. (12), leads to a direct correlation function inside the core given in Eq. (21). Thus, there is no need to make an MSA ansatz, such as Eq. (11), again in solving the one-component OZ equation. We shall call this procedure of using the direct correlation function, Eq. (43), a GOCM theory to distinguish it from OCM theory, which uses Eq. (20) as a direct correlation function between macroions. It is now clear why the OCM theory of Hayter and Penfold gave satisfactory $S(Q)$ with charge Z_0 in Eq. (20) as an adjustable parameter. When using Z_0 in OCM theory as an adjustable parameter, one essentially compensates for the deficiency of $V_R(r)$, which is an approximate form of Eq. (43). We thus conclude this section by emphasizing that the so-called DLVO interaction used in the current SANS literature is not an effective interaction potential between macroions, but is the effective direct correlation in the dilute limit, which is a thermodynamic quantity. Because OCM theory is a limiting case of GOCM theory, the charge extracted from a fit of OCM theory to SANS data would be inaccurate at a finite concentration. Figure 1.3 shows the theoretical calculation of Z_{OCM}/Z_{GOCM}, as one would extract through SANS data analysis, as a function of volume fraction η. One can see from Fig. 1.3 that the OCM charge is always higher than the GOCM charge, and at a volume fraction of 10%, Z_{OCM} is about 6% higher than Z_{GOCM}.

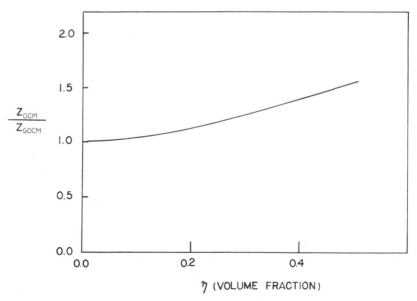

FIGURE 1.3. Theoretical comparison of the macroion charges one would extract from analysis of SANS data using one-component macroion (OCM) theory and generalized one-component macroion (GOCM) theory. Z_{OCM} is in general higher than Z_{GOCM}, since in OCM theory the γ factor [Eq. (16)], which ranges from 0.78 to 1, is assumed to be unity, and in GOCM theory, the screening of the surrounding macroions, besides that of the counterions, is also taken into account, which would reduce the extracted macroions' charge. As one can see, Z_{OCM} is about 6% higher than Z_{GOCM} when the volume fraction is 0.1.

V. The Charge and Structure Factor of Globular Proteins in Aqueous Solutions

In order to test the accuracy of the effective one-component direct correlation function we derived in Eq. (43), or equivalently the generalized DLVO interaction

$$U_{GDLVO}(r) = - k_B T c_{00}^{eff}(r), \qquad r > \sigma_o, \tag{48}$$

we shall apply the GOCM theory to calculate the structure factors of concentrated protein solutions. Protein solutions are ideal testing cases for the theory because the protein charge Z_o can be estimated independently from a titration experiment. In a titration experiment, one adds a known amount of salt, such as sodium acetate (NaAc), to a protein solution of known concentration $[p]$. Then, one adds a suitable amount of weak acid, such as acetic acid (AcA), to control the pH value of the solution. The pKa value of the acetic acid is precisely known to be 4.75. The pH value of the solution, or equivalently the hydrogen ion concentration $[H^+]$, can be accurately measured by an electrode. Then, it can be easily shown that the protein charge Z_o is given by [17]

$$Z_o = \frac{\Delta[AcA] - [NaAc]}{[p](1 + \Delta)}, \tag{49}$$

where

$$\Delta = 10^{(pH - pKa)}. \tag{50}$$

We shall illustrate this with a series of SANS measurements made on horse heart cytochrome-C [17]. The shape of cytochrome-C protein molecule is an oblate spheroid with a dimension $a \times b \times b = 15 \times 17 \times 17$ Å. The molecular weight is equal to 12,384 dalton [17,32]. Since it is nearly spherical, we can take it to be a hard sphere of diameter $\sigma_0 = 32.6$ Å. Wu and Chen [17] have performed a separate contrast variation measurement using a protein solution of volume fraction $\eta = 9.06\%$ to obtain the two unknown parameters in Eq. (8), namely, N_{ex} and v_d. In D_2O, $N_{ex} = 165$, and $v_d = 14808$ Å^3, which gives $A_p = 1.257$ cm^{-1}, knowing that $\sum b_i = 259.5 \times 10^{-12}$ cm from the amino-acid sequence. The hydrated diameter σ_H was obtained by a Guinier plot of SANS data at $\eta = 0.48\%$ solution with 10 mM acetate. $R_g = 12.7$ Å was, obtained which is fully consistent with the crystallographic dimension of the protein molecule. They thus used $\sigma_H = \sigma_o = 32.6$ Å to calculate the particle structure factor $P(Q)$ using the equation

$$P(Q) = \left[\frac{3j_1(Q\sigma_o/2)}{(Q\sigma_o/2)} \right]^2. \tag{51}$$

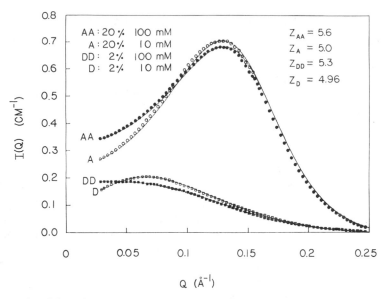

FIGURE 1.4. SANS intensity distributions and the fitted (by GOCM) curves (solid lines) from four protein solutions. AA and A are of high protein concentration (20%) in 100 and 10 mM of sodium acetate, respectively. DD and D are of low protein concentration in respective salts. The protein charges so obtained agree reasonably well with those measured by titration.

Computation of $S(Q)$ using GOCM requires an input of the volume fraction $\eta = \pi n_o \sigma_o^3 / 6$, the charge Z_o, and the Debye screening constant κ. Since the particle volume fraction is known for a given system and κ can be computed from the known ionic strength of the solution, only one adjustable parameter, Z_o, is needed to fit the entire distribution $I(Q)$ [17]. The results of such a fit for four cases are shown in Fig. 1.4. Curves AA and A represent two high-concentration (20%) protein solution in 100 and 10 mM of NaAc, respectively. Curves DD and D are the lower concentration cases, representing 2% protein solutions in the respective salts. As one can see from Fig. 1.4, GOCM theory fits the experimental data very well. The resultant fitted charges, Z_o, are also indicated in Fig. 1.4. Table 1.1 compares the charges from the titration experiment and that from the SANS experiment obtained by the GOCM theory. Two striking features can be readily observed: at a constant pH = 6.8, the protein charge is independent of the volume fraction, but it nevertheless depends on the ionic strength of the solution. From this comparison, we can conclude that the charge Z_o, which appears in Eq. (43), is equal to that obtained from the titration measurement. Figure 1.5 gives the structure factor $S(Q)$ computed by GOCM theory for the cases AA and DD. One notes from Fig. 1.5 that for a protein solution at a concentration of 2% with 100 mM added salt the structure factor is nearly unity and the solution is

TABLE 1.1. GOCM fitted parameters of cytochrome-C in acetate buffers.[a]

[NaAc] (mM)	[p] (mM)	η (%)	I (mM)	k	G	Z_p	$Z_{titration}$
100	15.72	18.13	139.8	4.08	4.32	5.6 ± 0.6	5.1 ± 0.1
	7.86	9.06	120.1	3.78	3.95	5.4 ± 0.5	5.1 ± 0.1
	3.93	4.53	110.3	3.63	3.72	5.3 ± 0.5	5.2 ± 0.1
	1.57	1.81	104.1	3.52	3.57	5.3 ± 0.5	5.2 ± 0.1
	0.786	0.91	102.2	3.49	3.51	5.3 ± 0.5	5.6 ± 0.1
	0.393	0.45	101.1	3.47	3.48	5.3 ± 0.5	5.6 ± 0.1
10	15.72	18.13	49.5	2.43	2.89	5.0 ± 0.5	5.0 ± 0.1
	8.06	9.30	30.3	1.90	2.34	4.9 ± 0.5	5.0 ± 0.1
	4.08	4.71	20.0	1.54	1.80	5.0 ± 0.5	4.9 ± 0.1
	1.64	1.89	14.0	1.29	1.56	5.0 ± 0.5	4.9 ± 0.1
	0.824	0.95	12.1	1.20	1.35	4.9 ± 0.5	5.0 ± 0.1
	0.412	0.48	11.0	1.14	1.24	5.0 ± 0.5	4.9 ± 0.1

[a] Input parameters are $\sigma_H = 33.2$ Å, $m = 112$, and $N_{ex} = 165$.

ideal. On the other hand, at 20% concentration, the structure factor is strongly depressed at small Q and hence the solution is very nonideal.

Study of the charge and structure factor of cytochrome-C was further extended to other pH values by Wu and Chen recently [33], using both SANS and SAXS. The only difference in the analysis of SAXS data from that

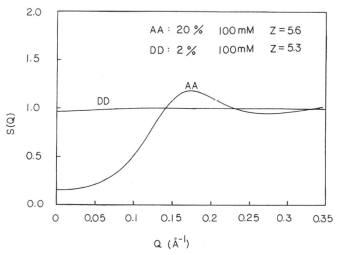

FIGURE 1.5. Extracted structure factors, $S(Q)$, of protein solutions, AA (20%) and DD (2%). $S(Q)$ of DD is nearly unity for all Q, and the solution is nearly ideal, while solution AA shows some local ordering due to interprotein interactions.

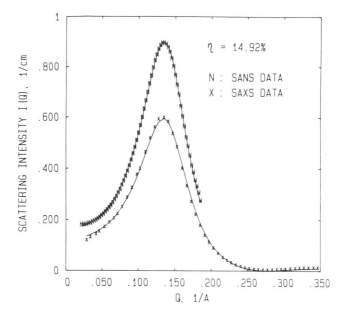

FIGURE 1.6. Comparison of SANS (N) and SAXS (X) intensity distributions for cytochrome-C solution at pH = 2.9 and a volume fraction of 14.9%. Solid lines are GOCM fits.

of SANS data is the computation of contrast. In the SAXS measurement, the contrast is, instead of the difference of scattering length densities between the particle and the solvent, the electron densities between them. Figure 1.6 gives the scattering intensities from both SANS and SAXS and the fitted curves for $\eta = 14.9\%$ and pH = 2.9. The agreement between them is excellent in absolute scale. It was found that protein surface charges extracted from SANS data and SAXS data are 15.6 ± 0.4 and 14.5 ± 0.4, respectively, which were considered to be consistent within experimental errors. However, the titration measurement gives a higher charge (20.3 ± 0.4). This is attributed to a pileup of the counterions along the protein surface because of high protein surface charge density, as we mentioned in Section I. This charge renormalization phenomenon was also observed in polymer latex particles at low ionic strength solutions [34]. As the macroion surface charge density becomes sufficiently high, the counterions would be attracted to the surface and would form a very steep distribution. As far as the intermacroion interaction is concerned, part of the condensed counterions look as if they belong to the macroion. Thus, the surface potential energy would be reduced to the order of $k_B T$, where the linear theory, such as GOCM theory, is applicable. It is therefore expected that the surface charge obtained from both SANS and SAXS would be consistent with each other, but smaller than that from the titration measurement.

VI. The Renormalized Charges of SDS Micelles

In comparison to globular proteins, micellar solutions pose a greater challenge to the theory for SANS data analysis. The reason is that the structure of micelles depends sensitively on the thermodynamic state of the solution. This implies that the two functions $P(Q)$ and $S(Q)$ in Eq. (7) are necessarily coupled. More precisely, the mean aggregation number of the micelles depends on the interaction between micelles and thus on the charge of the micelles. Because of this situation, we find it convenient to express the micellar charge in terms of the ratio $\alpha = Z_0/\bar{n}$. This seems to be a quantity that depends on the thermodynamic state of the solution. In spite of this complication, we found that the analysis of SANS data according to GOCM theory can still lead to a unique determination of the two factors $P(Q)$ and $S(Q)$ [4,11,35]. This fortunate situation arises from the fact that $\bar{n}$ is uniquely determined from the position of the interaction peak, as was pointed out in Section II. We have, in the past, done many experiments on ionic micelles, including lithium dodecyl sulfate [14,35,36], but the data analyses were performed using OCM theory [20], in which the effective double-layer interactions between micelles were taken to be the DLVO interactions. In these analyses, the effective charges of the micelles we extracted were subject to some errors because of the inaccuracy of the DLVO interactions. It can be shown that the OCM charge Z_{DLVO} is always higher than the GOCM charge Z_{GOCM} (see Fig. 1.3) by about 5 to 30% at high volume fractions. In this section, we show the results of our SANS data analyses with GOCM theory for a series of SDS micellar solutions at $T = 40°C$. The volume fraction ranges from 1.25 to 20%. The method of analysis parallels those of Refs. 29 and 30, except $c_{00}^{eff}(r)$ given in Eq. (43) was used to replace the DLVO interaction. The results of the analyses are presented in Table 1.2. Figure 1.7 gives the structure factors so obtained for solutions of different volume fractions. Figure 1.8 plots the fractional charge α as a function of volume fraction. Two striking features can be observed from Fig. 1.8: the fractional charge saturates at 0.3 in the dilute limit, and it drops down to a much lower level and saturates at 0.23 at the high volume fraction limit.

TABLE 1.2. Parameters extracted from SANS data analysis of SDS micellar solutions using GOCM.[a]

Volume fraction (η)	Micellar diameter σ_0 (Å)	$\bar{n}$	α
0.2	49	92	0.23
0.10	47	82	0.28
0.05	46	73	0.30
0.025	44	64	0.30
0.0125	42	53	0.30

[a] Experiments were performed at $T = 40°C$.

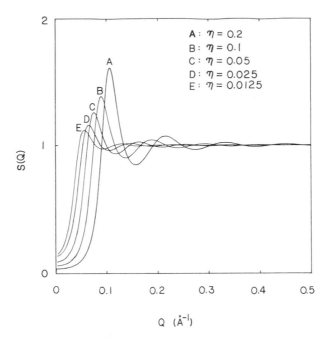

FIGURE 1.7. Structure factors extracted from SDS micellar solutions of various micellar volume fractions. Both peak height and peak position consistently increase as a function of volume fraction.

Even though one expects that SDS is fully dissociated at this temperature because it is a strong salt, one nevertheless observes an effective micellar charge that is significantly lower than the full dissociation ($\alpha = 1$). Physically, this means some parts of the counterions (Na^+) are always closely associated with the micelle in such a way that as far as the intermicellar interaction is concerned they should be considered as part of the micelle. Thus, the concept of a dressed micelle should be valid in the dilute limit where the double-layer free energy of a micelle–counterion system can be computed reasonably well by solving a Poisson–Boltzmann equation [35,37]. This kind of approach always predicts the dressed charge of the micelle to be significantly lower than the full charge, as seen in Fig. 1.8, at low volume fractions. The feature of the curve in Fig. 1.8 at the high volume fraction is however a more interesting one.

Consider a small layer of thickess, σ_1 (of the order of counterion size), around the micelle of radius a. Call this layer region I. Next, draw a sphere around the micelle with a radius R, which is half the average distance between two micelles in solution. Call the space between the two spheres of radii $(a + \sigma_1)$ and R region II. Let the number of counterions in region I and II be denoted by n_1 and n_2. Then, $\bar{n} = n_1 + n_2$. If we consider the counterions in region I to be part of the micelle when viewed from the other micelle, then one can define the fractional charge of the micelle as $\alpha = n_2/\bar{n} = (\bar{n} - n_1)/\bar{n}$. From

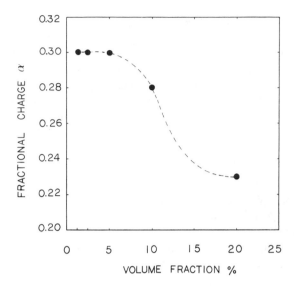

FIGURE 1.8 The fraction charges, $\alpha = Z_o/\bar{n}$, of SDS micelles extracted from SANS data analysis using GOCM theory. Saturation of α at 0.3 in the dilute limit indicates that considerable amounts of counterions are closely associated with the micelle in such a way that as far as the intermicellar interaction is concerned, they are part of the micelle. Thus, the charge so obtained is a renormalized charge.

this definition and a consideration of the Poisson–Boltzmann equation, one can derive, following Oozawa [38], the simple formula

$$\frac{1-\alpha}{\alpha} = (3\sigma_1/a) \frac{\eta}{1-\eta} \exp[\alpha P(1 - \eta^{1/3})], \qquad (52)$$

where P is a coulomb coupling parameter defined as

$$P = \frac{\bar{n}e^2}{\varepsilon k_B T(a + \sigma_1)}. \qquad (53)$$

The parameter P can be estimated to be between 16 and 25. This simple formula predicts that α would start from unity at a low volume fraction and saturate at a high volume fraction above $\eta = 0.15$ to a lower level determined by the parameter P. Thus, the saturation at the high volume fraction is simply due to the micellar–micellar interaction. The latter is taken into account in this shell model approach by a decrease in R, or equivalently an increase in η.

VII. Limitation of Generalized One-Component Macroion Theory

It can be seen from examples given in Sections V and VI that GOCM theory is rather successful in describing the structure factor $S(Q)$ of macroions. In the case of cytochrome-C at pH = 6.8 in 100 mM of sodium acetate, the surface charge density of the protein is determined by a titration experiment to be

2.5 μC per cm^2. Under this condition, the analysis of SANS data by GOCM theory gives the correct protein charge $Z_o/\bar{n} = 5.5$. On the other hand, in the case of SDS micellar solution at 1 wt.% concentration, the surface charge density of the micelle, based on the measured value of the aggregation number, is 15.3 μC per cm^2, but the fitted charge came out to be 4.6 μC per cm^2, corresponding to only 30% ionization of the monomers in the micelle. Also, only 70% of the titration charge was found from SANS analysis of a 14.9 wt.% cytochrome-C protein solution at pH $= 2.9$. We would like to interpret these results as an indication of substantial counterion condensation on the surface of the macroions. This phenomenon of counterion condensation has recently also been observed in a computer simulation by Watanabe et al. [39]. It should thus be qualified here that GOCM theory is a linearized theory and is not capable of explaining the phenomenon of the counterion condensation. Instead, it provides a reduced charge as a result of the linearization approximation. In order to treat the counterion condensation realistically, one must use a nonlinear theory, such as the multicomponent hypernetted chain approximation in the primitive model. This latter approach is capable of describing the counterion–micelle correlations in a more accurate way. Unfortunately, the multicomponent HNCA is known to be poorly convergent for a highly asymmetric polyelectrolyte solution with a high macroion charge.

We have recently shown that the HNCA can be solved for a micelle–counterion system at 0.04 micellar volume fraction with the micelle–counterion size ratio of 20 to 1 and a charge ratio 18 to 1. The theory can be fitted well to the measured structure factor of the sodium dodecyl orthoxylene sulfonate micellar solution at the corresponding condition [40]. Nevertheless, the micelle at this condition has an actual aggregation number of 50, which means that the fitted charge of 18 still indicates the counterion condensation phenomenon. We also fitted the same data with a multicomponent MSA, which gives an equally good fit with a micellar charge of 10. Based on this calculation, we can infer that both HNCA and MSA in their multicomponent versions are not sufficient to treat the micelle–counterion correlation accurately near the micellar surface, and the reduced micellar charges so obtained reflect the inadequacy of these theories. The same applied to GOCM theory, which is nothing but a contracted description of the multicomponent MSA. In practical terms, however, both the multicomponent MSA [30,40,41] and the generalized one-component macroion theory are analytical theories and are excellent tools for analysis of small-angle scattering data from strongly interacting colloidal solutions.

VIII. Perspective

We have shown that the generalized one-component macroion theory is an adequate analytical theory for the calculation of the intermacroion structure factor is charge colloidal solutions. It is a generalization of the one-component macroion theory of Hayter, Penfold, and Hansen [20,22] based on the DLVO interaction [6]. In GOCM theory, the OCM appears naturally as a limiting case where the volume fraction of the macroion approaches zero. GOCM theory is an accurate theory for obtaining the charge of proteins in aqueous solution at high concentrations when the protein surface charge density is of the order of 2.5 μC per cm^2. However, for proteins with higher surface charge densities [23] or for micellar solutions where the surface charge density is of the order of 15 μC per cm^2, the application of GOCM theory to analyses of SANS data generally gives rise to a reduced charge, or what we prefer to call a renormalized charge, because of charge condensation effect. The precise interpretation of this renormalized charges requires further development of an accurate theory of a multicomponent coulomb fluid. The multicomponent hypernetted chain approximation is shown to be insufficient for this purpose [40,42].

There is a further complication in the case of micellar solutions. Owing to the fact that micelles are in chemical equilibrium with surfactant monomers in solution, the micellar aggregation number is necessarily distributed according to the principle of chemical equilibrium. We have recently shown that for sodium dodecyl sulfate micellar solutions the size fluctuation as a result of the aggregation number distribution can be as much as 15% [7,43]. Under this condition, GOCM theory can be applied to calculate only the average intermicellar structure factor [7,43] up to a polydispersity of about 15%. In principle, to allow for micellar size distribution, one must use a multicomponent theory to calculate the partial structure factors. This approach, however, is feasible only in the case of multicomponent hard-sphere systems [44,45]. It is in principle also possible to calculate the partial structure factors in the multicomponent MSA [30]. A practical scheme for the computation has yet to be implemented.

Acknowledgments. The authors acknowledge a great deal of assistance from C. F. Wu in computational work related to GOCM theory, especially, with regard to the analyses of cytochrome-C and SDS data. For the theoretical formulation of multi-component coulomb fluid (MCCF), we owe much to Dr. L. Blum and Dr. D. Bratko. This research is supported by the National Science Foundation.

References

1. See a review by P.N. Pusey and J.A. Tough in "*Dynamic Light Scattering and Velocimetry*," ed. by R. Pecora, Academic Press, New York, 1982.
2. D.J. Cebula, J.W. Goodwin, G.C. Jeffrey, R.H. Ottewill, A. Parentich, and R.A. Richardson, *Faraday Discussion Chem. Soc.* **76**, 37 (1983).

3. See several review articles in "*Physics of Amphiphiles*: *Micelles, Vesicles and Microemulsions*," ed. by V. Degiorgio and M. Corti, North-Holland, Amsterdam (1985).
4. S.-H. Chen, *Ann. Rev. Phys. Chem.* **37**, 351 (1986).
5. S.-H. Chen and D. Bendedouch, Structure and interactions of protein in solution studied by small angle neutron scattering, in "*Methods in Enzymology*," ed. by C.H. Hirs and S.N. Timasheff, Academic Press, New York, 1986.
6. E.J. Verway and J.Th.G. Overbeek, "*Theory of The Stability of Lyophobic Colloids*," Elsevier Publ. Co. Inc., New York, 1948.
7. E.Y. Sheu and S.-H. Chen, *J. Phys. Chem.* **92**, 4466 (1988).
8. H.L. Friedman, "*A Course in Statistical Mechanics*," Prentice-Hall, Englewood Cliffs, New Jersey, 1985.
9. N.H. March and M.P. Tosi, "*Coulomb Liquids*," Academic Press, New York, 1984.
10. J. Schneider, W. Hess, and R. Klein, *J. Phys.* **A18**, 1221–1228 (1985).
11. C.F. Wu, E.Y. Sheu, D. Bendedouch, and S.-H. Chen, Studies of double layer interaction in micelle and protein solution by small angle neutron scattering, *Kinam*, Vol. 8, Series A, 37–61, Mexico, 1987.
12. J.A. Prins and H. Peterson, *Physica* **3**, 147 (1936).
13. L.W. Bahe, *J. Phys. Chem.* **76**, 1062 (1972), and unpublished results.
14. D. Bendedouch, S.-H. Chen, and W.C. Koehler, *J. Phys. Chem.* **87**, 2621 (1983). S.-H. Chen is grateful to Professor Lowell W. Bahe for communicating the unpublished results used.
15. E.Y. Sheu, S.-H. Chen, and J.S. Huang, *J. Phys. Chem.* **91**, 3306–3310 (1987).
16. E.Y. Sheu, C.F. Wu, and S.-H. Chen, *J. Phys. Chem.* **90**, 4179 (1986).
17. C.F. Wu and S.-H. Chen, *J. Chem. Phys.* **87**, 6199–6205 (1987).
18. J.P. Hansen and I.R. McDonald, "*Theory of Simple Liquids*," 2nd ed., Academic Press, New York, 1987.
19. E. Waisman, *Mol. Phys.* **25**, 45–48 (1973).
20. J.B. Hayter and J. Penfold, *Mol. Phys.* **42**, 109 (1981).
21. J.B. Hayter and J. Penfold, *J. Chem. Soc. Faraday Trans.* 1 **77**, 1851 (1981).
22. J.P. Hansen and J.B. Hayter, *Mol. Phys.* **46**, 651 (1982).
23. M.J. Gillan, *J. Phys.* **C7**, L1 (1974).
24. B. Svensson and B. Jonsson, *Mol. Phys.* **50**, 489 (1983).
25. B. Beresford-Smith and D.Y.C. Chan, *Chem. Phys. Lett.* **92**, 474 (1982).
26. E.Y. Sheu, "*Theoretical Development and Experimental Verification of a Primitive Model for the Inter-micellar Interactions*," Ph.D. thesis, M.I.T., Cambridge, Massachusetts, 1987.
27. L. Belloni, *J. Chem. Phys.* **5**, 519 (1986).
28. B. Beresford-Smith, D.Y.C. Chan, and D.J. Mitchell, *J. Coll. Int. Sci.* **105**, 216 (1985).
29. K. Hiroike, *Mol. Phys.* **33**, 1195 (1977).
30. G. Senatore and L. Blum, *J. Phys. Chem.* **89**, 2676 (1985).
31. M. Medina-Neyola and D.A. McQuarrie, *J. Chem. Phys.* **73**, 6279 (1980).
32. R.E. Dickson, T. Takano, E. Eisenbery, O.B. Kallar, L. Samson, A. Cooper, and E. Margoliash, *J. Bio. Chem.* **246**, 1511 (1971).
33. C.F. Wu and S.-H. Chen, *Biopolymers.* **27**, 1065 (1988).
34. S. Alexander, P.M. Chaikin, P. Grant, G.J. Morales, and P. Pincus, *J. Chem. Phys.* **80**, 5776–5781 (1984).
35. Y.S. Chao, E.Y. Sheu, and S.-H. Chen, *J. Phys. Chem.* **89**, 4395 (1985).

36. Y.S. Chao, E.Y. Sheu, and S.-H. Chen, *J. Phys. Chem.* **89**, 4862 (1985).
37. D.F. Evan, D.J. Mitchell, and B.W. Ninham, *J. Phys. Chem.* **88**, 6344 (1984).
38. F. Oozawa, "*Polyelectrolytes*," Marcel Dekker, New York, 1971.
39. K. Watanabe, M. Ferrario, and M. Klein, *J. Phys. Chem.* **92**, 819 (1988).
40. D. Bratko, E.Y. Sheu, and S.-H. Chen, *Phys. Rev. A* **35**, 4359 (1987).
41. E.Y. Sheu, C.F. Wu, S.-H. Chen, and L. Blum, *Physe. Rev. A* **32**, 3807 (1985).
42. S. Khan and D. Ronis, *Phys. Rev. A* **35**, 4295 (1987).
43. S.-H. Chen and E.Y. Sheu, *Makromol. Chem. Macromol. Symp.* **15**, 275–294 (1988); *J. Phys. Chem.* **92**, 4466 (1988).
44. P. van Beurten and A. Vrij, *J. Chem. Phys.* **74**, 2744 (1981).
45. W.L. Griffiths, R. Triolo, and A.L. Compere, *Phys. Rev. A* **35**, 2200 (1987).

2
Statistical Mechanics of Interacting Double Layers

M. Lozada-Cassou and D. Henderson

The force between interacting double layers is examined using two methods. In the first method, which is fairly conventional but useful in many applications, the colloidal particle (or in electrochemical applications, the electrode) is considered to be a giant particle, present in dilute concentration in a fluid mixture. The second method, similar to the first but much more original and potentially more powerful, treats a pair of particles as a single dumbbell, again present in dilute concentration in a fluid mixture. Triplet correlations are included in the second method. The two methods are identical for noninteracting double layers, but differ for interacting double layers. These methods are applied to noninteracting flat, spherical, and cylindrical double layers and to interacting spherical and flat double layers.

I. Introduction

In this chapter, we briefly survey two approaches based on liquid-state theory integral equations to calculate the electrostatic force due to the interaction of the double layers between colloidal particles in an electrolyte. We also include a discussion of noninteracting double layers.

The first method is quite conventional. The colloidal particles are regarded as giant particles, present in dilute concentration in a fluid mixture. Although conceptually very simple, this method has proven useful at least for noninteracting double layers. The second method is new and promising. A pair of particles is treated as a single dumbbell molecule, again present in dilute concentration, and the conventional liquid-state theory techniques are applied. By treating a pair of particles as a single dumbbell, three-body correlations are included even though only a pair calculation formalism is employed. For noninteracting double layers, the two methods are identical. This method is applied here solely to the problem of the interaction of the double layers of two charged plates in an electrolyte. However, the method is general and potentially very powerful. It can be applied to other different problems, such as the calculation of triplet correlation functions.

The model of the electrolyte employed here is the *primitive model* in which the solvent is represented by a dielectric continuum whose dielectric constant, ε, is a parameter and is equal to that of the actual electrolyte, and the ions and colloidal particles are represented as charged hard spheres. Obviously, it is preferable to represent the solvent as a fluid of discrete molecules. Although conceptually straightforward, a molecular model of the solvent greatly complicates the numerical solution of the equations. As a result, only a few calculations based on molecular models of the solvent have been made.

In the following sections, we will develop the two methods. We will apply the first (conventional) method to noninteracting flat, spherical, and cylindrical double layers and to the interaction of two spherical double layers. The article will conclude with an application of the second (new) method to the interaction of two flat double layers. In this article, some knowledge of liquid-state theory is presumed. Detailed treatments of liquid-state theory are available [1].

II. Conventional Method

The basis of most integral equation approaches to the double layer is the Ornstein–Zernike (OZ) relation

$$h_{ij}(r_{12}) = c_{ij}(r_{12}) + \sum_k \rho_k \int h_{ik}(r_{13})c_{kj}(r_{23})\,d\underline{r}_3$$

$$= c_{ij}(r_{12}) + \sum_k \rho_k h_{ik} * c_{kj}, \tag{1}$$

where $\underline{r}_i$ is the position of particle i, $r_{12} = |\underline{r}_1 - \underline{r}_2|$, $g_{ij}(r) = h_{ij}(r) + 1 = g_{ji}(r)$ is the *radial distribution function* for a pair of particles of species i and j, $c_{ij}(r) = c_{ji}(r)$ is the *direct correlation function* for a pair of particles of species i and j, $\rho_k = N_k/V$ is the number N_k of particles of species k in the volume V of the system, and the asterisk indicates a *convolution* [i.e., an integral of the type given in the first line of Eq. (1)].

Equation (1) is a definition of the direct correlation function. When coupled with some ansatz, it becomes an approximate integral equation for $h(r)$. One ansatz is the hypernetted chain approximation (HNCA),

$$h_{ij}(r) - c_{ij}(r) = \ln g_{ij}(r) + \beta u_{ij}(r), \tag{2}$$

where $u_{ij}(r)$ is the potential energy function for a pair of particles of species i and j, and $\beta = 1/kT$, k is the Boltzmann constant and T is the temperature.

Linearizing the right-hand side (RHS) of Eq. (2), $\ln g_{ij}(r) \simeq h_{ij}(r)$ yields a second ansatz, the mean spherical approximation (MSA).

Following Henderson et al. [2], a theory of interfacial systems can be developed by treating the surface (or colloid particle) as one of the spheres present in dilute concentration in a fluid mixture. Lozada-Cassou [3] has obtained these equations in a more direct manner by assuming one of the species to have the geometry of the external field, thus avoiding limiting processes that are often part of the approach of Henderson et al. Assuming the supporting electrolyte has two components, the positive and negative ions, and labeling the surface (or colloidal) particle species 3, the Ornstein–Zernike relations become, in the limit $\rho_3 \to 0$,

$$h_{11} = c_{11} + \rho_1 h_{11} * c_{11} + \rho_2 h_{12} * c_{21}, \tag{3a}$$

$$h_{12} = c_{12} + \rho_1 h_{11} * c_{12} + \rho_2 h_{12} * c_{22}, \tag{3b}$$

$$h_{22} = c_{22} + \rho_1 h_{21} * c_{12} + \rho_2 h_{22} * c_{22}, \tag{3c}$$

$$h_{31} = c_{31} + \rho_1 h_{31} * c_{11} + \rho_2 h_{32} * c_{21}, \tag{4a}$$

$$h_{32} = c_{32} + \rho_1 h_{31} * c_{12} + \rho_2 h_{32} * c_{22}, \tag{4b}$$

and

$$h_{33} = c_{33} + \rho_1 h_{31} * c_{13} + \rho_2 h_{32} * c_{23}. \tag{5}$$

Equations (3)–(5) are appropriate for a binary salt using the primitive model. If a molecular model of the solvent is employed, a fourth index and additional equations are required. On the other hand, in a colloidal system with only counterions in the supporting electrolyte, only one index is needed for the fluid and a second for the colloid, yielding three rather than six equations.

Equations (3) describe the electrolyte and can be solved without reference to the large spheres. They are just the pure electrolyte equations. Equations (4) describe the double layers around noninteracting large spheres. Equations (4) require c_{11}, c_{12}, and c_{22} as input. These need not come from the solution of Eqs. (3). They could come from computer simulations. Also, they could come from Eqs. (3) but using a different ansatz than that used to solve Eqs. (4). The functions h_{13} and h_{23} give the density profiles for the ions in the noninteracting double layers. From these can be calculated the charge and potential profiles of these noninteracting double layers.

Equation (5) describes the correlation between two interacting double layers. All that is required as input is h_{13}, h_{23}, c_{13}, and c_{23}. These need not come from Eqs. (4). They could come from computer simulation. Even if they come from Eqs. (4), the ansatz used in Eqs. (4) need not be the same as that used in Eq. (5). The potential energy of the interacting double layers is obtained from

$$w(r) = -kT \ln[1 + h_{33}(r)]. \tag{6}$$

The Poisson–Boltzmann (PB) theory is the standard theory of the double layer. The PB equations for the bulk electrolyte, and noninteracting double layer, can be obtained from the HNC versions of Eqs. (3) and (4), respectively, by setting $c_{ij}(r)$ equal to the coulomb expression $Q_i Q_j / \varepsilon r_{ij}$, where Q_i is the charge on spheres of species i. The PB theory should be most reliable at low concentrations where the noncoulombic terms in $c_{ij}(r)$ are least important because the coulomb interaction is poorly screened and very long ranged.

In principle, the PB equation for colloidal particles should result from Eq. (5) if some appropriate expression for the direct correlation function is used. However, the situation is more complex than for Eqs. (3) and (4) since, as is made clear in the next section, the HNC version of Eq. (5) does not yield the PB equation when only the coulomb term in $c_{33}(r)$ is retained. Some extension of the HNCA, including bridge diagrams, is evidently required. Until the set of diagrams needed to produce the PB equation is known, which is not presently the case, the connection between the PB theory for colloidal interactions and Eq. (5) is unclear.

III. New Method

One of us [4] has suggested an alternative method that is rather exciting and potentially very powerful. It is too early to know the accuracy of this method. However, this method reduces to the PB theory for colloids and, therefore, it warrants further study. The idea of this method is to regard some pair of particles as a single molecule, a dumbbell. The dumbbell could be a pair of ions, an ion and a large particle, or two large spheres (or cylinders or plates). Here, we will consider only the case of two flat plates.

Hence, taking the indices 1 and 2 for the ions and the index 3 for the dumbbell, we need only consider Eqs. (3) and (4). Equations (3) again describe the electrolyte. Equations (4) describe the ions in the presence of the dumbbell. Again, the c_{ij} in Eqs. (4) need not be obtained from Eqs. (3) using the same ansatz as that used in Eqs. (4) or, for that matter, even from Eqs. (4). The results for noninteracting double layers can be obtained from the new method in a number of ways, for example, by considering the pair of particles to be very far apart. The equations for noninteracting double layers are identical in the conventional and new methods.

IV. Noninteracting Double Layers

A. Planar Double Layer

Most calculations for noninteracting double layers have been made for planar double layers, where one solves Eqs. (4) in the limit of infinitely large spheres. However, as noted in the previous section, the same equations can be obtained directly without using this limit.

For example, if the MSA is used for both Eqs. (3) and (4), one can obtain the result [5]

$$g_{ij}(r) = g_{ij}^0(r) - \frac{\beta Q_i Q_j}{\varepsilon(1 + \Gamma a_i)(1 + \Gamma a_j)} \frac{f\left(r - \dfrac{a_i + a_j}{2}\right)}{r}, \tag{7}$$

where $g_{ij}^0(r)$ is the radial distribution function for an uncharged ij pair, Q_i and Q_j are the charges on the spheres, $a_1 = a_2 = a$ is the ion diameter, a_3 is the diameter of the large sphere (species 3 is infinitely dilute), and Γ is related to the Debye inverse screening parameter

$$\kappa = \left[\frac{4\pi\beta e^2}{\varepsilon} \sum_{i=1}^{2} \rho_i z_i^2\right]^{1/2}, \tag{8}$$

where e is the electronic charge and ρ_i and z_i are the number density and valence of the ions of spheres i by

$$\kappa = 2\Gamma(1 + \Gamma a). \tag{9}$$

Thus, $\Gamma = \kappa/2$ at low concentrations. The function $f(x)$ in Eq. (7) is complex. Analytical expressions are available [6]. However, for low concentrations,

$$f(x) \simeq e^{-\kappa x}. \tag{10}$$

Thus, for i and j equal to 1 or 2,

$$g_{ij}(r) = g^0(r) - \frac{\beta z_i z_j e^2}{\varepsilon(1 + \Gamma a)^2} \frac{f(r - a)}{r}, \tag{11}$$

where z_i includes the sign, as well as the magnitude, of the charge on an ion of species i, and $g^0(r)$ is the hard-sphere radial distribution function.

For the ion–large sphere correlation function,

$$g_{3i}(r) = g_{3i}^0(r) - \frac{\beta z_i e Q_3}{\varepsilon(1 + \Gamma a)(1 + \Gamma a_3)} \frac{f\left(r - \dfrac{a + a_3}{2}\right)}{r}, \tag{12}$$

where $i = 1, 2$. Setting

$$r = \frac{a_3}{2} + x \tag{13}$$

and

$$Q_3 = \frac{E a_3^2}{4}, \tag{14}$$

that is, E is the electric field at the surface of the sphere, we have, on taking the limit $a_3 \to \infty$,

$$g_i(x) = g_i^0(x) - \frac{\beta z_i eE}{\varepsilon \kappa} f(x - \sigma/2),$$ (15)

where $i = 1, 2$, and $g_i^0(x)$ is the profile for hard spheres near a hard wall [6]. The index 3 has been dropped for simplicity. From Eq. (15), the potential difference across the double layer is

$$V = \frac{E}{\varepsilon(2\Gamma)},$$ (16)

and the differential capacitance of the double layer is

$$C = \frac{\varepsilon}{4\pi}(2\Gamma).$$ (17)

The linearized PB results,

$$V = \frac{E}{\varepsilon \kappa}$$ (18)

and

$$C = \frac{\varepsilon \kappa}{4\pi}$$ (19)

are obtained by setting $a = 0$ (i.e., neglecting the noncoulombic interactions among the ions).

Note that the MSA is a weak coupling theory, valid for large T or ε or small z_i or E, and so at best, it can describe only linear effects in the double layer. However, nonlinear effects can be described by the HNCA. In contrast to the MSA, the HNCA does not yield analytical results. One must solve numerically Eq. (5), which in the limit of large a_3 becomes for the HNCA

$$\ln g_i(x) = -\beta z_i e\psi(x) + 2\pi \sum_j \rho_i \int_{-\infty}^{\infty} h_j(y)\mathscr{C}_{ij}(x, y)\, dt,$$ (20)

where $\psi(x)$ is the electrostatic potential, and $\mathscr{C}_{ij}(x, y)$ is an integral over the electrolyte direct correlation function and depends only on the approximation used in the theory of the bulk electrolyte. The simplest version of the HNCA is that obtained by using the MSA $\mathscr{C}_{ij}(x, y)$, the so called HNC/MSA approach. The HNC/MSA works well if a is not too small. However, if a is very small, it seems better [7] to use an HNC/HNCA approach using the HNCA for both Eqs. (3) and (4), that is, using the HNCA $\mathscr{C}_{ij}(x, y)$ in Eq. (20).

Some HNC/MSA results [8] for the diffuse layer potential, that is, the potential at the distance of closest approach,

$$\psi(a/2) = V - \frac{Ea}{2\varepsilon}, \tag{21}$$

are compared with computer simulation results [9] in Fig. 2.1. The agreement is very good. The nonlinear PB results, obtained by setting $\mathscr{C}_{ij}(x, y) = 0$ in Eq. (20), are much larger in magnitude, especially when the counterions are divalent. Note that the PB approach predicts a monotonic dependence of $\psi(a/2)$ on charge, which is incorrect.

Results similar to those of the HNCA have been obtained from the modified Poisson–Boltzmann [10] and Born–Green [11] approximations. There has also been some interest in improvements to the HNCA for the double layer near a flat electrode.

We close this section with a brief account of recent work involving the molecular nature of the solvent and the contribution of the electrons in the metal electrode. Using the MSA, Carnie and Chan [12] and Blum and Henderson [13] have calculated the potential difference across the double

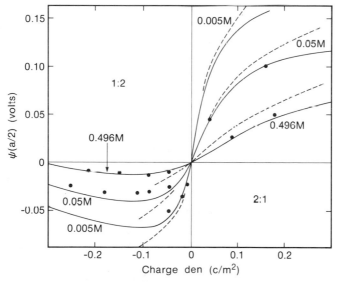

FIGURE 2.1. Electrostatic potential at the distance of closest approach, $x = a/2$, for an asymmetric electrolyte ($z_1 = 1$, $z_2 = 2$, $a = 4.25$ Å, $T = 298$ K, $\varepsilon = 78.5$) near a charged electrode as a function of the charge density on the electrode for three concentrations. The points give the computer simulation results (Ref. 9) and the solid and broken curves give the HNC/MSA and nonlinear PB results, respectively. The arrow indicates the point ($L = a$) at which p changes discontinuously because all the ions have been squeezed out. The value of the broken curve at $L = a$ is slightly less than that of the solid curve.

layer when the solvent is represented by hard spheres with point dipoles. They obtained a general expression for V (within the MSA). However, useful explicit results were obtained only for low concentrations. They found that

$$V = \frac{E}{\varepsilon\kappa} + \frac{Ea_i}{2\varepsilon} + \frac{\varepsilon - 1}{\varepsilon}\frac{Ea_s}{2\lambda}, \tag{22}$$

where a_i and a_s are the hard-sphere diameters of the ions and solvent molecules, respectively, and λ is related to ε by

$$16\varepsilon = \lambda^2(\lambda + 1)^4. \tag{23}$$

The first two terms can be obtained from Eq. (16) by expansion of Γ in powers of κ. The third term is the new term, resulting from the molecular nature of the solvent molecules. This new third term is important and, except at very low concentrations, is the dominant term. The second term, involving a_i, is relatively unimportant, which accounts for the experimental observation that the size of the ions is not a significant parameter in determining double-layer properties.

Badiali et al. [14] and Schmickler and Henderson [15] have considered the contribution of the metal electrons to the potential difference. This contribution arises because the conduction electrons penetrate slightly into the double layer. This spill over of electrons has only recently been postulated. Previously, the dependence of V and C on the nature of the electrode was assumed to be due to some unspecified chemical interaction between the metal and electrolyte. Using the MSA to describe electrolyte, Schmickler and Henderson obtained the result

$$V = \frac{E}{\varepsilon\kappa} + \frac{Ea_i}{2\varepsilon} + \frac{\varepsilon - 1}{\varepsilon}\frac{Ea_s}{2\lambda} + \frac{4\pi ne}{\alpha^2}, \tag{24}$$

when n is the electronic density of the metal, and α is a parameter, determined by minimizing the surface energy, describing the extent of the spill over. The reciprocal of α is a measure of the spill over, so that $\alpha = \infty$ means that the electrons do not penetrate at all into the solution.

Using Eq. (24), Schmickler and Henderson have calculated the capacitance of a planar double layer at small electrode charge and have obtained excellent agreement with experiment for a wide variety of metals and solvents. Results for aqueous systems are displayed in Fig. 2.2. Equation (24) explains simply the dependence of the capacitance on the metal. The more conduction electrons in the metal (i.e., the greater n), the greater the capacitance.

The main difficulty with Eqs. (22) and (24) is that, because of the MSA, they are useful only for small charge. Including solvent and metal effects in some nonlinear theory is clearly required.

B. Spherical and Cylindrical Double Layers

MSA results for a spherical double layer can be obtained from Eq. (12). Results for the HNCA, or some other approximation, could be obtained from Eq. (4). However, this has not yet been done.

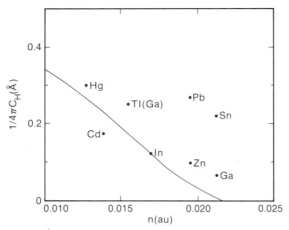

FIGURE 2.2. Inverse differential capacitance at the point of zero charge for sp metals immersed in aqueous electrolytes at high concentration as a function of the density of free electrons in the metals.

HNCA/MSA results have been obtained for cylindrical double layers by Lozada-Cassou et al. [16]. These calculations are of biological interest since a cylinder is a crude but plausible model of DNA. Some of their results are plotted in Fig. 2.3. The results are similar to the planar HNCA results. The potential at the distance of closest approach, often called the zeta potential in electrophoresis experiments, can be nonmonotonic. The nonlinear PB approximation seems a little more accurate for small-cylinder radii than for large-cylinder radii. In addition, the HNCA curves for different cylindrical radii can cross each other. For the concentrations shown, cylinders whose radii exceed 100 Å can be approximated as flat plates. At very low concentrations, it is probably necessary to go to a larger radius before the cylinder can be approximated as a flat plate since the range of the effective interaction, κ^{-1}, is greater at low concentrations. It is interesting to note that maxima in the zeta potential (i.e., the potential at the distance of closest approach) as functions of charge density have been seen experimentally by Carroll and Haydon [17] for large spherical molecules.

The potential at the distance of closest approach at the fixed charge density or the total potential difference is a monotonic function of the inverse cylinder radius, R^{-1}, in the PB theory. This need not be the case in the HNC/MSA. The potential at the distance of closest approach at both the fixed charge density and fixed total potential difference tends to zero as R^{-1} tends to zero. However, the approach is much faster at the fixed charge density.

There has been speculation that some form of charge condensation occurs when the linear charge density nears unity. The HNC/MSA calculations give no support to the conjecture.

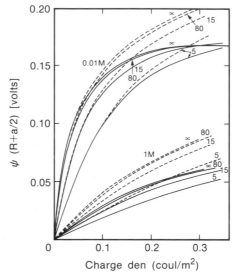

FIGURE 2.3. Electrostatic potential at the distance of closest approach, $r = R + a/2$, for a 1:1 electrolyte ($a = 4.25$ Å, $T = 298$ K, $\varepsilon = 78.5$) near a charged cylinder as a function of the charge density on the cylinder for two concentrations. The solid and broken curves give the HNCA/MSA and nonlinear PB results, respectively. The number by each curve gives the value of the cylinder radius, R, used.

V. Interacting Double Layers

There have been a few recent exact studies that should be useful in testing approximate theories. Blum et al. [18] have studied the interaction of two charged plates in a two-dimensional, one-component plasma. Jönsson et al. [19] have made computer simulation of the force between two charged plates in an electrolyte. This latter study is interesting because they find that for strong coupling (high charge, multivalent ions, low ε or T) the force between like, charged colloidal particles can be attractive, a result of considerable potential interest in colloid chemistry.

A. Conventional Method

Medina-Noyola and McQuarrie [20] have solved Eqs. (3) to (5) for the interaction of charge spheres in an electrolyte using the MSA. Their work is really a rederivation of Blum's results since Blum [5] has solved the MSA for charged hard spheres for an arbitrary number of spheres of any charge, size, and concentration.

Using Eq. (7), the colloid particle–colloid particle radial distribution function is (with $c = 3$)

$$g_{cc}(r) = g_{cc}^0(r) - \frac{\beta Q_c^2}{\varepsilon(1 + \Gamma a_c)^2} \frac{f(r - a_c)}{r} \tag{25}$$

if the colloid particles are present in vanishing concentration and where Γ is given by Eq. (8). In general, the interaction between the colloid particles is given by the logarithm of $g_{cc}(r)$. However, because of the linearization inherent in the MSA, the colloidal interaction is $kT[1 - g_{cc}(r)]$. Thus,

$$w_{cc}(r) = kT[1 - g_{cc}^0(r)] + \frac{Q_c^2}{\varepsilon(1 + \Gamma a_c)^2} \frac{f(r - a_c)}{r}. \tag{26}$$

For large spheres, this becomes

$$\frac{w_{cc}(r)}{a_c} = \frac{kT}{a_c}[1 - g_{cc}^0(r)] + \frac{\pi^2 \sigma_c^2}{\varepsilon \Gamma^2} f(r - a_c), \tag{27}$$

where σ_c is the charge per unit area of the colloidal particles. Equation (27) illustrates the fact that for large spheres $w_{cc}(r)$ becomes proportional to the diameter (or radius) in accord with the Derjaguin approximation,

$$w_{cc}(r) = \frac{\pi a_c}{2} p, \tag{28}$$

where p is the pressure (force per unit area) on a parallel set of charged flat plates.

At low concentrations, where $\Gamma \to \kappa/2$ and $g_{cc}^0 \to 1$,

$$\frac{w_{cc}(r)}{a_c} = \frac{4\pi^2 \sigma_c^2}{\varepsilon \kappa^2} e^{-\kappa(r - a_c)}. \tag{29}$$

Equation (29) is the linearized PB result for two *weakly* interacting double layers. The full *linearized* PB result for the interaction potential of two charged spheres is

$$\frac{w_{cc}(r)}{a_c} = -\frac{4\pi^2 \sigma_c^2}{\varepsilon \kappa^2} \ln[1 - e^{-\kappa(r - a_c)}] \tag{30}$$

if the two colloidal particles are kept at constant charge σ_c, or

$$\frac{\omega_{cc}(r)}{a_c} = \frac{1}{4} \varepsilon \psi_c \ln[1 + e^{-\kappa(r - a_c)}], \tag{31}$$

if the two colloidal particles are kept at constant potential ψ_c. Equation (29) is the large separation limit of Eqs. (30) and (31) since there is no distinction between constant charge and constant potential for weakly interacting spheres.

The fact that, at low concentrations, where noncoulombic interactions among the ions can be neglected, Eq. (27) yields only the linearized PB result for weakly interacting spheres, and not the full linearized PB result, presumably means that Eq. (27) is valid only for weakly interacting spheres. This is not overly surprising since, because $\rho_c = 0$, the convolutions in Eq. (5) involve only the h's and c's for noninteracting spheres.

Patey [21] has solved Eqs. (3) to (5) using the hypernetted chain approximation for all three sets of equations. In all of Patey's calculations, the charge on the colloid spheres, rather than the potential, was held constant. Some of Patey's results for $w_{cc}(r)$ are shown in Fig. 2.4. At high charge density and small separations, the HNCA *electrostatic* interaction between like, charged spheres can be attractive. Potentially, this is a very important result since the standard Derjaguin–Landau–Verwey–Overbeek (DLVO) theory is based on a balance of an attractive van der Waals interaction and a repulsive electrostatic (PB) interaction. Should there be an attrative component to the electrostatic interaction, the conventional ideas about colloid stability would have to be reexamined.

Teubner [22] criticized the calculations of Patey stating that the HNCA was correct only at large separations. In view of our comments about the

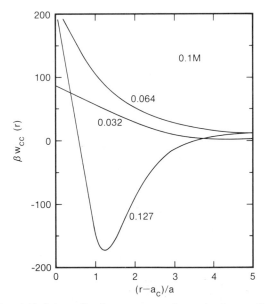

FIGURE 2.4. Electrostatic interaction between two charged spheres of radius $200a$ in an $0.1\ M$ electrolyte ($z = 1$, $a = 4.25$ Å, $T = 298$ K, $\varepsilon = 78.5$) as a function of the separation of the spheres as calculated by Patey (Ref. 20) from the HNCA. The number by each curve gives the value of $\sigma a^2/e$, where σ is the charge density, and e is the electronic charge.

MSA expression for w_{cc}, we believe that Teubner is correct. Both the MSA and HNCA are correct only at large separations. The PB theory for overlapping double layers is not a limiting case of either of these approximations. The problem does not lie with the OZ relations nor with the limit $\rho_c = 0$, since Eqs. (3) to (5) are really only definitions of the $c_{ij}(r)$. If an exact result for the $c_{ij}(r)$ were employed in Eqs. (3) to (5), exact results would be obtained. The problem lies with the MSA and HNCA which both only involve repeated convolutions of noninteracting double-layer h's and c's. Some more complex approximation involving so-called bridge diagrams would be more satisfactory.

In any case, whatever the defects of the HNCA, an attractive electrostatic force between like, charged spheres seems real since it has been seen in the simulations of Jönsson et al. [19]. It is reasonable to assume that the HNCA calculations of Patey exaggerate the tendency toward an attractive interaction at small separation but that the effect is real. Whether more accurate results than those of Patey can be obtained without great labor from Eqs. (3) to (5) is doubtful because the calculation of bridge diagrams is a notoriously difficult computational problem.

Presuming, as we do, that the attractive interaction for small separations and strong couplings is real, a reexamination of the DLVO theory is in order. Clearly, much needs to be done. More simulations and further HNCA calculations are required. In addition, the formulation of more elaborate theories is required.

We close this section by noting that a system of charged hard spheres in which the large spheres are present in nonvanishing concentration has been considered by Beresford-Smith et al. [23]. If Q_c and ρ_c are, respectively, the charge and density of the colloidal spheres, and the m components of the electrolyte have charge and density $z_i e$ and ρ_i, then electroneutrality requires that

$$Q_c \rho_c + e \sum_{i=1}^{m} z_i \rho_i = 0. \tag{32}$$

The OZ relations are

$$h_{ij} = c_{ij} + \rho_c h_i * c_{cj} + \sum_{k=1}^{m} \rho_k h_{ik} * c_{kj}, \tag{33}$$

where we have separated out the convolution involving h_{ic} and c_{cj} to emphasize that $\rho_c \neq 0$, which differs from the case in Eqs. (3) to (5). By considering the long-range behavior of the $c_{ij}(r)$, Beresford-Smith et al. show that to a good approximation $h_{cc}(r)$ can be calculated from the single equation

$$h_{cc} = c_{cc}^{\text{eff}} + \rho_c h_{cc} * c_{cc}^{\text{eff}}, \tag{34}$$

where c_{cc}^{eff} is defined in terms of the *effective* colloid–colloid pair potential.

$$U_{cc}^{eff}(r) = -\frac{Q_c^2}{\varepsilon r} e^{-\kappa_{eff}(r)}, \tag{35}$$

where

$$\kappa_{eff}^2 = \frac{4\pi\beta e^2}{\varepsilon} \sum_{i=1}^m z_i^2 \rho_i, \tag{36}$$

that is, only the electrolyte z_i and ρ_i appear explicitly in the definition of the effective screening parameter, κ_c^{eff}. Note, however, that κ_c^{eff} is a function of ρ_c because of Eq. (34).

The advantage of this approach is that only one equation, Eq. (34), need be solved to obtain h_{cc}. All of the other equations in the OZ relations have been included in an average way by means of the effective pair potential, Eq. (35). We note that

$$g_{cc}(r) \simeq \frac{e^{-\kappa r}}{r}, \tag{37}$$

where

$$\kappa^2 = \kappa_{eff}^2 + \frac{4\pi\beta}{\varepsilon} Q_c^2 \rho_c, \tag{38}$$

as we would expect. As $\rho_c \to 0$, $\kappa_{eff} \to \kappa$, as would be expected.

B. New Method

We have already mentioned that the conventional method (Section V.A) seems appropriate only for double layers at large separation. Lozada-Cassou [4] has attempted to overcome this problem by assuming the two colloidal particles form a single dumbbell molecule. Thus, one need only solve Eqs. (4) with species 3 being the dumbbell formed by the two colloidal particles, which are present in vanishing concentration.

The method is quite general and can be applied to systems of any geometry. However, numerical results are available only for interacting planar double layers. In this article, we confine ourselves to this case. The geometry is displayed in Fig. 2.5.

Using the HNCA, Eqs. (4) are

$$\ln g_{\alpha i}(r_{12}) + \beta u_{\alpha i}(r_{12}) = \sum_j \rho_j \int h_{\alpha j}(r_{23}) c_{ij}(r_{13}) \, dr_3, \tag{39}$$

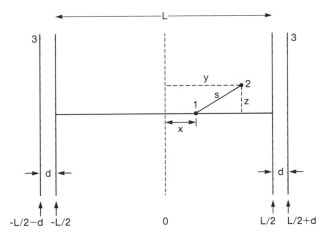

FIGURE 2.5. Geometry for two interacting charged plates.

where $\alpha = 3$ is the dumbbell composed of two plates. Following Lozada-Cassou and using cylindrical coordinates,

$$\ln g_{\alpha i}(x) + \beta u_{\alpha i}(x) = 2\pi \sum_j \rho_i \int_{-\infty}^{\infty} h_{\alpha j}(y)\,dy \int_0^{\infty} c_{ij}(s)z\,dz. \qquad (40)$$

We can write

$$c_{ij}(s) = -\frac{\beta z_i z_j e^2}{\varepsilon s} + \Delta c_{ij}^{sr}(s), \qquad (41)$$

where $\Delta c_{ij}^{sr}(s)$ is the noncoulombic part of $c_{ij}(s)$. The coulombic terms can be combined with $u_{\alpha i}(x)$ to give

$$\ln g_{\alpha i}(x) + \beta z_i e \psi(x) = 2\pi \sum_j \rho_i \left\{ \int_{-\infty}^{0} h_{\alpha j}(y)\,dy \int_{x+|y|}^{\infty} s \Delta c_{ij}^{sr}(s)\,ds \right.$$

$$\left. + \int_0^{\infty} h_{\alpha j}(y)\,dy \int_{|x-y|}^{\infty} s\,\Delta c_{ij}^{sr}(s)\,ds \right\}, \qquad (42)$$

where we have made use of the fact that $z\,dz = s\,ds$ and where $\psi(x)$ is the electrostatic potential

$$\psi(x) = \psi(L/2) + \frac{4\pi e}{\varepsilon}\left(\frac{L}{2} - x\right)\sum_j z_j \rho_j \int_0^{(L-a)/2} h_{\alpha j}(y)\,dy$$

$$+ \frac{4\pi e}{\varepsilon}\sum_j z_j \rho_j \int_x^{(L-a)/2} (x-y)h_{\alpha j}(y)\,dy, \qquad |x| \leq L/2, \quad (43a)$$

$$\psi(x) = \frac{4\pi e}{\varepsilon}\sum_j z_j \rho_j \int_x^{\infty} (x-y)h_{\alpha j}(y)\,dy, \qquad |x| \geq L/2 + d, \quad (43b)$$

where it is understood that the second integral in Eq. (43a) vanishes if $x \geq (L - 2)/2$.

Electroneutrality requires that

$$\sigma\left(-\frac{L}{2} - d\right) + \sigma\left(-\frac{L}{2}\right) + \sigma\left(\frac{L}{2}\right) + \sigma\left(\frac{L}{2} + d\right) = -e \sum_j z_j \rho_j \int_{-\infty}^{\infty} h_{\alpha j}(y)\, dy,$$

(44)

where the distance in the parentheses means that the charge density σ on the surface of the plate at that distance is to be used.

If we assume that the plates are symmetrically charged, we may make use of the symmetry about $y = 0$ and obtain

$$\ln g_{\alpha i}(x) + \beta z_i e \psi(x)$$

$$= 2\pi \sum_j \rho_j \int_0^{\infty} h_{\alpha j}(y)\, dy \left\{ \int_{x+y}^{\infty} s\, \Delta c_{ij}^{sr}(s)\, ds + \int_{|x-y|}^{\infty} s\, \Delta c_{ij}^{sr}(s)\, ds \right\}.$$

(45)

There are two cases to consider, $x \leq L/2$ and $x \geq (L/2) + d$. For $x \leq L/2$, Eq. (45) applies. If the two plates are far apart, then for $x \leq L/2$,

$$\ln g_{\alpha i}(x) + \beta z_i e \psi(x) = 2\pi \sum_j \rho_j \int_0^{\infty} h_{\alpha j}(y)\, dy \int_{|x-y|}^{\infty} s\, \Delta c_{ij}^{sr}(s)\, ds,$$

(46)

which, apart form a change of variables, is the HNCA for noninteracting double layers. For $x \geq (L/2) + d$, Eq. (46) applies at all separations.

In general, one must solve Eqs. (45) and (46) simultaneously and charge neutrality is satisfied only over all space, that is, Eq. (44). However, if the potential on both sides of the plate are the same, that is, $\psi(L/2) = \psi(L/2) + d$, the two regions, $|x| \leq L/2$ and $|x| \geq (L/2) + d$, decouple and Eqs. (45) and (46) can be solved separately. Also, for this case, the thickness of the plates is irrelevant. Furthermore, charge neutrality is satisfied separately in each of these regions if $\psi(L/2) = \psi(L/2) + d$.

If the two plates have equal potentials on both sides at infinite separation and both sides are kept at constant potential as the separation is varied, then automatically $\psi(L/2) = \psi(L/2) + d$. This situation is referred to as *constant potential*. As the plates are brought together at constant potential, the charge density on the inside of the plates tends to zero whereas that on the outside is constant. The charge density on the inside and outside of the plates need not be equal.

If the two plates are kept at constant charge density as the separation is varied, $\psi(L/2) = \psi(L/2) + d$ need not be satisfied. However, one could demand that the charge density on the inner side of the plates, $|x| = L/2$, be held constant as the separation is varied and that the charge density on the outer side of plates, $|x| = (L/2) + d$, vary so that $\psi(L/2) = \psi(L/2) + d$. Again, the charge density on the inside and outside surfaces of the plates need not be equal. This is the situation that is generally referred to as *constant charge*.

In the DLVO theory, in which the electrostatic force is obtained using the assumption $\Delta c_{ij}^{sr}(r) = 0$, the thickness and the outside of the plates are, in general, neglected. This is permissible only if $\psi(L/2) = \psi(L/2) + d$. Prior to Lozada-Cassou's work, only in the papers of Oshima [24] were the thickness and the outside of the plates taken into account. Thus, Eq. (45) is more general than the DLVO theory both because $\Delta c_{ij}^{sr}(r)$ can be nonzero and because the thickness and the outside of the plates are included, giving the possibilities of $\psi(L/2)$ and $\psi(L/2) + d$ being different and of charge transfer from one side of the plates to the other.

It is to be noted that in the new method the polarizability of the colloid particles is taken into account in a natural way, whereas it is more difficult to account for polarizability in the conventional method.

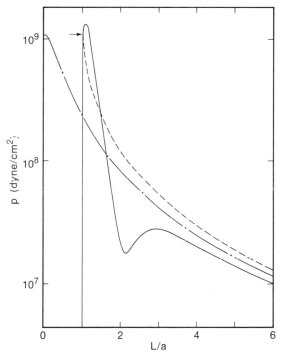

FIGURE 2.6. Force per unit area between two interacting charged plates held at a constant potential of 0.215 V immersed in a 0.01 M electrolyte ($z = 1$, $a = 4.25$ Å, $T = 298$ K, $\varepsilon = 78.5$) as a function of separation. The solid, broken, and dot–dash curves give the HNC/MSA results obtained using the approach of Lozada-Cassou, the PB results using an ionic distance of the closest approach of $a/2$, and the PB results using an ionic distance of the closest approach of zero, respectively. The arrow indicates the point ($L = a$) at which p changes discontinuously because all the ions have been squeezed out. The value of the broken curve at $L = a$ is slightly less than that of the solid curve.

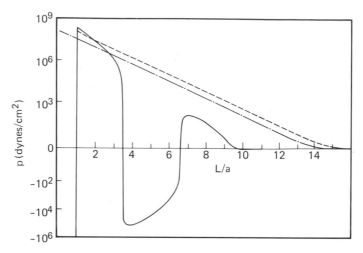

FIGURE 2.7. Force per unit area between two interacting charged plates held at a constant potential of 0.05 V in a 1 M electrolyte ($z = 1$, $a = 4.25$ Å, $T = 298$ K, $\varepsilon = 78.5$) as a function of separation. The solid, broken, and dot–dash curves have the same meaning as in Fig. 2.6.

The net pressure between the two plates is obtained by taking the difference between the pressure in the region between the plates and the pressure outside. Thus,

$$p(L) = -\frac{2\pi e^2}{\varepsilon}\left\{\left[\sum_j z_j \rho_j \int_0^{L/2} g_{\alpha j}(y)\, dy\right]^2 - \left[\sum_j z_j \rho_j \int_{L/2+d}^{\infty} g_{\alpha j}(y)\, dy\right]^2\right\}$$
$$+ kT\sum_j \rho_j\left[g_{\alpha j}\left(\frac{L-a}{2}\right) - g_{\alpha j}\left(\frac{L+a}{2} + d\right)\right], \qquad (47)$$

where for simplicity we have assumed symmetry about $y = 0$.

As was the case for the noninteracting double layers, there is a choice as to which results for Δc_{ij}^{sr} are used. The DLVO theory results from the assumption $\Delta c_{ij}^{sr} = 0$. We have made calculations using MSA results for Δc_{ij}^{sr}, which we refer to as the HNC/MSA approach. Some of our results for the HNC/MSA net pressure between the plates are displayed in Figs. 2.6 and 2.7. The HNC/MSA pressure is non-monotonic, which contrasts with the monotonic behavior of the PB theory. Two versions of the PB theory are plotted in Figs. 2.6 and 2.7: one in which the ions are point ions not only in their interactions with each other, but also in their interaction with the plates, and one in which the ions are point ions in their interactions with each other, but have a nonzero distance of closest approach to the plates. The force in the second version is more repulsive than in the first version, as expected.

The HNC/MSA pressure exhibits oscillations and is nonmonotonic. At higher concentrations, regions of attractive electrostatic forces between the plates do appear. At a concentration of 0.01 M, the pressure at a has a negative slope.

The charge density of the plates (held at constant potential) is plotted as a function of separation in Fig. 2.8. In constrast to the PB theory, the charge density is nonmonotonic, exhibiting a maximum near a at low concentrations and, although not very clear in the figure, a maximum near $3a$ at high concentrations. Clearly, these maxima are related to the minima in the pressure seen in Figs. 2.6 and 2.7.

Grimson et al. [25] have obtained results for the pressure using a different formalism. Their procedure is to write the free-energy density as a function of the density profiles using plausibility arguments. The actual profiles are found by minimizing the free-energy density. They consider only the region between the plates and ignore the outside. In particular, they assume that there is charge neutrality between the plates. We have already pointed out that this is equivalent to assuming that $\psi(L/2) = \psi(L/2) + d$. Their results are nonmonotonic and have some similarity to those shown in Figs. 2.6 and 2.7. However, for some cases, they find a minimum in the pressure at $L = 3a$ (their $h = L - a$), whereas we find a maximum at this distance. Their minimum at $L = 2a$ agrees with our results. The origin of the differences between their result and ours is puzzling. It is difficult to understand a minimum at $3a$.

Recently, Augousti and Rickayzen [26] have considered the problem of two plates immersed in a fluid containing ions and dipoles.

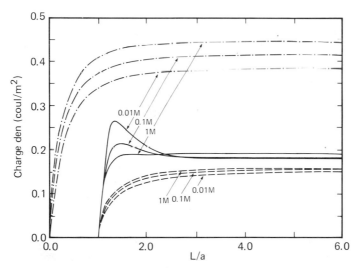

FIGURE 2.8. Charge density on two interacting charged plates held at a constant potential and immersed in an electrolyte ($z = 1$, $a = 4.25$ Å, $T = 298$ K, $\varepsilon = 78.5$) as a function of separation for three concentrations. The solid, broken, and dot–dash curves have the same meaning as in Fig. 2.6. The potentials used for 0.01 M, 0.1 M, and 1 M are 0.215 V, 0.160 V, and 0.105 V, respectively.

VI. Summary

We have outlined two approaches based on integral equations for describing interacting double layers. When the double layers are far apart and noninteracting, the two methods are identical. This new method can be regarded as an extension of the PB theory since this theory can be obtained as a special case. In contrast, the conventional method is, in a sense, less general than the PB theory since it yields the PB theory as a special case only at large separation. Both methods predict that the electrostatic interaction between colloidal particles can be attractive. Recent simulation studies [19] show that this attraction is real. There is some experimental evidence of attractive electrostatic forces [27].

In as much as the standard DLVO theory of collodial stability is based on an assumed balance between a repulsive electrostatic interaction (calculated using the PB theory) and an attractive van der Waals interaction, the possibility of attractive electrostatic interactions has potentially far-reaching implications in colloidal chemistry.

The pressure calculated from the new method shows much more structure than that calculated from the conventional method even when ion size is included. The pressure based on the new method shows oscillations, adhesive and nonadhesive minimums in the electrostatic pressure, and a negative slope in the electrostatic pressure at small separations, without the existence of van der Waals forces and discrete solvent molecules.

Many of these features have been seen in the recent experimental measurements at the Australian National University [28] of the forces between crossed cylinders immersed in fluids. It is uncertain whether the calculations reported here have any relation to these measurements since these calculations are purely electrostatic, whereas van der Waals and solvent molecule interactions are present in the real system. However, further study is warranted.

Computer simulations of the pressure between plates and spheres in a primitive electrolyte would be useful in determining the accuracy of the calculations outlined here.

Acknowledgments. This work arose out of discussions at the CECAM workshop on the electric double in Bellvue, France, in April 1983. The support of Dr. C. Moser and CECAM is gratefully acknowledged. M.L.C. is grateful to Dr. A. Haymet and the Department of Chemistry, University of California, Berkeley, for their hospitality during his stay in Berkeley in 1985. This work was supported in part by the Universidad Autonoma Metropolitana and CONACYT, Mexico.

References

1. See, for example, Barker J.A., and Henderson, D., *Rev. Mod. Phys.* **48**, 588 (1976); Hensen, J.P., and McDonald, I.R., *"Theory of Simple Fluids,"* Academic Press, New York, 1976.
2. Henderson, D., Abraham, F.F., and Barker, J.A., *Mol. Phys.* **31**, 1291 (1976).
3. Lozada-Cassou, M., *J. Chem. Phys.* **75**, 1412 (1981); **77**, 5258 (1982).
4. Lozada-Cassou, M., *J. Chem. Phys.* **80**, 3344 (1984).
5. Blum, L., *Mol. Phys.* **30**, 1529 (1975); Blum., L., and Høye, J., *J. Phys. Chem.* **81**, 1311 (1977).
6. Henderson, D., and Smith, W.R., *J. Stat. Phys.* **19**, 191 (1978).
7. Carnie, S.L., Torrie, G.M., and Valleau, J.P., *Mol. Phys.* **53**, 252 (1984).
8. Lozada-Cassou, M., Saavedra-Barrea, R., and Henderson, D., *J. Chem. Phys.* **77**, 5150 (1982); Lozada-Cassou, M., and Henderson, D., *J. Phys. Chem.* **87**, 2821 (1983); Borojas, J., Henderson, D., and Lozada-Cassou, M., *J. Phys. Chem.*, **87**, 4547 (1983).
9. Torrie, G.M., and Valleau, J.P., *J. Chem. Phys.* **73**, 5807 (1980); *J. Phys. Chem.* **86**, 3251 (1982).
10. Levine, S., Outhwaite, C.W., and Bhuiyan, L.B., *J. Electroanal. Chem.* **123**, 105 (1981).
11. Croxton, T.L., and McQuarrie, D.A., *Mol. Phys.* **42**, 141 (1981).
12. Carnie, S.L., and Chan, D.Y.C., *J. Chem. Phys.* **73**, 2949 (1980).
13. Blum, L., and Henderson, D., *J. Chem. Phys.* **74**, 1902 (1981).
14. Badiali, J.P., Rosenberg, M.L., Vericat, F., and Blum, L., *J. Electroanal. Chem.* **158**, 253 (1983).
15. Schmickler, W., and Henderson, D., *J. Chem. Phys.* **80**, 3381 (1984).
16. Lozada-Cassou, M., *J. Phys. Chem.* **87**, 3729 (1983); Gonzales-Tovar, E., Lozada-Cassou, M., and Henderson, D., *J. Chem. Phys.*, **83**, 361 (1985).
17. Carroll, B.J., and Haydon, D.A., *J. Chem. Soc. Faraday Trans I* **71**, 361 (1975).
18. Blum, L., and Jancovici, B., *J. Phys. Chem.* **88**, 2294 (1984); Blum, L., *J. Chem. Phys.* **80**, 2953 (1984).
19. Svensson, B., and Jönsson, B., *Chem. Phys. Letters* **108**, 580 (1984); Guldbrand, L., Jönsson, B., Wennerström, H., and Linse, P., *J. Chem. Phys.* **80**, 2221 (1984).
20. Medina-Noyola, M., and McQuarrie, D.A., *J. Chem. Phys.* **73**, 6279 (1980).
21. Patey, G.N., *J. Chem. Phys.* **72**, 5673 (1980).
22. Teubner, M., *J. Chem. Phys.* **75**, 1907 (1981).
23. Beresford-Smith, B., and Chan, D.Y.C., *Chem. Phys. Letters* **92**, 474 (1982); Beresford-Smith, B., Chan, D.Y.C., and Mitchell, D.J., *J. Colloid Interface Sci.* **105**, 216 (1985).
24. Oshima, H., *Colloid and Polymer Sci.* **252**, 158 (1974).
25. Grimson, M.J., and Rickayzen, G., *Mol. Phys.* **45**, 221 (1982).
26. Augousti, A.T., and Rickayzen, G., *J. Chem. Soc. Faraday Trans II* **80**, 141 (1984).
27. Ise, N., and Okubo, T., *J. Phys. Chem.* **70**, 1930 (1966); Ise, N., Ito, K., Okubo, T., Dosho, S., and Sogami, I., *J. Amer. Chem. Soc.* **107**, 8074 (1985).
28. Israelachvili, J.N., *Chimica Scripta* **25**, 7 (1985); Christenson, H.K., and Horn, R.G., *Chimica Scripta*, **25**, 37 (1985) and references therein.

3
Computer Simulation Studies of the Electrical Double Layer

B. Jönsson and H. Wennerström

Computer simulation techniques have been used to study the electric double layer with respect to both equilibrium and transport properties. Comparisons have been made between simulated results and results obtained from the Poisson–Boltzmann (PB) equation. The mean-field approximation underlying the PB equation turns out to be qualitatively correct for weakly coupled systems, but becomes less accurate for strongly coupled systems. For a system with high surface charge density and divalent counterions, the PB equation predicts a repulsive interaction between overlapping double layers, while the Monte Carlo simulations show a net attraction due to the strong ion–ion correlation. The inclusion of dielectric discontinuities into the basic primitive model does not seem to alter these conclusions.

Stochastic dynamics simulations make it possible to investigate not only equilibrium but also dynamic properties, such as counterion diffusion. The simulations show that the lateral diffusion is only marginally affected by the electrostatic interactions, while the transverse diffusion is completely governed by the inhomogeneous potential set up by the counterions. The simulations can also be used to test the validity of the Smoluchowski equation, and it is found to be an excellent approximation provided an accurate potential of mean force is used.

I. Introduction

In aqueous colloidal systems, surfactant aggregates and macromolecules, charged particles are common. The charges can be due to covalently bound ionic groups, as in sulfonated polystyrene latex particles, they can arise from self-association of charged molecules, as in surfactant systems, or they can be created by specific ion adsorption on the particle surface, as is the case in AgI sols.

Because of the long-range character of the electrostatic interactions, the charges often have a profound influence on the physicochemical properties of the system. The electrostatic effects are particularly important for two types of problems. The interparticle force has a strong and long-ranged electric component, which, for example, plays a crucial role in the celebrated Derjaguin–Landau–Verwey–Overbeek DLVO theory [1] of colloidal stability. Second, the chemistry of other charged species is strongly affected by the presence of the charged aggregate. There are numerous examples of the importance of such effects in biological systems, but they are also found, for example, in micellar catalysis.

These two different aspects of the relevance of electrostatic interactions also relate to different problems in the theoretical description of the electric double layer. The electrostatic force between two charged particles or surfaces is directly related to the derivative of the free energy of the system, while the chemistry at the surface is mainly determined by the ion distribution. Although closely related, the descriptions of these two properties turn out to be two rather different problems.

The traditional theoretical approach to the electric double layer is (1) to use the primitive model of electrolyte solutions and (2) to solve the statistical mechanical problem by a mean field approximation leading to the Poisson–Boltzmann (PB) equation. This approach has turned out to be extremely useful in a wide range of applications. Apparently, the primitive model is adequate to describe the long-range part of the ionic interactions, where the influence of the molecular nature of the solvent is small. In our opinion, the deeper relation between the primitive model and a full molecular description is yet to be unravelled. The surprising versatility of the approach implies that its success cannot be explained by a mere cancellation of errors.

In spite of the success of the PB approach, there are apparently situations where it fails, and there is a need for a deeper theoretical understanding of the electric double layer. One possible improvement is to abandon the continuum model and to devise a more realistic physical model, which explicitly takes the molecular nature of the solvent into account. Attempts along this line have been made using liquid-state theories. For example, a mixture of hard-sphere ions and hard-sphere dipoles has been treated within the mean spherical approximation by Carnie and Chan [2]. However, when attempting to describe a system with strong interactions, where a linearization is impossible, it seems that computer simulation techniques are the only means at hand. Simulations with such ambitions are in progress in several laboratories.

In this article, we will, however, concentrate on the second basic ingredient in the PB approach and that is the mean field approximation (MFA). This problem is certainly more clearcut and simpler to investigate, and considerable progress has been made during the last few years. Once the primitive model is accepted and the Hamiltonian defined, it is a straightforward exercise to test the validity of the MFA. Various improved analytical approximations have been used in connection with this problem, but it has turned out, as with many other problems in liquid-state theory, that computer simulation is a superior technique to test the accuracy of different approximate theories. Simulations provide, within statistical errors, the exact answer for a given physical model.

In the following sections, we will attempt to summarize a number of studies of the electric double layer performed in our laboratories using different simulation techniques. In Section II, we give a short description of the physical system, a lamellar liquid crystal, that is modelled using the concepts of the primitive model. A discussion of different simulation methods with a particular emphasis on problems encountered in simulations of inhomogeneous systems with long-range interactions, is found in Section III. In Section IV, we discuss ion distribution functions, and in Section V, the main section we deal with interparticle forces. Dynamic properties of mobile ions in the electric double layer are discussed in Section VI. This is followed by some concluding remarks in Section VII.

II. Model System

When mixing ionic amphiphlies, water and long-chain alcohols, a rich variety of different phases are formed [3]. At a sufficiently high amphiphile concentration, one usually finds a lamellar liquid crystalline region. The lamella is built up by amphiphiles with intercalating water containing the mobile counterions (see Fig. 3.1). In the majority of the simulations reported here, we have chosen parameters relevant for the conditions found in such liquid crystals. There are a number of reason for this choice:

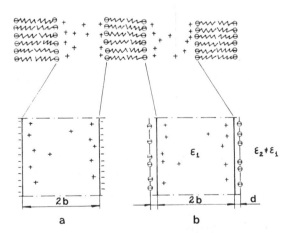

FIGURE 3.1. Schematic representation of a lamellar liquid crystal: (a) represents the basic model with uniformly charged surfaces, uniform dielectric permittivity, and point charge ions; (b) is an extended model with distinct wall charges and dielectric discontinuities taken into account.

1. We and others in our department have been working with these systems for a number of years, accumulating considerable knowledge and experience of lamellar liquid crystals,
2. The systems are amenable to detailed experimental studies of phase equilibria, water activities, and ion binding, as well as counterion, amphiphile, and water diffusion,
3. The charged surface of the lamella is molecularly homogeneous with a usually high charge density ($\approx 0.2 \, Cm^{-2}$).
4. In an idealized system, one can assume a planar geometry, and it is often sufficient to include only one species of mobile ions (the counterions), while the head group charges are constrained to a plane or uniformly smeared out on the walls. For a smeared-out wall charge density (σ), the statistical mechanical problem is characterized by only two dimensionless parameters,

$$S_1 = \frac{Ze\sigma b}{(kT\varepsilon_r\varepsilon_0)},$$ (1a)

and

$$S_2 = \frac{\sigma Z^5 e^5}{[b(kT\varepsilon_r\varepsilon_0)^3]},$$ (1b)

where kT is the Boltzmann factor, ε_r and ε_0 are the relative permittivity and the permittivity of the vacuum, respectively, and Ze is the ionic charge. The parameter b is defined in Fig. 3.1. This model system shows the unusual combination of being the simplest possible and yet probably the most thoroughly studied type of system that has an electrical double layer.

The system shown in Fig. 3.1 represents the basic model used in most of our simulations. In the majority of the simulations presented here, we have used the following parameter values: $\varepsilon_r = 77.3$, $T = 301 \, K$, $2b = 21.0 \, \text{Å}$ and $\sigma = 0.224 \, Cm^{-2}$. In addition, we have, in different contexts, introduced complications into the model, in attempts to make it more realistic. These include finite ion size effects, distinct wall charges, and changes in the dielectric permittivity at the surface. The extended models all have in common that they tend to introduce additional parameters. The numerical value of these parameters is not always an obvious choice, and yet they can have a substantial influence on the final result. Hence, in the following, we will, concentrate on systems with only point charge counterions and a uniformly smeared-out surface charge density.

III. Simulation Methods

In the study of equilibrium properties, such as (free) energies, forces, and distribution functions, the standard Monte Carlo (MC) method with the Metropolis algorithm [4] has been used. Compared to MC studies of simple liquids, there are two main complications arising from the inhomogeneity of the system and from the long-range character of electrostatic interactions.

The dimensions of the MC box are determined by the chosen distance, $2b$, between the charged walls and by the number of particles included in the simulation. The system is in principle infinite in the two directions parallel to the walls, and this is mimicked by the use of the minimum image convention. A remarkable consequence of the planar geometry is that the field from the two charged walls is zero in between the walls, and there is formally no effect of the wall charges on the ion distribution. However, in the simulations it is preferable to split the zero field from the infinite system into nonzero components, one due to the field from the walls inside the MC box and another due to the walls outside the box. This latter field is combined with a mean field from the ions outside the MC box, thus adding up to a constant and reasonably small external potential that takes care of the long-range correction. The mean ion distribution can be obtained either from the PB solution or in an iterative way using previous steps in the simulation. Both procedures seem to work as long as the correction term is small. The former procedure has been tested by successively increasing the number of particles in the simulation and looking for convergence in different physical properties. Table 3.1 shows the results from such a convergence test. For this specific case, there seems to be qualitative agreement with only two particles in the simulation, requiring less than a second of computation time! In most cases, however, we have used about 50 particles, and a typical run then takes an hour on a VAX 780-type computer.

Formally, the external potential can be written as a three-dimensional integral, which can be analytically integrated over the two lateral coordinates. The integration over the z-coordinate has to be done numerically. Before the MC simulation is started, this integral is tabulated for subsequent rapid use in the energy evaluation. This leads to an efficient MC procedure, where most of the computer time is spent in handling the minimum image convention and calculating the inverse of the square root of the distance

TABLE 3.1. Convergence of the MC simulations with long-range corrections.[a]

N	E_{ii}	E_{tot}	$-S$	$\rho(b)$	p_b
2	0.85	−0.21	5.76	14.2	0.781
6	4.41	−0.30	5.27	14.0	0.758
20	12.74	−0.39	5.03	14.5	0.746
80	33.94	−0.48	4.95	13.9	0.743

[a] E_{ii} is the ion–ion interaction energy, E_{tot} is the total energy, S is the one-particle excess entropy, and p_b is the fraction of ions within 3 Å from the charged wall. Units are kJ/mol, J/mol K, and mol/dm^3.

between the ions. This part of the MC program is but a few lines, and substantial improvements can be obtained by coding this part in assembler language.

Monte Carlo simulation techniques have been used by several other groups in studies of the electric double layer. Torrie and Valleau [5] and van Megen and Snook [6] have used the grand canonical ensemble approach, and cylindrical aggregates in the cell model have been studied by Bratko and Vlachy [7] and by Murthy et al. [8], just to mention a few examples. The double layer around a spherical particle has been investigated by Linse and Jönsson [9], in both the cell model and an isotropic solution.

To obtain information about dynamical properties, the usual molecular dynamics procedure does not work simply because the solvent is not included explicitly in the calculations. Within the primitive model, the influence of the solvent enters as a random disturbance, and this effect can be described in a Langevin equation as

$$\mathbf{p}_k(t) = -\xi \mathbf{p}_k(t) + \mathbf{F}_k(t). \tag{2}$$

Here, $\mathbf{p}_k(t)$ is the linear momentum of particle k, ξ is a friction factor, and $\mathbf{F}_k(t)$ is the force acting on particle k. This force has in the present case both a random, rapidly varying component $\mathbf{R}_k(t)$ due to the solvent and a deterministic component $\mathbf{K}_k(\mathbf{r}^N, t)$, where we have indicated that the electric force depends on the coordinates of all N particles in the system. In an electrolyte solution, the time variation in the deterministic force is slow relative to the velocity relaxation time ξ^{-1}, and the amplitude of $\mathbf{K}$ is small relative to the amplitude of $\mathbf{R}$. When these two conditions are fulfilled, the velocities can be eliminated from Eq. (2), and one obtains instead a stochastic differential equation involving particle positions only [10,11]:

$$\mathbf{r}_k(t) - \mathbf{r}_k(0) = (m\xi)^{-1} \int_0^t dt' \, [\mathbf{R}_k(t') + \mathbf{K}_k(\mathbf{r}^N, t')]. \tag{3}$$

The solution of Eq. (3) leads to a particle distribution function satisfying the Smoluchowski equation,

$$f_S(\mathbf{r}, t|r_0) = D_0 \nabla \cdot (\nabla - \beta \langle \mathbf{K} \rangle) f_S(\mathbf{r}, t|r_0), \tag{4}$$

where $\beta = (kT)^{-1}$, $D_0 = kT/m\xi$, $\langle \mathbf{K} \rangle$ is the mean force, and f_S is the self-propagator. The diffusion coefficient D_0 is obtained from experiment on dilute electrolytes and is assumed constant independent of the ion concentration. This means that the ion–solvent interaction, in the present model contained in $\mathbf{R}_k$ and ξ, is concentration independent. One can show that this assumption is consistent with the two other assumptions concerning the amplitude and time variation of the deterministic force [11]. When solving Eq. (3) by simulation techniques, one can use time steps on the order of 10^{-12} s, and the system can be followed over times on the order of 10^{-7} s.

The stochastic dynamics (SD) simulations provide the same information on equilibrium properties as MC simulations, but in addition, dynamic aspects of the ion distribution are revealed. The SD algorithm is slightly more complicated to work with and somewhat more time-consuming than the MC procedure. The same technique for handling long-range interactions as described previously can be used in the SD simulations.

IV. Distribution Functions

Because of the high symmetry of our model system, the one-particle density is a function of the transverse coordinate only, $\rho(\mathbf{r}) = \rho(z)$. It can thus be obtained rather easily from a simulation. Figure 3.2 shows some typical results obtained for monovalent and divalent counterions. The comparison with the PB results reveals that the mean field approximation works very well except close to the midplane, where the error, albeit small in absolute terms, is large on a relative scale. As will be apparent in the following sections, this is an important observation when considering the force between the walls. Near the wall, the agreement is extremely good. The reason is that the contact value theorem [12,13]

$$kT\rho(b) = \sigma^2/(2\varepsilon_r\varepsilon_0) + p_{osm} \tag{5}$$

is valid in both cases, and the first term on the right-hand side (rhs) dominates for the present case. As a consequence, it becomes numerically difficult to calculate the osmotic pressure from Eq. (5), since it is obtained as a difference between two large numbers.

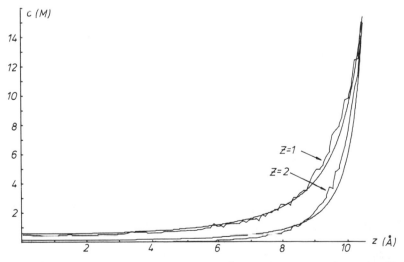

FIGURE 3.2. Concentration profiles (in mol/dm^3) between the two charged walls for monovalent and divalent ions. The smooth curves are from the PB equation, and the two others are from MC simulations. The charged walls are placed at ± 10.5 Å.

In most of the simulations presented here, we have used uniformly charged surfaces. One argument for this choice is simplicity; further, as advocated above, the introduction of, for example, distinct wall charges introduces additional parameters. As an example, Fig. 3.3 shows concentration profiles obtained from simulations with distinct charges with varying ionic radii. First, we note that the contact theorem is no longer valid in its simple form, Eq. (5), and second, that one can obtain almost any concentration profile by changing the ionic radius.

A complication of considerable importance in liquid crystals, and in many other systems, is the fact that the system has a dielectric discontinuity [14]. This feature is not taken into account in the basic model used to obtain the results in Fig. 3.2. A dielectric inhomogeneity affects the one-particle distribution both directly and indirectly by changing the interparticle interactions. The effects can be easily incorporated using the image charge technique. With image charges, it is essential to make the distinction between a uniform charge distribution and distinct charges on the walls. In the former case, there will be a repulsion between a counterion and its self-image, while in the latter case, there will also be an interaction with the images of the wall charges, which will give rise to a net attraction. However, this is only a qualitative argument since there is also an interaction with the images of all the other counterions and wall ions. With respect to interactions between the mobile ions, the discontinuity leads to a stronger coupling if one assumes that the nonpolar region has a lower permittivity. Within the PB approximation, the images will be uniformly smeared out: hence, they will give rise to a zero electric field between the charged walls. Clearly, one would expect larger deviations between PB results and simulation results in the presence of image

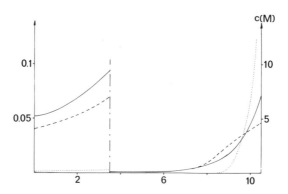

FIGURE 3.3. Concentration profiles with varying wall charge radii, but with $d = 0$ Å. The curves are obtained with divalent counterions, a surface charge density of 0.16 Cm^{-2}, and only one double layer. The left part of the diagram is magnified by a factor of 100. The solid line is the uniform surface charge density, the dashed line is the wall ion radius = 2 Å, and the dotted line is the wall ion radius = 1 Å. The charged wall is placed at 10.5 Å.

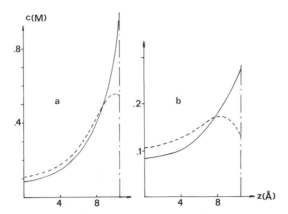

FIGURE 3.4. Concentration profiles with divalent counterions and image charge effects taken into account. The dielectric discontinuity is displaced a distance of 3 Å from the uniformly charged walls, which are placed at ± 10.5 Å. The surface charge density is (a) 0.056 and (b) 0.028 Cm^{-2}, and $\varepsilon_2 = 5$. Solid line, no images; dashed line, images.

charges. However, the image effects turn out to depend on a number of factors among which is the surface charge density. Figure 3.4 shows the effects of image charges on the one-particle distribution for two different surface charge densities. The most noticeable feature is that the self-image term seems to be less important at higher surface charge densities.

It is considerably more difficult to study two-particle distribution functions in simulations of inhomogeneous systems. With the given symmetry, the two-particle correlation function $g^{(2)}(\mathbf{r}, \mathbf{r}')$ is a function of three variables, z, z', and $r = [(x - x')^2 + (y - y')^2]^{1/2}$. It is very time-consuming to map the full three-dimensional function $g^{(2)}(z, z', r)$, but because of the high density close to the wall, one can obtain an estimate of $g^{(2)}(\delta, \delta, r)$, where δ denotes a 1-Å-thick slice close to the wall. This function is illustrated in Ref. 15. There is a distinct hole around the ion due to the Coulomb repulsion. The radius of the hole gives an indication of when the finite size of a counterion is of importance. In a simulation with uniformly charged walls and only counterions, there is no effect of ionic radii less than 2 Å. It is only when the counterion radius exceeds 4 Å that there is any noticeable effect [9]. In neither of our simulations are there any signs of the theoretically predicted [16,17] algebraic decay of $g^{(2)}(z, z', r)$. This could be due to numerical uncertainties or to the fact that the MC box is too small to pick up the effect.

V. Forces

The most important application of double-layer theory in colloid chemistry is in the evaluation of interparticle forces. The basic model system used in calculating the force is the case with two interacting charged surfaces [1], as in Fig. 3.1. In the canonical ensemble, the effective force F is given by the derivative of the free energy with respect to the separation,

$$F = -\{\partial A/\partial(2b)\}_{N,T}. \tag{6}$$

The osmotic pressure is then simply the force per unit area.

To understand the significance of simulations in determining the force, it is constructive to first identify the different contributions to the force. When two surfaces approach each other, the counterion concentration in between increases, thereby leading to a decrease in the entropy of mixing and to a repulsive force. In both the exact description and the PB approximation, this change in entropy is directly related to the concentration of ions in the midplane [13,18]. In the PB case, this is the only contribution to the force, since there is no direct interaction between the two halves of the system. The mean electric field does not penetrate from one half to the other. When fluctuations are included, the ion distributions on either side correlate to give an attractive component of interaction. To investigate this effect, it is necessary to go beyond the MFA, and computer simulation is one way of achieving this.

The most straightforward way to calculate the osmotic pressure is, of course, to use Eq. (5). However, since the p_{osm} term is usually smaller than the accuracy of $\rho(b)$, this is not a recommendable way. A more mechanistic approach, separating the entropic and energetic components of the force, was used in one of our earlier studies and gave acceptable numerical accuracy. It was also shown that a numerical derivation of the free energy, obtained through a charge integration procedure, was less reliable [19]. Another approach is based on the relation

$$p_{osm} = p_{osm,id} - (\partial U_{ex}/\partial V), \tag{7}$$

where the derivative is evaluated at constant excess entropy [20]. There exist several other methods for calculating the force, some of which were explored in Ref. 21. It seems clear that the optimal method to use depends on the specific system under study.

When discussing the simulation results, it is useful to recall that the basic model system is completely defined through the two dimensionless parameters S_1 and S_2, while the PB solution depends only on S_1. For a large region of parameter space, the PB prediction of the force is accurate, and in the simulations, we have been particularly interested in situations when this is not true, that is, when ion–ion correlations become important. These correlations are more pronounced when the coupling between the ions are stronger, which implies that S_2 should be large. From the expressions for the dimensionless parameters, one sees that, for example, multivalent counterions and high surface charge density lead to a stronger coupling.

As an illustration, Figs. 3.5 and 3.6 show how the force varies with σ for parameters relevant to a lamellar liquid crystal. For monovalent counterions, the deviation from the PB prediction is moderate up to the highest realistic

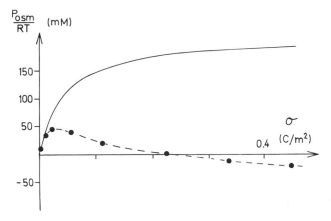

FIGURE 3.5. The osmotic pressure as a function of surface charge density for divalent counterions. The solid curve is from the PB equation, and the dashed curve shows the simulation result. $2b = 21$ Å.

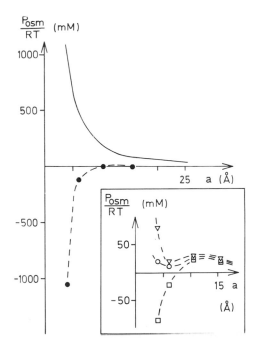

FIGURE 3.6. The osmotic pressure as a function of the distance between the charged walls with divalent ions. The solid curve is the PB result, and the dashed one MC result. The insert shows three simulated curves with varying surface charge density. From top to bottom, it is 0.07, 0.08, and 0.09 Cm^{-2}. $2b = 21$ Å.

charge densities, while for $Z = 2$, there are dramatic effects, and the traditional double-layer repulsion changes to a double-layer attraction at high σ! This surprising result is due to a combination of two effects. First, the attractive component due to ion–ion correlations between the two halves of the system becomes substantial, thereby allowing the possibility of a net attraction. The second important effect is that the correlations make the double layer contract, and the concentration of ions in the midplane is considerably lowered relative to a mean field description [13]. This latter effect does not exist only for short separations; the PB theory will overestimate the repulsive component at all separations.

The presence of dielectric discontinuities close to the charged surfaces influences the force between the surfaces essentially through two mechanisms. With uniformly charged surfaces, the counterions will feel an extra repulsion close to the wall, which, as discussed in the previous section, leads to an increase in ion density at the midplane and hence an increased repulsion. In addition, the image charges give rise to stronger correlations between counterions, thus making it easier for the attractive force component to develop. The increased correlations also reduce the midplane concentration, and it is the balance between these effects that determines the final net force. However, simulations show that the latter mechanisms dominate at large separations, while there is an additional repulsive component appearing at short separations because of the first mechanism. Similar effects have also been shown to exist in zwitterionic systems [14].

In summary, the simulations show that when counterions are weakly coupled, that is, when S_2 is small, the PB approximation usually gives a qualitatively correct description of the electrostatic force acting between charged aggregates. However, when counterions couple strongly, the ion–ion correlations have important implications, and under some realistic conditions, the traditional double-layer repulsion turns into a double-layer attraction, implying that the PB description is qualitatively wrong under these conditions.

VI. Ion Diffusion

Most theoretical studies of the electric double layer have been concerned with equilibrium rather than dynamic aspects. This merely reflects the difficulties encountered when describing time-dependent phenomena in a statistical mechanical framework and not the relative importance of dynamic properties in double-layer context. An early contribution in this area was made by Lifson and Jackson [22], who solved the one-particle Smoluchowski equation (S-equation). Stochastic simulation techniques were later introduced by Ermak [23].

It is constructive to discuss the lateral (parallel to the walls) and transverse (perpendicular to the walls) diffusion separately. In the first case, the average lateral force will be zero and the one-particle S-equation will be one of free diffusion. However, since the instantaneous force is nonzero, there will be an effect on the lateral diffusion from the particle interactions in the simulations. Table 3.2 shows how the lateral diffusion varies with surface charge density, for both the basic model and in an extended one with distinct charges on the walls. The general observation is that electrostatic interactions have only a minor effect on the lateral motion [24] in analogy with what is found for uniform electrolytes.

The transverse motion provides a more interesting case from a theoretical point of view. Since the mean force is nonzero, we can now use the S-equation and compare with the simulation results. Before going on with this comparison, it may be instructive to consider the approximations underlying the one-particle S-equation. Formally, one can write an N-particle S-equation, which is the counterpart to Eq. (3), but is formulated as a partial differential equation. It is straightforward to reduce the N-particle equation, and one obtains a hierarchy of coupled n-particle equations ($1 \leq n \leq N$). By assuming that the relaxation of the ionic correlations are instantaneous (IRA, instantaneous relaxation approximation), the one- and two-particle equations decouple, and we arrive at Eq. (4) for the self-propagator. This derivation is given in detail in Ref. 25.

We can now test the IRA by solving the S-equation using a mean force obtained from simulations, and we can compare the result with the corresponding quantity calculated directly in the simulation. This will give information about the relative importance of what may be referred to as dynamic correlations. In a uniform electrolyte, this is usually called the relaxation effect. There is also the possibility of substituting for the mean force the corresponding PB quantity. So, we are left with three different alternatives in order to describe the counterion dynamics: (1) the SD simulations, which are formally exact; (2) the S-equation with the exact mean

TABLE 3.2. The lateral diffusion coefficients for counterions (D_i) and wall charges (D_w) as a function of surface charge density.

σ (Cm^{-2})	Basic model,[a] D_i/D_{i0}	Extended model ($d = 1$ Å) D_i/D_{i0}	D_w/D_{w0}
0.080	0.93(0.77)	0.80	0.93
0.200	0.90(0.67)	0.70	0.86
0.224	0.87(0.65)		
0.267	0.85(0.63)	0.67	0.86
0.401	0.82(0.60)	0.62	0.85

[a] The values within parentheses are for divalent ions; $2b = 26$ Å, $D_{i0} = 2.10^{-9}$ m^2 s^{-1}, and $D_{w0} = 10^{-10}$ m^{2}s^{-1}.

force (referred to below as the IRA); and (3) the S-equation with the PB force, referred to as SPB. In the last case, not only the dynamic correlations but also the static correlations (MFA) have been neglected. Figure 3.7 shows a comparison of the three propagators $f_s(\delta, t|\delta)$, where δ denotes a layer of thickness 1 Å close to the wall. The propagators give the probability of finding an ion in this shell at time t provided it was there at $t = 0$. The general conclusion is that the IRA is an excellent approximation giving results indistinguishable from the simulation results. The SPB approach seems to be at least qualitatively correct judging from Fig. 3.7. However, it turns out that the propagator $f_s(\delta, t|\delta)$ is a poor measure of the different approximations. A much more critical test is to compare the mean first passage times (MFPT) for a counterion, formally obtained by integrating twice over the self-propagator,

$$\tau(z_A|z_0) = \int_0^\infty dt \int_{-b}^{z_A} dz\, f_s(z, t|z_0). \tag{8}$$

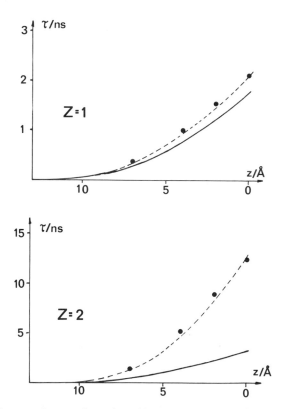

FIGURE 3.7. The two figures show the self-propagator $f_s(\delta, t|\delta)$ for monovalent and divalent counterions. The solid line is the SPB result, the dashed line the IRA, and the filled circles represent the simulated result. The unit of time is 10^{-12} s and $D_0 = 2 \cdot 10^{-9}$ m^2 s^{-1}.

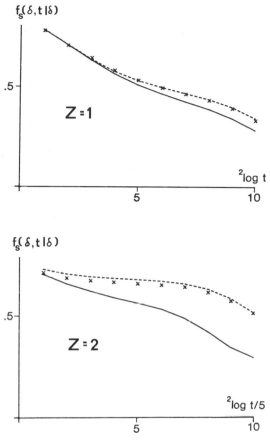

FIGURE 3.8. The mean first passage time with monovalent and divalent counterions; $2b = 26$ Å. The solid line is the SPB result, the dashed line is the IRA result, and crosses represent the simulated result.

The MFPT is the average time it takes an ion to diffuse from z-coordinate z_0 to z_A. Figure 3.8 shows a comparison of the three different approaches for both monovalent and divalent counterions. Again, we find that the IRA is an excellent approximation, while it is obvious that the neglect of static correlations inherent in the PB approach also affects the dynamics.

The insignificant effect of dynamic correlations on the transverse diffusion may be contrasted with the reduction of the lateral diffusion coefficient, as evidenced in Table 3.2. The difference may be qualitatively understood by noting that the outward transverse diffusion starts in a potential well several kT deep. This means that this diffusion process is much slower than the free diffusion; hence, the assumption of an instantaneous relaxation of the ionic correlations around the diffusing ion is valid.

VII. Conclusions

In the previous sections, we have shown how static and dynamic properties of a model system of an electric double layer can be obtained from computer simulations. There is a twofold use of these results. On the one hand, they are important guides when developing approximate analytical theories. For example, the force results have been reproduced by Kjellander and Marcelja [26,27] using an integral equation approach based on the HNC closure. The other use of the simulation results is for a direct interpretation of experimental results. In such applications, there is always an element of uncertainty due to questions about the validity of the basic model. However, it has been found in a number of cases that the PB description gives a good account of electrostatics in colloidal systems. It is then natural to assume that when parameters are changed to a region where the PB approximation breaks down the basic model is still valid.

Clearcut experimental evidence of attractive double-layer forces is found from the swelling behavior of liquid crystalline lamellar phases of ionic amphiphiles. The much studied surfactant Aerosol-OT readily forms a lamellar phase when mixed with water. The regular counterion is sodium, and in this case, the lamellar phase exists in the range of 10 to 60 % surfactant. The swelling behavior can be quantitatively accounted for using the PB approximation. However, when the sodium ion is exchanged for a divalent ion, such as Ca^{+2}, the lamellar phase does not swell in contrast to the PB prediction [28]. From the simulations, it is clear that there exists an attractive electrostatic force in this system. That this is the basic explanation of the experimental results is further substantiated by experiments on systems where the ratio Na^+/Ca^{+2} is varied. When the Na^+ content is relatively high, the PB treatment is adequate, but around a 1:1 ratio there is a sudden change in the swelling properties, and a transition to a regime with water-poor lamellae occurs [29]. In the transition region, one observes an equilibrium between two lamellar phases with different water contents. These observations are qualitatively in agreement with what one would predict from the simulations (see Fig. 3.6), and there is also a quantitative agreement to the extent such a comparison can be made.

Studies of other Ca^{+2}-containing surfactant systems show that the existence of attractive electrostatic forces is a general phenomenon, but is particularly clearly demonstrated in the AOT system. There are other situations where the appearance of attractive double-layer forces is the most likely explanation of the experimental results, and it may be fruitful to reinvestigate some of these old problems considering the possible occurrence of attractive double-layer forces.

References

1. E.J.W. Verwey and J.Th.G. Overbeek, "*Theory of the Stability of Lyophobic Colloids*," Elsevier, Amsterdam, 1948.
2. S.L. Carnie and D.Y.C. Chan, *J. Chem. Phys.* **73**, 2949 (1980).
3. T. Tadros, ed., "*Surfactants*," Academic Press, London 1984.
4. N. Metropolis, A.W. Rosenbluth, M.N. Rosenbluth, A.H. Teller, and E. Teller, *J. Chem. Phys.* **21**, 1083 (1953).
5. G.M. Torrie and J.P. Valleau, *Chem. Phys. Lett.* **65**, 343 (1979).
6. W. Megen and I.J. Snook, *J. Chem. Phys.* **73**, 4656 (1980).
7. D. Bratko and V. Vlachy, *Chem. Phys. Lett.* **90**, 434 (1982).
8. C.S. Murthy, R.J. Bacquet, and P.J. Rossky, *J. Phys. Chem.* **89**, 701 (1985).
9. P. Linse, G. Gunnarsson, and B. Jönsson, *J. Phys. Chem.* **86**, 413 (1982); P. Linse and B. Jönsson, *J. Chem. Phys.* **78**, 3167 (1983).
10. S. Chandrasekhar, *Rev. Mod. Phys.* **15**, 1 (1943).
11. T. Åkesson and B. Jönsson, *Mol. Phys.* **54**, 369 (1985).
12. D. Henderson and L. Blum, *J. Chem. Phys.* **69**, 5441 (1978).
13. H. Wennerström, B. Jönsson, and P. Linse, *J. Chem. Phys.* **76**, 4665 (1982).
14. B. Jönsson and H. Wennerström, *J.C.S. Faraday Trans.* 2 **79**, 19 (1983).
15. B. Jönsson, H. Wennerström, and B. Halle, *J. Phys. Chem.* **84**, 2179 (1980).
16. A.L. Nichols and L.R. Pratt, *J. Chem. Phys.* **77**, 1070 (1982).
17. S.L. Carnie and D.Y.C. Chan, *Mol. Phys.* **51**, 1047 (1984).
18. R.A. Marcus, *J. Chem. Phys.* **23**, 1057 (1955).
19. L. Guldbrand, B. Jönsson, H. Wennerström, and P. Linse, *J. Chem. Phys.* **80**, 2221 (1984).
20. V. Vlachy and D. Bratko, *J. Chem. Phys.* **75**, 4612 (1981).
21. B. Svensson and B. Jönsson, *Chem. Phys. Lett.* **108**, 580 (1984).
22. S. Lifson and J.L. Jackson, *J. Chem. Phys.* **36**, 2410 (1962).
23. D.L. Ermak, *J. Chem. Phys.* **62**, 4189, 4197 (1975).
24. T. Åkesson and B. Jönsson, *J. Phys. Chem.* **89**, 2401 (1985).
25. T. Åkesson, B. Jönsson, B. Halle and D.Y.C. Chan, *Mol. Phys.* **57**, 1105 (1986).
26. R. Kjellander and S. Marcelja, *Chem. Phys. Lett.* **112**, 49 (1984).
27. R. Kjellander and S. Marcelja, *J. Chem. Phys.* **82**, 2122 (1985).
28. A. Khan, K. Fontell, G. Lindblom, and B. Lindman, *J. Phys. Chem.* **86**, 4266 (1982).
29. A. Khan, B. Jönsson, and H. Wennerström, *J. Phys. Chem.* **89**, 5180 (1985).

4
The Ionic Environment of Rod-like Polyelectrolytes

P.J. Rossky, C.S. Murthy, and R. Bacquet

Association between electrolyte ions and polyions is of fundamental importance. Here, we discuss the nature of the distribution of small ions around polyelectrolytes and examine the sensitivity of the distribution to salt concentration and polyion charge density. Two polyion models are considered, namely, a simplified one in which the polyion is taken to be a uniformly charged cylinder and an atomically more detailed one in which the smaller ions interact with the polyion on an atom by atom basis. The hypernetted chain integral equation and Monte Carlo computer experiments are used to develop the results for ionic distributions. The results confirm that the ionic environment around highly charged polyions is relatively insensitive to large changes in bulk electrolyte concentrations in the case of the simpler model of the polyion. Further, the concept of an effective net charge for the polyion as a determinant for the asymptotic electrostatic potential holds. The results for the more detailed model of the polyion indicate that charge association can be understood qualitatively using the simpler, uniformly charged model. Such a simplification is, however, not sufficient for determining the local ionic concentrations and the spatial extent of the association quantitatively.

I. Introduction

The importance of charge association between ions of a supporting electrolyte and macroions or ionic molecular assemblies in complex electrolyte solutions is widely appreciated. The resulting screening of repulsive electrostatic interactions has a fundamental influence on the stability of colloidal solutions, the stability and structure of micellar and lamellar aggregates, and the conformation of and interaction between polyelectrolytes in solution. Such influences include both traditional long-range Debye–Hückel-like screening and strong short-range interactions. In the latter case, these interactions can produce a high degree of counterion association with the macroion.

The significance of the latter type of association, which is characteristic of macroionic solutes, is twofold. First, the screening of the interaction between nearby charges within the macroion cannot be expected to be adequately represented by results, such as the Debye–Hückel expression, which are based on a diffuse ion atmosphere description incorporating the bulk concentration. Second, a naive diffuse atmosphere description of long-range screening is, in general, also inappropriate because of short-range association. Rather, in such cases, a more accurate view of the long-range interactions is often obtained by ascribing to the macroion an effective net charge that is composed of the bare charge less that of the closely associated counterions [1–4].

For the example of cylindrical, or rod-like, polyelectrolytes, such as DNA, it has been demonstrated [4] within the context of the mean field Poisson–Boltzmann theory that the long-range electrostatic potential is given rigorously by such an approach even in the absence of specific short-range interactions, as long as the macroionic charge density is not too low. It is this phenomenon that underlies Manning's formulation for polyelectrolyte behavior [2].

Our goal is to discuss the nature of the polyelectrolyte ionic atmosphere per se, from the point of view of the example of rod-like polyions. In particular, our interest is in characterizing the small ion distributions quantitatively and determining the sensitivity of these distributions to such quantities as salt concentration and polyion charge density. Of special interest is a determination of the spatial extent around the polyion that is required to incorporate a given fraction of the neutralizing counterion charge. It is this neutralization that underlies the usefulness of models based on effective net macroionic charges.

We first discuss selected results [5,6] obtained for a quite simplified polyion model in order to demonstrate the response of the ion atmosphere to polyion charge density and salt concentration. This polyion model includes only a uniform polyion charge density and no atomic detail in its structure. We then examine corresponding results [7] for an atomically detailed polyion model in order to examine the limitations imposed on our description of ionic distributions by such simplifications in polyion modeling.

Section II, we outline the methods used in these studies, including a general description of the models employed and the theoretical techniques used to evaluate the ionic distributions. In Section III, we present representative results for ionic distributions that manifest the association phenomena. The conclusions are presented in Section IV.

II. Methods

We first describe the models used and then the theoretical methods employed. Because of space limitations, we do not present these in full detail but refer to the literature for this information [5, 7].

We note at the outset that all of the models we consider treat the aqueous solvent at a dielectric continuum level. That is, all ionic interactions between solutes are given by the reduced value obtained by dividing the bare coulombic interaction by the bulk dielectric constant of water at 25°C, namely, 78.358. Correspondingly, no effort is made to account for differences in the dielectric response between the polyion interior and the solvent. Ionic sizes are taken as given by atomic radii. While this is perhaps not ideal, we avoid the arbitrary assignment of hydrated radii, which would, for example, expand the radius by a distance equal to the diameter of a water molecule. While the model we use treats the solvent in a simplified manner, there seems little question that the general behavior of interest here is adequately represented.

The simplified polyion solution model incorporates a single, infinitely long, uniformly charged, cylinder. The magnitude of the polyion linear charge density is expressed in reduced units by $\xi = e^2\beta/\varepsilon b$, where e is the magnitude of the electronic charge, β is the inverse of the product of Boltzmann's constant and the temperature, ε is the solvent bulk dielectric constant, and b is the polyion contour length incorporating unit polyion charge. Here, we emphasize the case of DNA for which $\xi \cong 4.2$ for water at 25°C. Correspondingly, we will assume throughout that the polyion is negatively charged and the counterions are positive.

The single polyion is immersed (at infinite dilution) in a simple electrolyte solution that is characterized by the bulk salt concentrations far from the polyion. Both electrostatic and short-ranged repulsive core interactions are included between small ions and between the small ions and the polyion. The form of the core repulsive interaction used is either an inverse ninth power law or an exponential; the precise parameterization of the potentials used can be found in Ref. 6, but this is not essential for the present purposes.

The second polyion model considered is an atomically detailed model of the B conformational form of double helical DNA in an idealized geometry [8]. We consider specifically a base sequence with alternating C and G bases. The small ions of the supporting electrolyte interact with this polyion on an atom by atom basis. The polyion atoms each carry a partial electrostatic charge and have a characteristic atomic radius, so that the ion–polyion atom interaction takes the same form as that between small ions in the electrolyte. That is, it consists of an electrostatic term and a short-ranged core interaction characterized by the atomic radii [7]. The atomic parameters for the polyion were based on the potential function developed by Kollman [9].

To evaluate the ionic distribution surrounding the polyion, we have used two approaches. For the uniformly charged cylindrical polyion, we can employ analytically based integral equation techniques [10,11]. In the present context, such approaches provide the radial distribution of electrolyte ions with respect to the polyion axis as the solution to a nonlinear integral equation. The particular approximation used here is the so-called hypernetted chain (HNC) approximation [11]. This approach is established as a very good one for electrolyte problems [12], although we have independently tested the accuracy of these results for particular cases [6], as will be noted. We do not describe the integral equation theory here, since excellent detailed discussions are available in the literature [5,11].

To carry out the tests of approximations referred to above or to study the more detailed model requires computer simulation. The simulations are carried out by standard Monte Carlo methods [13]. The basic simulation cell consists of a single polyion with its axis aligned with that of a cylindrical sample of salt solution. The solution includes both the neutralizing counterions and added salt. The infinite extent in the axial direction is mimicked by the usual periodic boundary conditions [13]. The radial boundary was chosen at a sufficiently large distance so that boundary effects are demonstrably absent [6]. Thus, the salt concentration corresponding to the bulk value of the HNC calculations is determined by the ion concentrations at this outer boundary.

The only subtle element in these simulations results from the strong inhomogeneity in the electrolyte distribution produced by the polyion field. As a result, the small ions experience a very long-ranged interaction in the axial direction produced by this average distribution. We account for this by including in the potential an additional term equal to the mean electrostatic potential produced by the infinite segments of electrolyte and polyion that extend beyond the basic simulation cell. This approach is completely analogous to the approach used by Valleau and co-workers [14] in their treatment of the charged planar interface. For the uniformly charged polyion, this external potential was evaluated using a self-consistent result for the small ion charge distribution, while for the detailed polyion model, we used as a convenient approximation a field evaluated from the ion distribution for a corresponding state of the uniformly charged polyion system. The full numerical details for the evaluation of this long-ranged potential have been given elsewhere [6].

We have also evaluated the ionic distributions in the Poisson–Boltzmann approximation [6] for the uniformly charged polyion, and we include those results for comparison in a few cases.

III. Results

The first results we consider are for the simplified polyion model immersed in an NaCl solution of molar stoichiometric concentration c_{ST}. The polyion radius is nominally that widely taken [2] for DNA, 10 Å, but we consider a variety of polyionic charge densities.

Before examining the results obtained from the HNC approximation, it is important to establish the reliability of the approach [6]. Representative comparisons of HNC results with simulation and Poisson–Boltzmann theory are shown in Figs. 4.1 and 4.2. In these figures, we show the *relative* counterion concentration $g(r)$ as a function of distance for three concentration conditions and a polyionic charge density equivalent to DNA. This relatively high charge density case is expected to be the most demanding [12], so that it provides a stringent test.

In Fig. 4.1, we show the Na^+ profile for systems with bulk NaCl concentrations of about 0.01 M in (a) and 0.001 M in (b). The crosses are simulation results, the solid curve is the HNC result, and the dashed curve is that predicted by the mean field Poisson–Boltzmann theory. Although all curves are similar, it is clear that the HNC theory is quantitatively significantly more accurate than is the mean field theory and that the inclusion of ionic correlations yields significantly increased local counterion concentrations (by $\sim 15\%$) over those predicted by the simpler mean field theory.

Figure 4.2 shows a similar comparison, but now for a mixed electrolyte that models a solution including both NaCl and $MgCl_2$, in concentrations of 0.155 M and 0.022 M, respectively. Here, the Mg^{2+} radial distributions are shown in (a), and the Na^+ radial distributions are shown in (b). Similar trends are observed; that is, all three curves are in general agreement, but the HNC approximation is in closer quantitative accord with the simulation. Both the simulation and HNC approximation predict higher local counterion concentrations than are predicted by mean field theory.

Having seen that the HNC approximation produces reliable results under a variety of conditions, one can readily examine the response of the charge distribution to widely varying polyion charge and bulk concentration conditions [5]. Such results are shown in Fig. 4.3. In this figure, we present Na^+ counterion distributions for bulk concentrations $c_{ST} = 10^{-1}\ M$, $10^{-2}\ M$, and $10^{-3}\ M$ and for reduced polyion charge densities of $\xi = 0.5, 0.9,$

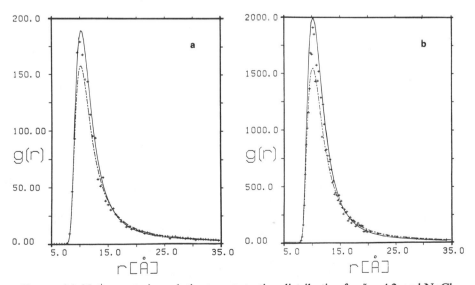

FIGURE 4.1. Na^+ counterion relative concentration distribution for $\xi = 4.2$, and NaCl concentrations (a) $c_{ST} = 0.01074\ M$ and (b) $c_{ST} = 0.00104\ M$. Monte Carlo, $+$; HNC, ——; and Poisson–Boltzmann, - - -.

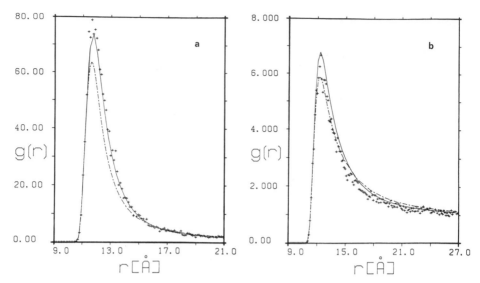

FIGURE 4.2. Counterion relative concentration distributions for (a) Mg^{2+} and (b) Na^+ for $\xi = 4.2$ and a bulk salt mixture concentration of 0.155 M NaCl and 0.022 M $MgCl_2$. Symbols as in Fig. 4.1.

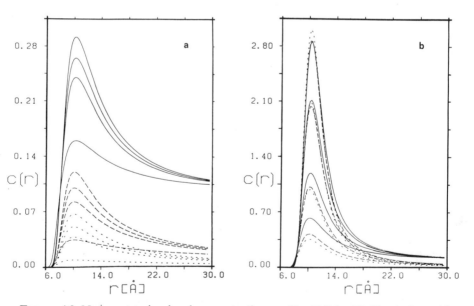

FIGURE 4.3. Na^+ counterion local concentration profiles (M) for NaCl solution with $c_{ST} = 10^{-1} M$ (——), $10^{-2} M$ (— —), and $10^{-3} M$ (- - -) and polyion charge densities of (a) $\xi = 0.5, 0.9, 1.0,$ and 1.1 and (b) $\xi = 2.0, 3.0, 4.2,$ and 5.0.

1.0, 1.1, 2.0, 3.0, 4.2, and 5.0. The results for smaller values of ξ are grouped in Fig. 4.3(a), while for larger values the results are collected in Fig. 4.3(b). In Fig. 4.3, we use concentration units to describe the distributions since this clearly manifests the magnitude of the counterion accumulation, as well as permitting more convenient comparison among systems of differing bulk concentration.

The asymptotic value of each curve is simply equal to the bulk concentration. However, in the region near the polyion, the behavior changes qualitatively as the polyion charge density increases. Although a strong bulk salt concentration dependence exists at small ξ, it diminishes as ξ increases. For high charge densities, comparable to DNA ($\xi \cong 4.2$), the Na^+ concentrations in the neighborhood of the polyion are essentially constant over a change in bulk concentration of two decades.

This type of behavior has been inferred from earlier theoretical and experimental studies [2-4, 15]. However, the apparently continuous nature of the onset of this insensitivity of local concentration to bulk concentration is in distinct constrast to popular treatments of the polyionic environment [2]. Furthermore, it is clear that the validity of the expression for the asymptotic electrostatic potential in terms of a definite *net* polyionic charge [4] stands independently of the physical picture of the short-ranged ionic distributions, which is manifest in the present results.

To further examine the concept of an effective net charge, it is instructive to evaluate the total small ion charge incorporated inside any given radius from the distribution function. We denote this quantity by $q(r)$ and express it as the fraction of polyion charge neutralized. Hence, $q(r)$ is initially zero at small r and is asymptotically equal to 1 at long distance. This result is shown in Fig. 4.4 as a function of polyion charge density for the particular NaCl concentration, $c_{ST} = 0.1\ M$.

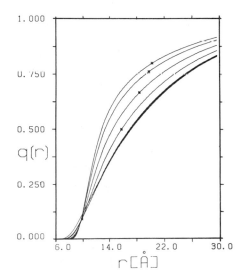

FIGURE 4.4. Total integrated small ion charge $q(r)$ for $c_{ST} = 10^{-1}\ M$ NaCl. The curves in order of increasing amplitude (beyond 10 Å) correspond to $\xi = 0.5, 0.9, 1.0, 1.1, 2.0, 3.0, 4.2,$ and 5.0. The asterisks mark the points at which $q(r) = 1 - \xi^{-1}$ for $\xi > 1$.

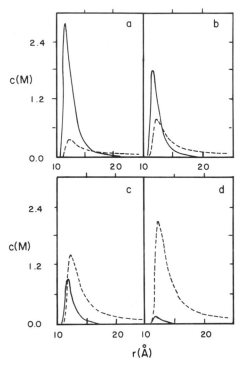

FIGURE 4.5. Counterion local concentration profiles for $\xi = 4.2$ and various NaCl/MgCl$_2$ mixture compositions; Mg^{2+}, ——; Na$^+$, – – –. For each frame, the molar ionic strength and ratio of Na$^+$ to Mg^{2+} bulk concentrations are, respectively, (a) 0.0178 M, 26.67; (b) 0.0166 M, 80.0; (c) 0.0151 M, 500.0; (d) 0.0150 M, 7500.

In formulations that focus on the long-range electrostatic potential [4], as well as those that assume an idealized definite spatial region for counterion accumulation [2], the fraction of polyion charge neutralized by counterions has the definite value $1 - \xi^{-1}$ for $\xi > 1$ and 0 for $\xi \leq 1$ for the case of cylindrical polyion geometry and univalent counterions. In Fig 4.4, we indicate on each curve for which $\xi > 1$ the distance at which this fraction is obtained.

While it is clear that the counterion charge is accounted for rather rapidly, it is also clear that this accumulation is gradual and that the distance enclosing the particular fraction $1 - \xi^{-1}$ is dependent on polyion charge density. However, in all cases, the distance enclosing this charge is $\lesssim 10$ Å in thickness.

An important phenomenon that is closely linked to the charge accumulation discussed previously is that of competitive ion association. It has been pointed out [2] that if one simply carried out an electrically neutral exchange of one multivalent ion in the bulk for an appropriate number of monovalent ions in the high concentration region near the polyion there would be a substantial gain in solution entropy resulting simply from entropy of mixing considerations. While the real process is necessarily more complex, the efficacy of, for example, Mg^{2+} in displacing Na$^+$ from the macroion environment is predictably large from this effect alone.

The size of the effect can be clearly seen in the results in Fig. 4.5 [16]. Here, we show one sequence of radial concentration profiles for the Na^+ and Mg^{2+} ions in $NaCl-MgCl_2$ mixtures near DNA. In the sequence from (a) to (d), the ionic strength is nearly constant at about 0.015 M, but the concentration ratio of Na^+ to Mg^{2+} varies from about 27 to 7500. As can be seen from these results, the Mg^{2+} concentration near the polyion always exceeds its bulk value by a substantially larger amount than does Na^+. Further, it is only under concentration conditions in the bulk that include Na^+ in great excess that one finds Na^+ as the majority species near the polyion. At higher ionic strength, the same phenomenon persists although it is predictably less dramatic.

We now turn to selected results for the more detailed polyion model. As indicated, these results are obtained using simulation; the basic cell included 20 base pairs spanning an axial length of 67.6 Å with an explicit inclusion of 820 polyion atoms [7].

The well-known duplex DNA structure [8] is significantly different from the simplified model considered previously in two important ways. First, the molecular surface in not a smooth cylinder, but rather it is characterized by two significant grooves, one of which approaches the cylindrical symmetry axis within about 4 Å. Second, the charge distribution is, of course, not uniform, but is distributed over specific atoms with the bulk of the net (negative) charge being localized on the phosphate groups that link the monomeric units.

Here, we consider a system corresponding to a bulk electrolyte concentration of about 0.1 M NaCl. We focus on the average radial distribution of Na^+ ions measured from the polyion axis. We compare this result, first, to the corresponding result for a uniformly charged perfectly cylindrical polyion and, second, to the result obtained with the detailed model of the polyionic *electrostatic* potential, but in which the detailed polyionic molecular shape has been enclosed by a cylinder of radius 11 Å, which completely excludes small ions from any closer approach. The latter comparison permits an evaluation of the relative significance of molecular shape and charge localization in the description of the ionic environment.

We should note that for the detailed polyion, the radial distribution of ions does not represent a complete description of the ionic distribution. The axial and angular correlations that are also present are, however, not of immediate concern here.

The results for the three cases described are shown in Fig. 4.6. The three curves are, from left to right, the fully detailed model, the detailed model with the exclusion radius, and the uniformly charged cylinder model. As is clear from the figure, the detailed model shows substantial radial penetration, a simple result of the macroionic shape. The peak at about 6 Å is traceable to small ions associated with the polyion within the deep groove referred to previously.

When the ions are artificially excluded from the grooves, the distribution becomes much more similar to that for the uniformly charged cylinder, also shown. However, there remain distinct differences. First, as a function of

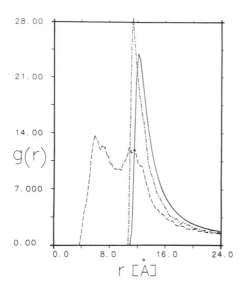

FIGURE 4.6. Na⁺ relative concentration distribution for various models of the DNA polyion in NaCl solution at $c_{ST} \cong 0.1\ M$. Detailed model ($c_{ST} = 0.114\ M$), – – –; detailed model with 11 Å exclusion radius ($c_{ST} = 0.117\ M$), – – – –; uniformly charged cylinder model ($c_{ST} = 0.1\ M$) ——.

radial distance, the distribution for the excluded radius case approaches that for the detailed model more rapidly than it approaches the uniformly charged model. Second, the distribution for the excluded radius case is distinctly more compact than that for the uniformly charged model. The interpretation of these results is that the localization of charge on the polyion, and the correspondingly larger local fields, causes stronger small ion association than is present for a dispersed polyion charge. The differences between the detailed and very simplified models are not solely attributable to molecular shape.

These differences in the structure of the ion atmosphere are also clear in the integrated charges $q(r)$. In Fig. 4.7, we plot the three results for $q(r)$, the total small ion charge enclosed within a radius r measured from the polyion axis (cf. Fig. 4.4). The curves are indicated by symbols corresponding to Fig. 4.6. As is clear from Fig. 4.7, the rise of the integrated charge densities for the polyion models with localized electrostatic charge is significantly more steep than that for the uniformly charged polyion model. Further, they merge into one another by about 30 Å, at a significantly higher value than that obtained for the uniformly charged case. (The small deviation of $q(r)$ above the value 1 between about 40 Å and 60 Å is presumably a result of the approximation made in treating the long-ranged mean field term in the simulation for this case; see Section II.)

The significant conclusion we draw from this analysis is that the phenomenon of charge association as a function of polyion charge and salt concentration can be qualitatively understood from a uniformly charged model, but that quantitative estimates of the local ionic concentrations and of the spatial extent of the association region cannot be well represented with such drastic simplification.

FIGURE 4.7. Total integrated small ion charge $q(r)$ for various models of the DNA polyion in NaCl solution. Models and labels as in Fig. 4.6.

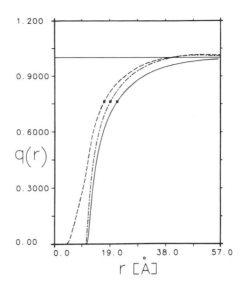

IV. Conclusions

The distribution of small ions around polyionic species is clearly an important feature for understanding the interaction among such species and the stability of polyionic assemblies. We have described these distributions for one class of such materials, namely, rod-like polyelectrolytes, within the limitations of accessible models. From these results, it follows that the spatial extent of the association region depends on the polyionic charge density and the bulk salt conditions. Further, any detailed description requires the detailed specification of the local electrostatic potentials. The results obtained from very simplified models provide only a generic picture that can only be valid for interactions at long range. Correspondingly, models that incorporate an effective net polyionic charge in the description of long-ranged interactions do not provide a basis for the formation of a physical picture of the short-range ionic structure.

An important area that remains to be fully explored is the interpretation of experimental probes of these small ion distributions, in particular, via small ion nuclear magnetic resonance (NMR) measurements [7,17,18]. As these results are sensitive to both the spatial distribution of ions and local electric field amplitudes, it appears clear that an adequate interpretation of such measurements requires a treatment employing a rather detailed model. Studies along these lines are now underway in our laboratory, and they promise to further elucidate the nature of these important ionic interactions.

Acknowledgments. This research has been supported by a grant from the National Institute of General Medical Sciences. P.J.R. is an Alfred P. Sloan Fellow and a recipient of the National Science Foundation Presidential

Young Investigator Award, the Camille and Henry Dreyfus Foundation Teacher–Scholar Award, and the NIH Research Career Development Award from the National Cancer Institute, Department of Health and Human Services.

References

1. D. Bendedouch, S.-H. Chen, W.C. Koehler, and J.S. Lin, *J. Chem. Phys.* **76**, 5022 (1982); M. Kotlarchyk and S.-H. Chen, *J. Chem. Phys.* **79**, 2461 (1983).
2. G.S. Manning, *Accounts of Chem. Res.* **12**, 443 (1979).
3. M. Guéron and G. Weisbuch, *Biopolymers* **19**, 353 (1980).
4. G.V. Ramanathan, *J. Chem. Phys.* **78**, 3223 (1983); B.H. Zimm and M. LeBret, *J. Biomolec. Struct. and Dyn.* **1**, 461 (1983).
5. R. Bacquet and P.J. Rossky, *J. Phys. Chem.* **88**, 2660 (1984).
6. C.S. Murthy, R. Bacquet, and P.J. Rossky, *J. Phys. Chem.* **89**, 701 (1985).
7. M. Rami Reddy, P.J. Rossky, and C.S. Murthy, *J. Phys. Chem.* **91**, 4923 (1987).
8. S. Arnott, P. Campbell-Smith, and P. Chandresekharan, "*Handbook of Biochemistry and Molecular Biology*," Vol. 2, p. 411 (Chemical Rubber Co., Cleveland, 1976).
9. S. Weiner, P. Kollman, D. Case, U.C. Singh, C. Ghio, G. Alagona, S. Profeta, and P. Weiner, *J. Am. Chem. Soc.* **106**, 765 (1984).
10. R.O. Watts in "*Specialist Periodical Reports, Statistical Mechanics*," Vol. 1, The Chemical Society, London, 1973.
11. H.L. Friedman and W.D.T. Dale, in "*Modern Theoretical Chemistry, Statistical Mechanics*," Part 5A, B.J. Berne, ed., Plenum, New York, 1977.
12. J.P. Valleau, L.K. Cohen, D.N. Card, *J. Chem. Phys.* **72**, 5942 (1980); M. Lozado-Cassou, R. Saavedra-Barrera, and D. Henderson, *J. Chem. Phys.* **77**, 5150 (1982).
13. J.P. Valleau and S.G. Whittington, in "*Modern Theoretical Chemistry, Statistical Mechanics*," Part 5A, B.J. Berne, ed., Plenum, New York, 1977.
14. G.M. Torrie and J.P. Valleau, *J. Chem. Phys.* **73**, 5807 (1980).
15. G.S. Mannins, *Accounts of Chem Res.* **12**, 443 (1979).
16. R. Bacquet, Ph.D. thesis, University of Texas at Austin, 1985; R. Bacquet and P.J. Rossky, *J. Phys. Chem.* **92**, 3604 (1988).
17. H. Wennerström, G. Lindblom, and B. Lindman, *Chemica Scripta* **6**, 97 (1974); B. Lindman, *J. Mag. Res.* **32**, 39 (1978).
18. M.L. Bleam, C.F. Anderson, and M.T. Record, Jr., *Biochemistry* **22**, 5418 (1983).

5
Experimental Study of Solvation Forces

R.G. HORN and B.W. NINHAM

In addition to the classical DLVO forces, well-known in colloid science, a number of other forces arising from the structural rearrangements of the intervening liquid molecules and ions may contribute to the interactions among colloidal species at separations of the order of a few molecular diameters. These forces go under a variety of names, such as structural forces, hydration forces, and solvation forces. First, we review the forces due to liquid structure, and we outline the circumstances under which they can be expected to be significant. Following this, recent results of direct measurements of these forces in a number of situations are examined, and their implications are discussed.

I. Introduction

In the Derjaguin–Landau–Verwey–Overbeek (DLVO) theory [1] of colloid particle interactions, there is a clear distinction between repulsive electrostatic (double-layer) forces [2] that keep charged particles apart and attractive van der Waals forces [3] that, depending on the balance between them, can lead to contact and coagulation. The repulsive forces decay exponentially at large distances with a range set by the concentration and valencies of an intervening electrolyte. They also depend on surface potential or charge. The opposing attractive forces have a much more complicated form, but to rough approximation have a power law dependence. In the net potential energy of interaction illustrated in Fig. 5.1, the maximum, which acts as a barrier to coagulation, varies in height with salt or surface potential. On the addition of salt, the barrier disappears and coagulation, now diffusion controlled, is rapid.

The theory developed from these ideas gives a good account of systems for which it was devised [e.g., solid (high Hamaker constant) gold or silver iodide sols] and underpins colloid science. It has not been so successful in dealing with properties of some colloidal aggregates, for example, surfactants, clay swelling, air bubble interactions in salt solutions, or the spacing of liquid

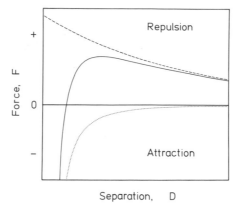

FIGURE 5.1. The classical DLVO theory of colloid stability [1] considers the net force
(——) between two surfaces in an electrolyte solution to be the sum of an electrostatic
double-layer repulsion (– – –) and a van der Waals attraction (· · ·). The strength of the
repulsion depends on surface charge, and its range depends on electrolyte concentration, whereas the van der Waals attraction is taken to be essentially independent of
these quantities. Since the attraction has an inverse power law dependence on surface
separation compared with the double-layer repulsion's quasiexponential decay, it
always dominates at short range. The surfaces should adhere if they overcome the
force barrier, whose height depends on the balance between attractive and repulsive
components.

crystal multibilayers. Thin films, adhesives, foams, emulsions, microemulsions, and many other systems of central industrial and biological importance
are not properly described by the older theory. Very different, very strong,
short-ranged stabilizing forces are operating here. They have been variously
termed structural components of disjoining pressure, structural forces, and
hydration or solvation forces, and their existence, range, and magnitude has
occasioned an emotional debate over at least 150 years [4].

What is new is that these structural forces, or forces due to surface-induced
liquid structure (and vice versa), have now been measured for a whole range
of systems. It is now beginning to be understood how the forces operating in
colloid science depend on the nature of the surface and on solution
properties.

II. Forces Due to Liquid Structure: Theoretical Expectation

It would be an easy matter to define solvation forces as anything additional
to those described by DLVO theory. To do so is rather like delineating the
nature of a dog's tail and defining a dog as anything else besides the tail. A

dog is a more complicated entity. Hence, before reporting measurements, and to fix ideas, we first attempt to lay out the shape of things that might be expected.

A. Solid Surfaces and Simple Liquids

Consider first two solute molecules immersed in a solvent made up of molecules that interact via a Lennard–Jones potential. The pair distribution function $g(r) \equiv h(r) + 1 = e^{-\beta W(r)}$, where $W(r)$ is the potential of mean force between the solute molecules. The distribution function [and $W(r)$] oscillates, reflecting the granularity of the solvent at very small distances of separation. At large distances, the discrete nature of the solvent is immaterial and $W(r)$ goes over asymptotically to a form $\sim 1/r^6$. That asymptotic form can be derived from Lifshitz theory of dispersion interactions (which includes all many-body forces), which treats the solvent as if it has continuum dielectric and other properties. The asymptotic long-range tail of $W(r)$ is an effective two-body interaction. The more important oscillatory part of the actual force operating can be called a microstructural force.

This is not the force operating between two solute molecules of nonzero size in a real constant pressure (Lewis–Randal) ensemble. Rather, it is the potential of mean force in the McMillan–Mayer (constant density) ensemble. The connection between the McMillan–Mayer, $W(r)$, and the real, constant pressure $W_P(r)$, requires a conversion that involves the single particle solute–solvent self-energy, also called the displacement term or solvation free energy, or in other words, partial molal quantities [5,6]. The actual force involves not just oscillations reflecting the bulk solvent granularity, but contributions reflecting adsorption excesses, that is, perturbations in the solvent density (and more generally orientation, hydrogen bonding, etc.) induced by the solute molecule—together with additional perturbation in the solute-induced solvent density profile due to interactions. Since real forces are always measured at constant pressure, we see that even at this simple level we expect microstructural forces to involve both bulk unperturbed solvent structure and solute-induced solvent structure. The latter depends on solute size, solute–solvent, solute–solute and solvent–solvent interactions.

With two solid surfaces, rather than solute molecules, we expect similar features. Asymptotically, continuum theories hold, but at some point the asymptotic forms break down. What is surprising at first sight is that, as will be seen, the microstructural forces between surfaces even in simple liquids oscillate strongly over at least 10 solvent molecular diameters. By contrast, for two solute molecules, the microstructure is not really an issue beyond about two molecular diameters. The reason, of course, is that the potential of mean force between a solvent molecule and a surface is of much larger range $[\mathcal{O}(1/r^3)]$ than that between a solvent molecule and a small solute particle $[\mathcal{O}(1/r^6)]$, leading to very different adsorption excesses.

B. Solid Surfaces: The Double Layer

Next, consider a primitive model electrolyte that separates the two charged surfaces of interacting colloidal particles. Suppose the surface charge or potential is fixed. In the primitive model, we treat the water as a continuum and the electrolyte ions as point charges. At a single charged surface, there will be a profile set up in ionic concentration, that is, a surface-induced liquid structure, the liquid here being the electrolyte. The overlap of two such profiles gives rise to the usual double-layer force [2]. It is par excellence the classic example of the existence of forces due to liquid structure.

At a higher level of approximation, we have to admit microstructure to the suspending solvent, water, and expect to measure a microstructural force which overlays the macrostructural double-layer force.

C. Solid Surfaces: Liquid-Mediated Surface Structure

Return now to our primitive-model point-ion electrolyte and suppose that the surface contains fixed dissociable groups, for example, $(AH) \rightleftarrows A^- + H^+$. Then, the degree of dissociation will be changed by both the added electrolyte and the proximity of another charged surface. The resulting double-layer interaction occurs neither at fixed surface charge nor at fixed potential, and the surface structure (charge density) regulates with distance [7,8]. This is a force due to the liquid (electrolyte, pH)-mediated surface structure.

D. Desolvation Forces

As an extension of that concept, suppose that the ions of our electrolyte, such as H^+ and Na^+, have different solvation energies (hydration). If the cations are present in sufficient excess in the reservoir electrolyte solution, when two charged surfaces are brought together, cations can be driven to a state where they give up water of hydration and bind to the interacting surfaces. There is here an additional secondary hydration force above and beyond the double-layer and microstructural forces.

E. Fluid, Nonplanar Surfaces

At a further level of complication, consider the surface of an ionic micelle. The head group area (i.e., surface charge) and curvature (size of the micelle) are set by a balance of hydrocarbon tail interactions, attractive surface tension due to exposed hydrocarbon at the hydrophobic core–water interface, and repulsive electrostatic interactions between the charged head groups. Even at the level of the primitive model, these head group interactions are set by the electrolyte concentration (and that of monomers and micelles) in a self-consistent way [9]. If two such micelles are brought together, there is a change in head group area because the double-layer interaction alters local ionic concentrations, and the surface rearranges in response; for example, with increased surfactant concentration or increased salt, micelles usually increase in size and grow through a sequence of intermediates through spheres → cylinders → vesicles → lamellar phase [10,11]. Sometimes, depending on counterion hydration, the sequence can be reversed [12]. While not solvation forces directly, these rearrangements, dissolution, or self-assembly processes clearly cannot be encompassed by DLVO theory, and they involve the self-energy of the surface and the monomers of which the surface is constituted.

F. Fluid, Planar Surfaces

For multilayers of liquid crystals, such as ionic double-chained surfactants, lecithins, galactolipids, or polyoxyethylene surfactants, constrained to a planar geometry, all of the previously discussed forces are involved to a greater or lesser extent. The additional complication for lecithins, for example, is that the head groups and their associated neighboring water molecules can be as bulky as the hydrocarbon tails. The interaction process involves, besides the entire self-assembly process, the adjustment of the head groups and tails in response to the additional chemical potential set up because of the proximity of an adjacent surface. Measured forces above and beyond van der Waals and double-layer forces are usually called hydration forces, or solvation forces if the solvent is nonaqueous. However, unlike that in Sections II.A through II.D, the deconvolution of measurements is clearly a very complex matter.

G. Hydrophobic Surfaces

Between hydrophobic surfaces immersed in a polar solvent, there are extraordinarily long-ranged strong attractive forces that have been recently measured. They reflect surface-induced reorientation of water in ways not understood.

H. Van der Waals Double-Layer Forces

A definition of solvation forces as anything beyond the continuum tail or DLVO theory is clearly not good enough. To pursue our metaphor further, even if it were acceptable for dogs, the class of animals with tails is broader than dogs. A classification can be made if the surfaces are smooth on a molecular scale and solid. In that case, we can distinguish forces due to the microstructure of an intervening liquid and those due to surface-induced liquid structure, both of which merge into the asymptotic forms derivable from continuum theory. But, for surfaces formed from molecules that spontaneously self-assemble, these distinctions are silly, especially since the surfaces are often ill defined over the operative distance scale.

Even at the level of continuum theory, the old distinction between van der Waals forces (considered to be independent of electrolyte concentration) and double-layer forces is blurred. Temperature-dependent van der Waals forces are important in oil–water systems. These forces are screened by salt with a decay length proportional to the Debye length of the intervening electrolyte [3]. (Such forces are due to fluctuation correlations in the Onsager–Samaris image profiles.) Between charged surfaces, correlations between fluctuations about the inhomogeneous double-layer ionic profile give additional forces not recognized by the standard theory [13,14]. These can, depending on surface charge and counterion type (and especially hydration radius), be an additional determinant of colloid stability.

III. Measurements

We now describe recent measurements that illustrate some of these forces. In doing so, we will follow the same logical sequence of increasing complexity set out in the previous section.

The results have been obtained using the force measurement apparatus developed by Israelachvili [15]. Two molecularly smooth mica surfaces are mounted in a crossed cylinder geometry and interact across an intervening liquid. The distance D between the surfaces is measured using an optical interference technique with a resolution of 0.1–0.2 nm. One of the micas is mounted on a cantilever spring whose deflection is measured to determine the force F, with an accuracy of 10^{-7} N. A more detailed description of the apparatus is given in Ref. 15. In some of the experiments, the mica provides a substrate for either a monolayer or a bilayer of surfactant.

In the Derjaguin approximation [16], the force between crossed cylinders of radius R is proportional to the interaction energy per unit area E_p between two flat plates at the same separation.

$$F(D) = 2\pi R E_p(D).$$

Force measurements scaled by the radius F/R can then be interpreted directly in terms of the energy between parallel plates.

A. Solid Surfaces and Simple Liquids

The force between mica surfaces in the nonpolar liquid octamethyl-cyclo-tetrasiloxane (OMCTS) oscillates between attraction and repulsion as the surface separation varies; see Fig. 5.2 [14]. The spatial periodicity of the oscillations corresponds to the molecular diameter of the liquid, $\sigma \approx 0.8$ nm. At short distances, the amplitude of the oscillations in the force law, that is,

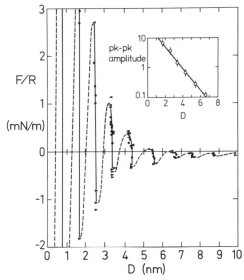

FIGURE 5.2. The force F (normalized by the radius of curvature R) between mica surfaces immersed in the simple nonpolar liquid OMCTS, whose molecular diameter $\sigma \approx 0.8$ nm [17]. Strong oscillations in the force result from the distribution of the solvent molecules in layers near the smooth, solid surfaces. Points represented by arrows facing right (▶) are measured at minima in the force curve, and those by arrows facing left (◀) at maxima. Because one mica surface is mounted on a spring, regions of the force law with positive slope (– – –) are unstable and cannot be measured with this technique. Points represented by solid circles (●) are measured on the stable regions (——).

the difference in force between a repulsive maximum and a neighboring attractive minimum, is large, and it decays as the separation increases. After about 10 oscillations, the force merges into its asymptotic tail: a van der Waals attraction.

This striking force law is an example of the microstructural force discussed in Section II.A. It results from a tendency of the liquid molecules to form layers next to the smooth, solid surface, so that the solvent density distribution function away from the surface has a series of peaks separated by approximately one molecular diameter. The modulation of the density profile extends much farther from a planar surface than the equivalent radial density distribution around a small solute molecule.

This force is measurable because both the surface roughness and distance resolution are finer than the size of the molecules of the liquid. Comparable force laws have also been measured between mica surfaces in cyclohexane [18], iso-octane, benzene, carbon tetrachloride [19], and other simple liquids, and also in the nematic liquid crystal 4'-n-pentyl-4-cyanobiphenyl [20]. In each case, the spatial periodicity seen in the force matches the molecular size.

B. *Electrical Double-Layer Forces*

Many measurements of electrical double-layer forces have been made between mica surfaces, with and without adsorbed surfactant layers, in aqueous electrolyte solutions. Some of these will be discussed below, but first we exhibit the double-layer force measured in propylene carbonate [21], a nonhydrogen-bonding liquid, free from some of the complications of water. This liquid is strongly polar (dipole moment 4.9D, dielectric constant 65). Electrolytes will dissolve and dissociate in it. Furthermore, cations (K$^+$) dissociate from the mica, leaving a negatively charged surface and an associated electrical double layer. The repulsive force measured when two double layers overlap is shown in Fig 5.3. The range of the force varies with electrolyte concentration exactly as predicted by classical theory.

At short distances, we again find a microstructural force due to the distribution of propylene carbonate molecules ($\sigma \approx 0.5$ nm) in layers next to the surfaces (Fig. 5.4). Beyond a separation of about 10 diameters, the force merges into the DLVO force: the sum of the double-layer repulsion and a van der Waals attraction.

It should be noted that the magnitude of the oscillatory part of the force is very similar to that measured in nonpolar liquids, but the oscillations are superimposed on the double-layer repulsion. When the repulsion is increased by changing the electrolyte concentration, the oscillatory part—both maxima and minima—is increased by the same amount. In other words, the net force can be approximately decomposed into the sum of a double-layer repulsion and an oscillatory microstructural component, which is comparable to that found in simple nonpolar liquids. This simple additivity suggests that the strong molecular dipoles have little effect on the ordering of the solvent near

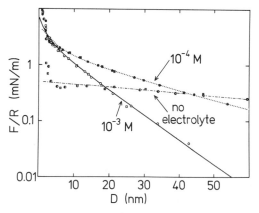

FIGURE 5.3. Double-layer repulsion measured between mica surfaces (which are negatively charged) in the polar liquid propylene carbonate, with different concentrations of added 1:1 electrolyte [21]. The lines show theoretical fits. The range of the repulsion decreases with electrolyte concentration, in accordance with theory.

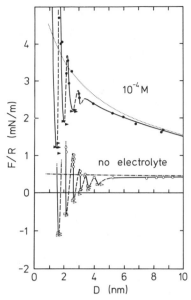

FIGURE 5.4. The short-range part of two of the force curves shown in Fig. 5.3. In this region, the force is oscillatory, with spatial periodicity equal to the molecular diameter of propylene carbonate, about 0.5 nm. The oscillations are superimposed on the double-layer repulsion extrapolated to short range (–·–·– in the absence of added electrolyte), and when that repulsion is increased by adding 10^{-4} M electrolyte ($\cdots$), the oscillatory part—both maxima and minima—is increased by the same amount. Symbols have the same meaning as in Fig. 5.2. [The outermost oscillations (beyond 3 nm) become difficult to detect when the underlying repulsion becomes very steep at 10^{-4} M.]

the surface. The dominant feature is still the hard-sphere packing, and dipole–dipole or dipole–surface charge interactions are relatively unimportant in determining the net force. (In water, the story can be different because the appropriate scaled parameter that measures the relative strength of dipole moment to hard-core size in determining bulk liquid properties is much larger.)

C. Liquid-Mediated Surface Structure: Charge Regulation

With mica surfaces interacting across aqueous solutions of alkali metal salts, it has been shown [22] that the measured double-layer force obeys neither constant surface charge nor constant surface potential boundary conditions. Rather, the surface charge depends (among other things) on the proximity of a second surface, so that it varies with separation. A simple surface dissociation model [22] gives a good account of this charge regulation and how it affects the double-layer force (Fig. 5.5).

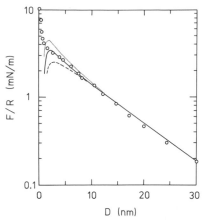

FIGURE 5.5. Surface charge regulation affects the double-layer force measured between mica surfaces in aqueous electrolyte solutions [22], 10^{-3} M CsCl in this case [39]. The top ($\cdots$) curve shows the theoretical force expected for surfaces interacting at constant surface charge; the bottom ($- - -$) curve shows the interaction at constant surface potential. The middle (——) curve, which fits the experimental points for $D > 2$ nm, is obtained when the surface charge is regulated according to the simple surface dissociation model presented by Pashley [22]. At short range, the force is not attractive, as the DLVO theory would predict, but strongly repulsive. This is because of the work required to remove water bound to Cs^+ ions adsorbed on the surface; hence it is called a secondary hydration force.

D. Desolvation Forces

In the same series of measurements, Pashley found that under certain solution conditions there is a strong short-range repulsion [23] between the mica surfaces. This contrasts strongly with the DLVO theory, which predicts that van der Waals attraction should dominate in this regime. With the same surface dissociation model, but allowing H^+ to compete with the other cations for negative surface sites on the mica, it was shown that the strong repulsion occurs when the surfaces are predominantly covered by cations [22,24] other than H^+. The cations that adsorb are hydrated, and it is the work required to remove water of hydration from the cations as they are brought together that gives rise to the repulsion, called a secondary hydration force (Fig. 5.5).

In a more recent experiment, remarkable for its accuracy, it was found that at very short range the hydration force between mica surfaces becomes oscillatory [25] because of the same microstructural effects discussed earlier (Fig. 5.6). Here, the spatial periodicity is only 0.25 nm, corresponding to the diameter of a water molecule. In this case, the oscillatory part of the force does not simply add to the double-layer repulsion as it does in propylene carbonate (Section III.B). There appears to be another component, mono-tonic and still of short range, that becomes increasingly repulsive as the

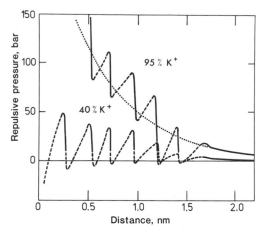

FIGURE 5.6. The hydration pressure between flattened mica surfaces in aqueous KCl solution [25]. Oscillations with spacing 0.25 nm—about the diameter of a water molecule—are evident, but the total force depends on how many K^+ ions are adsorbed to the mica surface, which in turn depends on the solution conditions. The upper curve shows a much stronger repulsive pressure when the coverage approaches 100% of the lattice sites. At zero coverage (not shown), there is no hydration repulsion.

electrolyte concentration increases (Fig. 5.6). Whether this is due to some subtle electrostatic effect not included in the standard theory of the double layer or whether it is a result of the extensive hydrogen-bonding properties of water is still an open question. Measurements in propylene carbonate [21] and ethylene glycol [26] give support to the latter idea, but the issue has not yet been settled unequivocally. The best simulation data so far also exclude the possibility of polarization orientation effects as the culprit.

E. Planar, Fluid Surfaces

Interactions between surfaces that are not so rigid and smooth as mica have also been measured, and as alluded to in Section II.F, their detailed interpretation in terms of fundamental mechanisms becomes problematic. Much interest was generated by the work of LeNeveu, Rand, Parsegian [27–29], and others, who measured the repeat spacing in multilamellar liquid crystal phases of lecithin and water as a function of osmotic stress. Equating this to mechanical pressure, they showed that there is a strong, short-ranged hydration repulsion between the polar head groups of lecithin across a water layer. This appears to be another desolvation force, as water of hydration resists the approach of opposing head groups. A short-range repulsion such as this explains why the lamellar phase forms with water layers between the (uncharged) lecithin bilayers and does not collapse under the influence of van der Waals forces.

The existence of a hydration repulsion in this system has been confirmed by measurements using mica surfaces coated with lecithin bilayers, formed either by adsorption from a vesicle dispersion [30] or by a Langmuir-Blodgett deposition [31]. Both experiments reveal a monotonic, exponentially decaying repulsion with a range of 2–3 nm. There are, however, slight differences between them that are indicative of the complexity of the system and the difficulty of isolating the underlying mechanisms. First, it is found that the area per molecule in the lecithin bilayer decreases as bilayers are forced closer together [28] in the multilamellar system, whereas it would tend to increase when two isolated bilayers (in an infinite reservoir of water) are pushed together. Second, the bilayers in a multilamellar system have more freedom to move (e.g., to undulate) than those deposited on mica. Steric hindrance effects between undulating bilayers probably contribute to the repulsion between them [32].

Neither the osmotic stress or the Israelachvili technique detects any oscillatory or microstructural component in this force. The reason for this is presumably that the lecithin head group is quite large, so on the scale of a water molecule, this surface looks quite rough. The surface is no longer like a mathematical plane, and water molecules are not found in well-defined layers. Furthermore, the head groups are quite mobile, and their motion would itself tend to average out any structure in the water density profile that might occur instantaneously.

Note that a very similar repulsion has been measured between lecithin bilayers in ethylene glycol [33], and that lecithin is known to form lamellar phases in this and other liquids [34,35], so the short-range hydration/solvation force is not unique to water.

Further measurements have been made on other double-chained surfactants adsorbed onto mica, and they show that the strength of the hydration repulsion varies considerably, depending on the particular head group involved. With phosphatidylethanolamine, closely related to the phosphatidylcholine head group of lecithin, the hydration repulsion extends only half as far (about 1.2 nm) [31]. The common plant lipids monogalactodiglyceride (MGDC) and digalactodiglyceride (DGDG) also have weaker hydration forces, extending about 0.6 and 1.2 nm, respectively [36].

With charged surfactants, the anionic phosphatidylglycerol [37] or cationic dialkyldimethlyammonium salts [38], very strong repulsions have been measured at short range. However, it turns out that these can be entirely attributed to the double-layer repulsion between these highly charged surfaces, and there is no need to invoke a hydration force at all (Fig. 5.7). For the cationics with hydroxide or acetate as the counterion, there is no binding, so the surface charge is known from the surface area per molecule. The theoretical double-layer force can be computed with no adjustable parameters, and it matches the measurements at all separations down to 0.5 nm (the smallest measured). Here, we have an apparently unequivocal test of the theory, and the test is passed with flying colors.

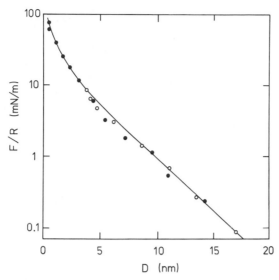

FIGURE 5.7. The force measured between mica surfaces coated with a bilayer of fully dissociated cationic surfactant, dihexadecyldimethylammonium ion in 10^{-2} M electrolyte [38]. In this case, the repulsion is very strong, but it can all be attributed to double-layer repulsion between these highly charged surfaces, without needing to invoke any additional hydration repulsion. The continuous line shows the theoretical fit, which is accurate at all distances down to a bilayer–bilayer separation of 0.5 nm. The strong repulsion prevented the surfaces from being forced closer than this.

F. The Hydrophobic Force

Perhaps the most interesting and important result comes from measurements between surfactant monolayers coated on mica surfaces. Here, the surfactant is absorbed with its head group next to the mica and its hydrocarbon tail exposed to form a hydrophobic surface. When two such surfaces are brought together in water, a strong attraction—significantly (10–100 times) stronger than the van der Waals force over most of its range—is measured (Fig. 5.8). This force was first measured by Israelachvili and Pashley [39] who worked with adsorbed monolayers of hexadecyltrimethylammonium bromide, and the result was greeted with disbelief. While the measured force itself could not be argued away as an artifact, the circumstances that the surfactant was soluble and the contact angle of water was only about $65°$ engendered disbelief in its interpretation as a hydrophobic effect. However, it has since been convincingly confirmed by several workers [40,41] using insoluble surfactants to form a monolayer with a contact angle $>90°$.

This force has been called the hydrophobic attraction, because it is presumed to have similar origins to the hydrophobic attraction frequently invoked in explaining oil agglomeration, surfactant aggregation, and many other phenomena. However, for the first time, it has been shown to be a force

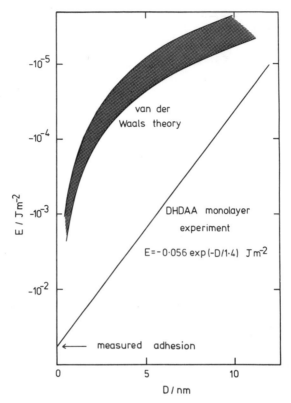

FIGURE 5.8. The attractive energy of interaction between two flat hydrophobic surfaces formed by depositing a monolayer of dihexadecyldimethylammonium acetate (DHDAA) on mica as a function of separation in water. Experimentally, the hydrophobic attraction is found to decay exponentially with distance, with a decay length of 1.4 nm [40]. It is much stronger than the power law attraction expected for van der Waals forces (shaded band), whose upper and lower bounds are given by values appropriate for the mica–water–mica and hydrocarbon–water–hydrocarbon interactions.

of considerable range (between large surfaces), decaying exponentially with a decay length of 1 or 2 nm. At distances as great as 15 nm, it is stronger than the van der Waals attraction given by continuum theory. The apparent attraction between small hydrophobic solutes in water is generally reckoned to result from the ability of water to form a hydrogen-bonded cage around the nonpolar molecule. It is difficult to conceive of such a cage around solutes whose radius of curvature is 1 cm. Nevertheless, it would seem that water next to the surfaces is being structured by their hydrophobic nature in some highly unfavorable way (compared with bulk water) [42], which leads to a long-range attractive force. Once again, the geometry has amplified the effect considerably, making the range over which water structure is affected surprisingly large.

G. Van der Waals Forces that Depend on Electrolyte Concentration

In measurements of the force between MGDG and DGDG bilayers deposited on mica, it has been found that the van der Waals attraction that was measurable in the range of 2–6 nm was reduced by a factor of 2 when the aqueous electrolyte concentration was increased from 10^{-5} M to 0.2 M [36]. The explanation for this is that about half of the van der Waals force in this system arises from the so-called zero-frequency term (or temperature-dependent term). This is a contribution from correlations in permanent dipole fluctuations. It is screened by the electrolyte in the same way as the double-layer force [3].

IV. Summary: New Forces and Old

The force between hydrophobic surfaces is probably the most startling of the new forces that have been measured, and the implications of these results both for biology and colloid science have not yet been fully comprehended. More subtle, but no less important, are the hydration forces that exist between charged surfaces in electrolyte or between lecithin bilayers. Just to complicate matters further, add to these the microstructural forces that are dominant at short range (between smooth, rigid surfaces). We see a rich plethora of solvation forces revealed by recent theoretical and experimental work. The details depend on many variables, such as the surface, liquid, electrolyte type, surface charge, and polarity; the list is long and our understanding is far from complete.

In the standard DLVO theory for most classical applications, the electrostatic potential barrier to flocculation occurs at a distance on the order of one Debye length, which is 3 nm in 10^{-2} M electrolyte. In biological systems, one works at an electrolyte concentration of 0.15 M, giving a Debye length of 0.8 nm, and it is clear from the results we have described that the classical theory simply cannot deal with the regime. Indeed, in some cases it will be inadequate at 3 nm. Yet, confusingly, there are situations where the nonlinear Poisson–Boltzmann equation gives an excellent account of double-layer interactions right down to very small distances (0.5 nm or less) [38], as shown in Fig. 5.7. Unlike the usual situation in which an ion-binding parameter is used to cover a multitude of sins, here there are no parameters, and the older theory appears to work well, and for the first time. It also appears to give a good account of the aggregation of ionic surfactants [9].

Even without considering specific solvation effects, newer theories recognize that the customary division of force into attractive van der Waals and repulsive double-layer components is blurred. It was shown a decade ago [3] that the zero-frequency component of the van der Waals force, because of fluctuation correlations in the Onsager–Samaris image profile (which gives rise to the change in surface tension at the air–water or oil–water interface in the presence of an electrolyte), is screened by the electrolyte. This has been confirmed by the experiment described in Section III.G. More recent theories [13,14] demonstrate that when the surfaces are charged corrresponding fluctuations in the inhomogeneous electrolyte profile give rise to an additional attractive force that, in the only case considered so far (counterions only), is substantial. The status of such forces is not entirely clear at this time. What is clear is that even asymptotically the old distinction between double-layer and van der Waals forces has gone. Forces are forces, and that is all.

References

1. Verwey, E.J.W., and Overbeek, J.Th.G., "*Theory of the Stability of Lyophobic Colloids,*" Elsevier, Amsterdam, 1948.
2. Chan, D.Y.C., and Mitchell, D.J., *J. Colloid Interface Sci.* **95**, 193 (1983).
3. Mahanty, J., and Ninham, B.W., "*Dispersion Forces,*" Academic Press, New York, 1976.
4. Ninham, B.W., *Adv. Colloid Interface Sci.* **16**, 3 (1982).
5. Chan, D.Y.C., Mitchell, D.J., Ninham, B.W., and Pailthorpe, B.A., in "*Water: A Comprehensive Treatise,*" Vol. 6, p. 239, F. Franks, ed., Plenum, New York, 1979.
6. Pailthorpe, B.A., Mitchell, D.J., and Ninham, B.W., *J. Chem. Soc. Faraday Trans. 2* **80**, 115 (1984).
7. Ninham, B.W., and Parsegian, V.A., *J. Theoret. Biol.* **31**, 405 (1971).
8. Healy, T.W., and White, L.R., *Adv. Colloid Interface Sci.* **9**, 303 (1978).
9. Evans, D.F., Mitchell, D.J., and Ninham, B.W., *J. Phys. Chem.* **88**, 6344 (1984).
10. Israelachvili, J.N., Mitchell, D.J., and Ninham, B.W., *J. Chem. Soc. Faraday Trans. 2* **72**, 1525 (1976).
11. Mitchell, D.J., and Ninham, B.W., *J. Chem. Soc. Faraday Trans. 2* **77**, 601 (1981).
12. Brady, J., Evans, D.F., Warr, G., Grieser, F., and Ninham, B.W., *J. Phys. Chem.* **90**, 1853 (1986).
13. Guldbrand, L., Jönsson, B., Wennerström, H., and Linse, P., *J. Chem. Phys.* **80**, 2221 (1984).
14. Kjellander, R., and Marcelja, S., *Chem. Phys. Lett.* **112**, 49 (1984); *Chemica Scripta* **25**, 112 (1985).
15. Israelachvili, J.N., and Adams, G.E., *J. Chem. Soc. Faraday Trans. 1* **74**, 975 (1978).
16. Derjaguin, B.V., *Kolloid Zh.* **69**, 155 (1934). For a more recent derivation, see White, L.R., *J. Colloid Interface Sci.* **95**, 286 (1983).
17. Horn, R.G., and Israelachvili, J.N., *J. Chem. Phys.* **75**, 1400 (1981).
18. Christenson, H.K., Horn, R.G., and Israelachvili, J.N., *J. Colloid Interface Sci.* **88**, 79 (1982).
19. Christenson, H.K., *J. Chem. Phys.* **78**, 6906 (1983).
20. Horn, R.G., Israelachvili, J.N., and Perez, E., *J. Phys. (Paris)* **42**, 39 (1981).
21. Christenson, H.K., and Horn, R.G., *Chem. Phys. Letters* **98**, 45 (1983).
22. Pashley, R.M., *J. Colloid Interface Sci.* **83**, 531 (1981).
23. Pashley, R.M., *J. Colloid Interface Sci.* **80**, 153 (1981).
24. Claesson, P.M., Herder, P., Stenius, P., Ericksson, J.C., and Pashley, R.M., *J. Colloid Interface Sci.* **109**, 31 (1986).

25. Pashley, R.M., and Israelachvili, J.N., *J. Colloid Interface Sci.* **101**, 511 (1984).
26. Christenson, H.K., and Horn, R.G., *J. Colloid Interface Sci.* **103**, 50 (1985).
27. LeNeveu, D.M., Rand, R.P., Parsegian, V.A., and Gingell, D., *Biophys. J.* **18**, 209 (1977).
28. Parsegian, V.A., Fuller, N., and Rand, R.P., *Proc. Natl. Acad. Sci. USA* **76**, 2650 (1979).
29. Lis, L.J., McAlister, M., Fuller, N., Rand, R.P., and Parsegian, V.A., *Biophys. J.* **37**, 657 (1982).
30. Horn, R.G., *Biochim. Biophys. Acta* **778**, 224 (1984).
31. Marra, J., and Israelachvili, J.N., *Biochemistry* **24**, 4608 (1985).
32. Evans, E.A., and Parsegian, V.A., *Proc. Natl. Acad. Sci. USA* **83**, 7132 (1986).
33. Persson, P.K.T., and Bergenstahl, B.A., *Biophys. J.* **47**, 743 (1985).
34. Moucharafieh, N., and Friberg, S.E., *Molec. Cryst. Liq. Cryst. Letters* **49**, 231 (1979).
35. Evans, D.F., Kaler, E.W., and Benton, W.J., *J. Phys. Chem.* **87**, 533 (1983).
36. Marra, J., *J. Colloid Interface Sci.* **107**, 446 (1985).
37. Marra, J., *Biophys. J.* **50**, 815 (1986).
38. Pashley, R., McGuiggan, P.M., Ninham, B.W., Evans, D.F., and Brady, J., *J. Phys. Chem.* **90**, 1637 (1986).
39. Israelachvili, J.N., and Pashley, R.M., *Nature* **300**, 341 (1982); *J. Colloid Interface Sci.* **98**, 500 (1984).
40. Pashley, R.M., McGuiggan, P.M., Ninham, B.W., and Evans, D.F., *Science* **229**, 1088 (1985).
41. Claesson, P.M., Blom, C., Herder, P., and Ninham, B.W. *J. Colloid Interface Sci.* **114**, 234 (1986).
42. Lee, C.Y., McCammon, J.A., and Rossky, P.J., *J. Chem. Phys.* **80**, 4448 (1984).

6
Statistical Mechanics of Confined Systems: The Solvent-Induced Force Between Smooth Parallel Plates

M.S. WERTHEIM, L. BLUM, and D. BRATKO

The force between two parallel plates immersed in a fluid of hard spheres is examined. Exact information about the model system is generated by Monte Carlo simulation of the density profile between two parallel, plane, hard walls. Theoretical calculations of two types are carried out and are found to agree very well with the simulations in certain regimes. The successful theory for very small gaps, barely exceeding one sphere diameter σ, incorporates the exact limiting law that the density approaches the fugacity as the gap approaches σ. The successful theory for wider gaps uses a superposition approximation of single-wall profiles, which are adequately approximated by the Percus shielding approximation. For high densities, we obtain oscillatory profiles similar to those obtained experimentally by Israelachvili and co-workers and in the computer simulations of Lane and Spurling.

I. Introduction

There is considerable interest in the statistical mechanics of confined systems such as (1) the state of water in small cavities, (2) the flow of solutions through narrow pores, and (3) the effect of the intercolloidal medium on the mean force between two large colloidal particles.

Recently, Israelachvili and co-workers have made direct measurements of the forces between two mica plates immersed in a liquid [1]. For liquids of neutral molecules, the force exhibited pronounced oscillations as a function of the distance between the plates. The observed behavior reflects the interplay of different forces present in real systems (see Ref. 2). In colloid science, they are traditionally classified into

1. repulsive core forces of very short range,
2. long-ranged electrostatic forces between charges and/or dipoles, and
3. van der Waals forces, of range intermediate between 1 and 2 above.

We take the view that it is essential to isolate the role of repulsive core forces by first studying purely repulsive models. The geometry chosen is that of a system confined between plane, parallel walls. This is, in fact, the traditional geometry used in studies of colloidal forces [3–5]. In adopting the hard-sphere model of solvent molecules, we depart from the precedent of Freasier and Nordholm [6], Lane and Spurling [7], and Chan et al. [8], who studied systems with soft repulsion. These papers and the review by Israelachvili [1] contain references to a large body of earlier work.

In Section II, we briefly summarize the theoretical basis of the approximations we have tested. In Section III, we discuss the Monte Carlo simulation method. The results are compared with theory in Section IV.

II. Limiting Laws

Quite different considerations apply to the two limiting cases for the distance between the hard walls, (1) a small gap barely exceeding one sphere diameter and (2) a large gap, that is, one of at least several diameters. Despite the conceptual simplicity of the hard-sphere fluids, the practical problem of bridging the gap between the two limiting laws is nontrivial.

A. Narrow Gap Limit

The appropriate relation for understanding the narrow gap limit is the fundamental equation for nonuniform systems [9,10]

$$\ln[\rho_A(1)/z_A(1)] = c_A(1). \tag{1}$$

Here, $\rho_A(1)$ is the density of particles of species A with an orientation of Ω_1 and a center located at $\underline{r}_1$. Similarly, $z_A(1)$ is a local fugacity, related to the constant fugacity z_A by

$$z_A(1) = z_A \exp[-\beta u_A(1)], \tag{2}$$

where $u_A(1)$ is the external potential for a particle of species A with coordinates $\underline{r}_1, \Omega_1$. The fugacity z_A is related to the chemical potential μ_A by

$$z_A = \Lambda_A \exp(\beta \mu_A), \tag{3}$$

where Λ_A is the usual factor arising from the kinetic energy partition function. For monatomics, we have $\Lambda_A = (2\pi M_A/\beta h^2)^{3/2}$. As usual, we define $\beta = 1/k_B T$, where T is the Kelvin temperature and k_B is Boltzmann's constant.

The function $c_A(1)$ can be represented by a graphical expansion in powers of either local density or local fugacity. The leading terms—one or two field points—are given by

$$c_A(1) = \mathord{\substack{\circ\\\bullet}} + [\mathord{\bigvee} + \tfrac{1}{2}\mathord{\bigvee}] + \cdots \tag{4}$$

and

$$c_A(1) = \mathord{\substack{\circ\\\bullet}} + \tfrac{1}{2}\mathord{\bigvee} + \cdots \tag{5}$$

A line between two points i, j represents a Mayer f-function:

$$f_{AB}(i, j) = \exp[-\beta u_{AB}(i, j)] - 1, \tag{6}$$

where $u_{AB}(i, j)$ is the pair potential. The black circles, so-called field points, carry factors

$$z_A(i) \text{ in Eq. (4)} \quad \text{and} \quad \rho_A(i) \text{ in Eq. (5)} \tag{7}$$

and imply integration over $\underline{r_i}$ and Ω_i and summation over species.

For our system of hard spheres between hard walls, all integrations are squeezed down to a two-dimensional region, namely, a plane, as the gap approaches the sphere diameter σ. As a result, all graphs vanish, and we have the limiting law

$$\rho_A = z_A. \tag{8}$$

Clearly, this result applies whenever the confining geometry reduces the accessible space to lower dimensionality.

This has an interesting corollary for a liquid in coexistence with its vapor. The equality of fugacities of the two phases implies that fluid in a confined space connected to the liquid takes on a density close to z, which is comparable to β times the vapor pressure. In other words, the liquid is vaporized in the gap, a possibility recently discussed by Massaldi and Borzi [11] in their analysis of flow through narrow pores.

It should be noted that the discussion leading to Eq. (8) does not apply to ionic systems, because the graphs in Eqs. (4) and (5) contain divergences. A correct discussion requires careful treatment of screening [12-14] and will be given elsewhere.

For very small gap distances L, that is, $(L - \sigma) \ll \sigma$, Eqs. (4) and (5) are effectively expansions in powers of $(L - \sigma)$. We have adopted the approximation of retaining only the leading graph in Eq. (5). For hard spheres, the resulting integral equation for $\rho(x)$ is

$$\ln\left[\frac{\rho(x)}{z}\right] = -\pi \int_{-(L-\sigma)/2}^{(L-\sigma)/2} \rho(x')[\sigma^2 - (x - x')^2] \, dx', \tag{9}$$

where x is the distance normal to the plates, measured from the center. Our theoretical narrow gap profiles are based on the analytic solution of this integral equation.

B. Wide Gap Limit

For a wide gap, one expects the density profile near each wall to be unaffected by the other wall and hence equal to the density profile near a single wall. Single-wall profiles are approximated adequately by the Percus shielding approximation [15]. By applying Eq. (4) to the two-wall system the two one-wall systems and the bulk, one can obtain a graphical expansion of the quantity

$$\ln\left[\frac{\rho_0\rho(x)}{\rho_L(x)\rho_R(x)}\right], \tag{10}$$

where ρ_0 is the bulk density and $\rho_L(x)$ and $\rho_R(x)$ are the densities in the presence of only the left and right wall, respectively. In the wide limit, all graphs become negligible, and we obtain

$$\rho(x) = \rho_L(x)\rho_R(x)/\rho_0, \tag{11}$$

an approximation already suggested by Percus [15].

In the narrow limit, however, this approximation leads to an asymptotic value of $\rho = (\beta P)^2/\rho_0$, which is far too small compared to z. Here, P denotes the pressure.

III. Monte Carlo Simulation

Computer simulations of idealized test systems, such as hard-sphere fluids, provide an unbiased test of theoretical models. Previous simulations of systems confined to a channel [7,16,17] deal with the regime where the channel is not too narrow. However, the very narrow limit may be relevant to some interesting phenomena, such as the coagulation of colloids.

Our system consists of a hard-sphere fluid confined between two plane, parallel hard walls of infinite extent. Physically, the system is in equilibrium with a reservoir of fluid at prescribed bulk density ρ_0, or equivalently, prescribed value of $\beta\mu$, where μ is the chemical potential. Such an open system is simulated by the grand canonical ensemble method [18], in which μ and the volume V of the system are fixed, while the number of particles N is allowed to fluctuate. Periodic boundary conditions were used in the two directions parallel to the walls. The Monte Carlo cell was a rectangular box of dimensions $a \times b \times L$, where L is the normal distance between the walls. The actual distance accessible to the fluids is $L - \sigma$.

At the begining of the run, N particles were placed randomly in the box. The run consisted of attempted additions and removals of particles, with equal a priori probability p, and attempted moves of randomly chosen particles, with probability $1-2p$. In most of the runs, the value chosen was $p = 0.15$. Each attempted move was made whenever the moved particle did not overlap with another particle or a wall. The acceptance probability of an attempted addition f_{ij} or deletion f_{ji}, where $N_j = N_i + 1$, was determined by the expressions [19]

$$f_{ij} = r, \quad f_{ji} = 1, \quad \text{for} \quad r < 1,$$

and

$$f_{ij} = 1, \quad f_{ji} = 1/r, \quad \text{for} \quad r > 1. \tag{12}$$

Here, r is the dimensionless quantity

$$r = (V\Lambda/N_j) \exp[\beta(\mu - U_j + U_i)], \tag{13}$$

where U_i and U_j are total configurational potential energies of states i and j. The average of r over ensemble states and positions of the inserted (deleted) particle is 1. When μ is expressed in terms of the excess chemical potential $\Delta\mu$ of the bulk system, using

$$\exp(\beta\mu) = (\bar{N}/V\Lambda) \exp(\beta\Delta\mu), \tag{14}$$

we have

$$r = (\bar{N}/N_j) \exp[\beta(\mu - U_j + U_i)], \tag{15}$$

where $\bar{N}$ is the mean number of particles in the Monte Carlo box at the bulk density.

The relation between ρ_0 and $\Delta\mu$ is given to sufficient accuracy by the expression [20]

$$\beta\Delta\mu = \eta(8 - 9\eta + 3\eta^2)/(1 - \eta)^3, \quad \eta = \pi\rho_0\sigma^3/6, \tag{16}$$

which is obtained by integrating the Carnahan–Starling equation of state [21]. The validity of Eq. (16) has been tested by direct simulation of $\beta\Delta\mu$ [22]. We have checked the validity of Eq. (16) by replacing the walls by periodic boundary conditions at the two densities at which experiments were run.

TABLE 6.1. Summary of simulations at $\rho_0\sigma^3 = 0.28731$.

	Box dimensions in units of σ		
			Extrapolated contact
$N_c{}^a$	Parallel	$L - \sigma$	density $\rho_c\sigma^3$
3.75	3.5 × 3.5	4	0.538 ± 2%
3	2.5 × 2.5	3	0.547 ± 2%
0.2	5 × 5	1.5	0.545 ± 2%
3	6 × 6	1	0.531
2.5	19.5 × 15.5	1	0.542 ± 1%
2.5	6 × 6.5	0.75	0.503 ± 1.5%
3	8 × 8	0.5	0.553
2.5	17.5 × 11.85	0.5	0.558 ± 0.2%
3	10 × 10	0.25	0.745
2.5	25 × 25.5	0.15	0.915 ± 0.2%
3	15 × 15	0.1	1.037
3	25 × 25	0.05	1.216
2	25 × 25	0.025	1.335
2	34 × 34.5	0.012	1.41

^{a} N_c, number of configurations/10^6.

Results were within 1% of the prediction of Eq. (16). Moreover, the correct bulk density was recovered as ρ far from the walls in the limit of large gaps.

The simulations were carried out on a Hewlett–Packard 9000 series 500 computer, using the built-in random number generator. The number of accepted configurations was $\sim 5 \times 10^5$ for the equilibration and 2 to 3×10^6 for the production runs. A run with about 100 particles and 2×10^6 configurations took about 10 hours. In some runs, the displacement lengths of moves were limited to assure an acceptance ratio of 0.4 to 0.5. In other runs, the new position was chosen randomly in the box. The latter procedure led to essentially the same final results as the former one, but with much slower convergence. Simulations were carried out for a number of gap widths at the two densities $\rho_0 \sigma^3 = 0.28731$ and $\rho_0 \sigma^3 = 0.6$. Monte Carlo box

TABLE 6.2. Summary of simulations at $\rho \sigma^3 = 0.6$.

N_c	Box dimensions in units of σ		Extrapolated contact density $\rho_c \sigma^3$
	Parallel	$L - \sigma$	
2	3.5×3.5	4	$2.56 \pm 1\%$
2.5	4.5×4.5	3.5	$2.48 \pm 3\%$
2.5	5×5	3.25	$2.62 \pm 2\%$
2.5	4.5×4.5	3	$2.40 \pm 3\%$
2.5	4.5×4.5	3	$2.50 \pm 3\%$
2.5	4.5×4.5	2.75	$2.54 \pm 3\%$
2.5	5×5	2.5	$2.54 \pm 1\%$
2.5	5×5	2.25	$2.66 \pm 3\%$
2.5	5×5	2	$2.73 \pm 2\%$
2.5	7×7	1.75	$2.51 \pm 2\%$
2.5	6×6	1.5	$2.24 \pm 1\%$
2.5	8×9	1.25	$2.66 \pm 2\%$
2.5	5.65×5.5	1.125	$3.10 \pm 2\%$
2.5	5×5	1	$3.32 \pm 2.5\%$
2.5	5.5×5.25	1	$3.40 \pm 1\%$
2.5	5×5	0.875	$2.90 \pm 2\%$
1.5	5×5	0.75	$2.06 \pm 4\%$
2.5	7.5×6.85	0.5	$1.66 \pm 1\%$
5	12×10.5	0.5	$1.67 \pm 2\%$
2.5	9×9	0.25	$2.43 \pm 1\%$
2.5	15.25×15.5	0.15	$3.74 \pm 1\%$
2.5	7.15×6.5	0.15	$3.70 \pm 1\%$
2.5	15.25×15	0.15	$3.74 \pm 1\%$
2.5	15×15	0.1	$5.20 \pm 1\%$
2.5	12.15×23.5	0.075	$6.68 \pm 2\%$
2.5	15×15	0.0555555	$8.55 \pm 1\%$
2.5	10×10	0.05	$9.60 \pm 2\%$
2.5	25×25	0.005	$47.5 \pm 1\%$
1	100×100	0.0025	$65.8 \pm 3\%$
2	100×100	1.25×10^{-5}	$122.6 \pm 0.5\%$
2	200×200	6.25×10^{-6}	$122.8 \pm 0.5\%$

dimensions, the number of configurations, and the results are summarized in Tables 6.1 and 6.2.

The density profiles $\rho(x)$ were obtained by counting particles in discrete parallel slices of the gap, as proposed by Karlström [16]. The mean number density of the slice was ascribed to the central plane. Since the outermost planes are half a slice thickness from the wall, extrapolation was used to obtain the contact densities ρ_c at the wall. The accuracy of this procedure was estimated by the method of subaverages [23] to be about 1–2% at the lower density and 1–3% at the higher density.

IV. Discussion of Results

The best test of approximations is the comparison of the complete density profile in the gap. Figures 6.1–6.5 compare Monte Carlo results with the profiles predicted by the wide gap approximation (W) of Section II.B. The single-wall profiles required for W were calculated from the analytic solution [24] of the Percus shielding approximation [15]. The agreement is quite

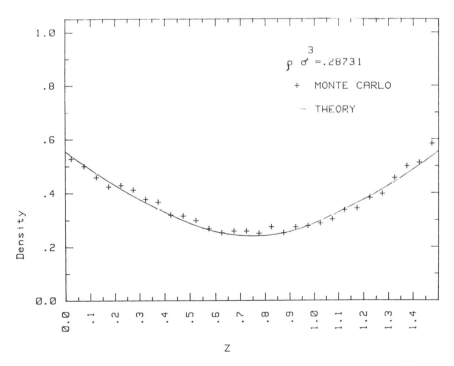

FIGURE 6.1. Density profile for bulk density $\rho_0^* = 0.28731$ and gap width $\hat{L} = 1.5$. Crosses are Monte Carlo results. The curve was calculated from the wide gap approximation W.

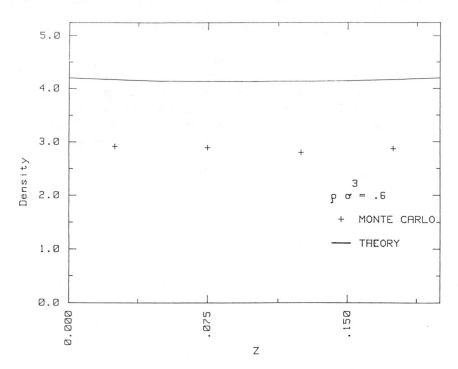

FIGURE 6.2. Same as Fig. 6.1, but with $\rho_0^* = 0.6$ and $\hat{L} = 0.2$.

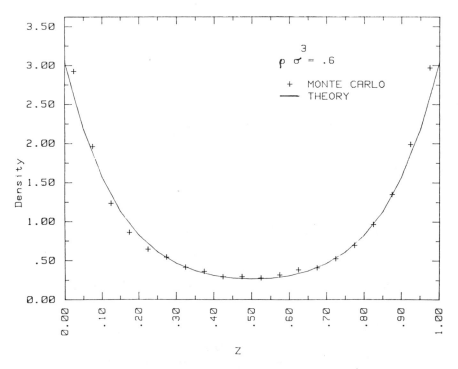

FIGURE 6.3. Same as Fig. 6.1, but with $\rho_0^* = 0.6$ and $\hat{L} = 1$.

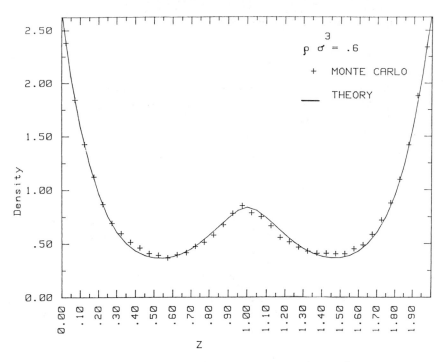

FIGURE 6.4. Same as Fig. 6.1, but with $\rho_0^* = 0.6$ and $\hat{L} = 2$.

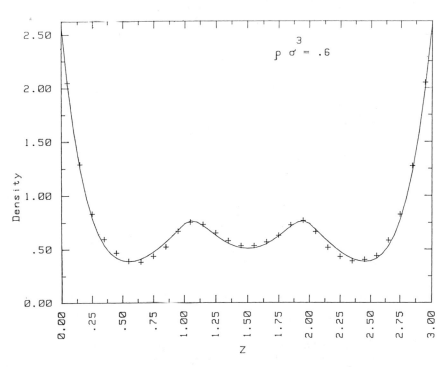

FIGURE 6.5. Same as Fig. 6.1, but with $\rho_0^* = 0.6$ and $\hat{L} = 3$.

good for values of $\hat{L} = (L - \sigma)/\sigma \geq 1$. For $\rho_0^* = \rho_0\sigma^3 = 0.6$ and $\hat{L} = 0.2$, both the simulation and W exhibit the flat profile characteristic of small gaps; the values, however, disagree. It is noteworthy that the values predicted by W are too large. Thus, we are still far from the small gap limit, where the values predicted by W are much too small.

An indication of the range of validity of the two limiting laws is contained in Figs. 6.6 and 6.7. Here, we compare the wall contact densities ρ_c^* as a function of gap width L. For $\rho_0^* = 0.28731$, the predictions of the narrow gap limit (N) and the wide gap limit (W) are parallel curves over a considerable range of L. The simulation results display the crossover from validity of N to validity of W in the range $\hat{L} = 0.07$ to 0.25. As ρ_0^* increases, the range of validity of N is rapidly squeezed down to very small gaps. This is so because the effective expansion parameter in Eq. (5) is

$$\pi(L - \sigma)\sigma^2\bar{\rho} = \pi\hat{L}\bar{\rho}^*, \tag{17}$$

where $\bar{\rho}$ is the mean number density in the gap. For small gaps, the increase in $\bar{\rho}$ with increasing ρ_0 is extremely rapid, similar to the behavior of the fugacity z. For this reason, only W is shown in the comparison for $\rho_0^* = 0.6$. Here, N takes over only for gaps of order $\hat{L} = 0.01$. The overall agreement between simulation and W is quite satisfactory for $\hat{L} > 1$. Except for very small $\hat{L}$, neither N nor W is satisfactory for $\hat{L} < 1$.

The force per unit area of the wall is $\rho_c k_B T$. Experimental observations [1] of the total force between two mica surfaces in the geometry of crossed

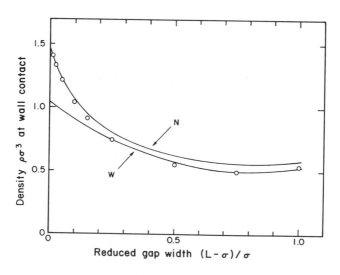

FIGURE 6.6. Wall contact density ρ_0^* as a function of gap width $\hat{L}$ at $\rho_0^* = 0.28731$. The circles are Monte Carlo results. The curves marked N and W are the narrow and wide gap approximations, respectively.

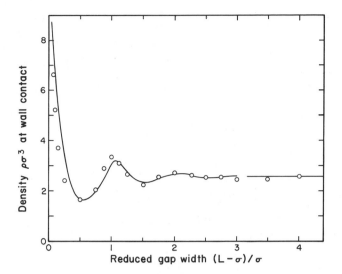

FIGURE 6.7. Wall contact density ρ_c^* as a function of gap width $\hat{L}$ at $\rho_0^* = 0.6$. The circles are Monte Carlo results. The curve was calculated from the approximation W. The straight line segment for $\hat{L} > 3$ is the asymptotic value of ρ_0^* for large $\hat{L}$.

cylinders show many large oscillations as a function of separation. This is qualitatively similar to our ρ_c. A direct comparison is impossible for two reasons.

1. Hard spheres and smooth hard walls are far from a faithful representation of the medium and the mica surfaces.
2. The experimental quantity is a total force, which depends on unknown features of surface geometry, while we calculate the force per unit area on plane parallel walls.

Acknowledgment. We thank Dr. G. Karlström for sending us a preprint of his work. M.S.W. was supported by the National Science Foundation under grant number CHE-8211236, L.B. by the Petroleum Research Fund under grant number 15473-AC7, and D.B. by the U.S. National Science Foundation through the U.S.-Yugoslavia Fund for Scientific and Technological Cooperation under grant number INT-8711845.

References

1. J.N. Israelachvili, *Phil. Mag.* **43**, 753 (1981).
2. R.G. Horn and B.W. Ninham, this volume.
3. B.V. Derjaguin, *Kolloid Z.* **69**, 155 (1934).
4. B.V. Derjaguin and L. Landau, *Acta Physicochim.* **14**, 633 (1941).

5. E.J.W. Verwey and J.T.G. Overbeek, "*Theory of the Stability of Lyophobic Colloids*," Elsevier, Amsterdam, 1948.
6. B.C. Freasier and S. Nordholm, *J. Chem. Phys.* **79**, 4431 (1983).
7. J.E. Lane and T.H. Spurling, *Aust. J. Chem.* **34**, 1529 (1981).
8. D.Y. Chan, D.J. Mitchell, B.W. Ninham, and B.A. Pailthorpe, in "*Water, a Comprehensive Treatise*," Vol. 6, Chapter 5, F. Franks, ed., Plenum, New York, 1979.
9. J.E. Mayer, *J. Chem. Phys.* **15**, 187 (1947).
10. F.H. Stillinger and F.P. Buff, *J. Chem. Phys.* **37**, 1 (1962). J.L. Lebowitz and J.K. Percus, *J. Math. Phys.* **4**, 116 and 1495 (1963).
11. H.A. Massaldi and C.H. Borzi, *J. Phys. Chem.* **85**, 1746 (1981).
12. C. Gruber, J.L. Lebowitz, and P.A. Martin, *J. Chem. Phys.* **75**, 944 (1981).
13. L. Blum, C. Gruber, J.L. Lebowitz, and P.A. Martin, *Phys. Rev. Lett.* **48**, 1769 (1982).
14. L. Blum, C. Gruber, D. Henderson, J.L. Lebowitz, and P.A. Martin, *J. Chem. Phys.* **78**, 3195 (1983).
15. J.K. Percus, *J. Stat. Phys.* **23**, 657 (1980).
16. G. Karlström, *Chem. Scripta* **25**, 89 (1985).
17. J.J. Magda, M. Tirrell, and H.T. Davis, *J. Chem. Phys.* **83**, 1888 (1985).
18. D.B. Chesnut and Z.W. Salsburg, *J. Chem. Phys.* **38**, 2861 (1963).
19. J.P. Valleau and G.M. Torrie, in "*Modern Theoretical Chemistry*," Vol. 5, B. Berne, ed., Plenum, New York, 1977.
20. D.E. Sullivan, *Phys. Rev. B.* **20**, 3991 (1979).
21. N.F. Carnahan and K.E. Starling, *J. Chem. Phys.* **51**, 635 (1969).
22. J.D. Adams, *Mol. Phys.* **28**, 1241 (1974).
23. W.W. Wood, in "*Physics of Simple Liquids*," H.N. Temperley, J.S. Rowlinson, and G.S. Rushbrooke, eds., North Holland, Amsterdam, 1968.
24. M.S. Wertheim, *J. Chem. Phys.* **84**, 2808 (1986).

7
Computer Experiments for Structure and Thermodynamic and Transport Properties of Colloidal Fluids

C.S. HIRTZEL and R. RAJAGOPALAN

The use of computer experiments for investigating equilibrium and nonequilibrium phenomena in supramolecular fluids is relatively recent. While in some respects dispersions of supramolecular and colloidal species can be treated as macroscopic analogues of the simpler atomic fluids, they differ substantially from the latter because of possible dependence of electrostatic and steric interactions on the density or volume fraction of the dispersions or the monomeric units that form the building blocks of the dispersed species. In the case of micellar solutions and microemulsions, the structure of the micelle or the microemulsion droplet itself may change with the above parameters. Additionally, in the case of dynamic properties (such as diffusion coefficients and rheological properties) of the dispersions, the accompanying electrohydrodynamic phenomena in the dispersions may need attention. Despite these complexities, even highly simplified treatments of colloidal dispersions as one-component or multicomponent supramolecular fluids offer considerable insights. This chapter presents an overview of Monte Carlo and Brownian dynamics experiments for colloidal fluids and a discussion of some of the results obtained from such experiments. In addition, based on equilibrium computer experiments, the conditions under which simpler theoretical techniques, such as integral equation theories and perturbation theories, are useful for estimating the structure and properties of model dispersions are outlined.

I. Introduction

It has been recognized for many years that dispersions of supramolecular species (of which the association colloids are perhaps one of the more complex examples) behave very much like a collection of simple atomic or molecular units in a statistical and structural sense (see Chapter 6 of Ref. 1 for a brief historical review and additional references). Nevertheless, the application of statistical mechanics and liquid-state physics to such dispersions has shown substantial growth only in recent years. While the entire task of describing the very complex hierarchy of equilibrium and nonequilibrium phenomena observed in supramolecular systems (particularly multicomponent ones) from first principles is certainly not within reach presently, some progress has been made in the last few years in understanding the interactions, structure, phase transitions, and critical phenomena in dispersions of species of simpler shapes. The other chapters in this book present numerous such examples of instances where statistical mechanics and liquid-state theory form the basis for interpreting interactions in dispersions, micellar solutions, and microemulsions. Additional examples may be found in other recent monographs [2,3].

This chapter will focus on one aspect of this area, namely, statistical mechanical computer experiments for studying properties of dispersions. Computational studies can take a number of forms depending on the specific phenomenon or the resolution of the length and time scales of interest. For instance, computer simulations have been used effectively to probe the structure of electrical double layers [4], ionic environments around polyelectrolytes [5], and the so-called structural forces [6], as illustrated in some of the other chapters in the present book. Also, at the molecular level, computer experiments can be designed to study the transient formation of pores, vesicles, or similar complexes in surfactant systems [7-10]. However, the objective of the present chapter will be to focus on interactions on a *coarser* scale. Specifically, we present a few classes of computer experiments that have been used for examining the static and dynamic structure of globular species of supramolecular dimensions (roughly 5 to 1000 nm) dispersed in a *structureless* continuum. The effect of the host fluid thus enters essentially in two ways: (1) as an effective dielectric and (2) as a continuum that serves to transmit the hydrodynamic stress field due to the motion of the species. In this sense, the dispersion is analogous to a fluid whose constituents (particles) are of colloidal dimensions (hence the name *supramolecular fluids*). The interparticle interactions in such a fluid, however, may be sensitive to the possible interactions between the electrolytes and surfaces of the colloidal particles. This latter aspect may sometimes be sufficiently important to make the study of supramolecular systems considerably more difficult. However, no attempt is made in this chapter to account for this complexity.

In addition, because of space limitations, we shall not go into the finer details of the computer simulation procedures. The objectives are to present the outlines of a few selected procedures that are useful and to summarize the major conclusions based on the available results. Some topics that need further attention are also suggested at the end. Despite these restrictions, a sufficient number of general and specific references is cited throughout the text to allow the readers to obtain more detailed information on the topics discussed.

Before we proceed, the following references are worth noting. Surveys of the statistical mechanical literature on colloids have been presented by a number of authors [11-13]. An overview of analytical approaches and results are available in Ref. 14. A general overview of research on interacting dispersions and many related references prior to 1985 are presented in the monograph cited earlier [1]. These references address association colloids only peripherally, although many of the topics discussed in these references are appropriate in the general context. However, a compendium that focuses specifically on the physics of micelles, vesicles, and microemulsions [15] may be consulted for issues of particular interest in association colloids. A detailed overview of scattering techniques as applied to supramolecular solutions has been recently presented by Chen [16].

II. Equilibrium Structure and Properties

An example of long-range statistical ordering in strongly interacting dispersions is illustrated by the experimental data from Brown et al. [17] shown in Fig. 7.1. These data were collected by light scattering experiments on dilute model dispersions of polystyrene latex particles in a monovalent aqueous electrolyte at an estimated concentration of approximately 10^{-6} M [18]. As evident from the figure, significant local structures can occur even at very low volume fractions; the highest particle concentration shown in Fig. 7.1

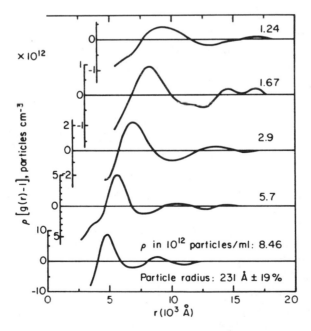

FIGURE 7.1. Development of structure as a function of particle concentration in a charged colloidal fluid, shown in terms of the radial distribution function (experimental data on polystyrene dispersions [17]).

corresponds to a volume fraction of less than about 5×10^{-4}, but liquid-like structure persists for distances on the order of 20 to 30 particle diameters. In this section of the chapter, we review some of the methods available for exact determination of the equilibrium structure, such as the one shown in Fig. 7.1, of colloidal fluids. This includes a discussion of general Monte Carlo procedures, the Metropolis (or Boltzmann sampling) Monte Carlo techniques, and non-Boltzmann computer experiments. Typical results for the structure and for some selected thermodynamic properties of colloidal fluids are presented. Last, a few remarks on the *transport* properties that are derivable from the knowledge of the equilibrium structure are given. Only brief descriptions of the above are presented here. Additional details on the computational methods may be found in the comprehensive monographs edited by Binder [19,20] and in Haile and Mansoori [21]. Details on computer experiments as applied to colloidal fluids are available in the numerous references cited in Ref. 1. Selected results on Monte Carlo studies of colloidal fluids are also presented in Ref. 14.

A. General Monte Carlo Procedures

The term Monte Carlo method refers to a general technique of estimating the value of a multidimensional integral by sampling the space of the integrand using random numbers. The Monte Carlo method represents the physical or mathematical system of interest in terms of the sampling procedure, which satisfies the same probability laws as the system being studied. For example, suppose that some function $f = f(x)$ for $x \in [0, 1]$. Let the variables X_i, $i = 1, 2, \ldots, n$, denote n uniform random variates. Then, the ensemble average of the function f, denoted by $\langle f \rangle$, is

$$\langle f \rangle = \frac{1}{n} \sum_{i=1}^{n} f(X_i) \tag{1a}$$

$$\approx \int_0^1 f(x) \, dx. \tag{1b}$$

A primary advantage of the Monte Carlo method is that the efficiency of evaluating integrals using the Monte Carlo technique does not depend on the dimensionality (number of dimensions) of the integral. In the case of crude Monte Carlo methods, the random variates $\mathbf{X}$ are assumed to be uniformly distributed on the interval $[0, 1]$.

More specifically, in the case of colloidal fluids, the multidimensional integrals to be evaluated are of the form

$$\langle f \rangle = \frac{\int f(\mathbf{r}^N) \exp[-U(\mathbf{r}^N)/k_B T] \, d\mathbf{r}^N}{\int \exp[-U(\mathbf{r}^N)/k_B T] \, d\mathbf{r}^N}, \tag{2}$$

where $\langle f \rangle$ denotes the ensemble average of function f (e.g., osmotic pressure or internal energy), and $\int g(\mathbf{r}^N)\, d\mathbf{r}^N$ represents the multidimensional integral

$$\int \cdots \int g(\mathbf{r}_1 \cdots \mathbf{r}_N)\, d\mathbf{r}_1 \cdots d\mathbf{r}_N. \tag{3}$$

The Monte Carlo method can be described briefly as follows. Some number (N) of particles, which interact through a specified pair potential, is placed in a box of known volume (V) in some initial configuration. Subsequently, a large number of configurations typical of the equilibrium state is generated. The structural and thermodynamic properties of interest are then evaluated, based on averaging of the appropriate quantities over these configurations, where equal weights are assigned to all configurations. For a system at equilibrium, the probability of the system being in a particular state associated with energy U is proportional to the Boltzmann factor, that is, $\exp(-\beta U)$, where $\beta = (k_B T)^{-1}$. The Monte Carlo method is then based on generating configurations in a manner that guarantees that any given configuration occurs with a probability proportional to $\exp(-\beta U)$ for that configuration.

B. Metropolis Monte Carlo Experiments

The Metropolis Monte Carlo technique used traditionally in statistical physics is one example of an importance sampling technique, a more sophisticated Monte Carlo technique. The major rationale underlying the motivation for more sophisticated techniques can be illustrated by considering Eq. (2). If the function $f(\mathbf{r}^N)$ is not a sensitive function of $\mathbf{r}^N$, then the integral in the numerator of Eq. (2), that is, $\int f(\mathbf{r}^N) \exp[-\beta U(\mathbf{r}^N)]\, d\mathbf{r}^N$, is dominated by low-energy regions. The importance sampling techniques are based upon generating configurations that have probability densities as follows:

$$h(\mathbf{r}^N) = \frac{\exp[-\beta U(\mathbf{r}^N)]}{\int \exp[-\beta U(\mathbf{r}^N)]\, d\mathbf{r}^N}. \tag{4}$$

The Metropolis Monte Carlo algorithm is based on a particular method (based on Markov chain theory) for generating configurations that satisfy Eq. (4). This algorithm was developed by Metropolis and co-workers for the study of the equilibrium behavior of dense gases and liquids; see, for example, Metropolis et al. [22] or Binder [19,20] for details of the Metropolis method and the monograph by Hammersley and Handscomb [23] for various related sampling techniques. Specific applications to colloidal fluids are available elsewhere [1,18,24–29].

Calculation of Structure and Thermodynamic Properties

The most direct piece of theoretical information on the equilibrium structure of an isotropic colloidal dispersion is the radial distribution function, $g(r)$. Physically, the radial distribution function is a measure of the local density as a function of distance from the center of an arbitrary particle. The product $\rho g(r)$ then represents the local number density of the dispersion relative to an arbitrary particle (ρ denotes the bulk-averaged number density of particles). The position of the first peak in $g(r)$ corresponds approximately to the average interparticle separation, and the height of the first peak is a measure of the number of nearest neighbors. One advantage of the radial distribution function is that it is also accessible through experiments using appropriate scattering techniques; the structure factor $S(K)$, which is obtained directly from such experiments, is related to $g(r)$ through Fourier transform.

The radial distribution function can be determined from the computer experiments by using the fact that the average number of particles separated by a distance that is in the interval $[r, r + \Delta r]$ is related to $g(r)$ as

$$\langle \eta(r, r + \Delta r)\rangle = 4\pi\rho r^2 g(r)\,\Delta r, \tag{5}$$

where ρ is the number density of particles. The function $\eta(r, r + \Delta r)$ is given by

$$\eta(r, r + \Delta r) = \sum_{i>j}^{N}\sum^{N} b(\xi_{ij}; r, r + \Delta r), \tag{6}$$

where ξ_{ij} denotes the distance between particles i and j and the function b is a binary function defined as

$$b(\xi; r, r + \Delta r) = \begin{cases} 1 & \text{if} \quad \xi \in [r, r + \Delta r] \\ 0 & \text{otherwise} \end{cases}. \tag{7}$$

The radial distribution function, together with the pair potential of interaction for the dispersion, provides the information required to obtain many of the equilibrium thermodynamic properties of the system (e.g., internal energies and osmotic pressures). A quantity M, which can be expressed in terms of the radial distribution function $g(r)$ as

$$M = 2\pi N\rho \int_0^\infty X(r)g(r)r^2\,dr, \tag{8}$$

can be obtained as a Monte Carlo average of the function $X(r)$. That is, M is determined as the ensemble average

$$M = \langle \omega \rangle, \tag{9}$$

where

$$\omega = \sum_{i>j}^{N}\sum^{N} X(r_{ij}). \tag{10}$$

(This method is known as direct Monte Carlo averaging.) Some specific examples of calculations of thermodynamic properties are given in Table 7.1. Other relationships are available in standard references on statistical mechanics.

Metropolis Monte Carlo Experiments on Colloidal Fluids: Results

Some typical results are presented for Metropolis Monte Carlo computer experiments on colloidal fluids. The pair potential of interaction $u(r)$ is assumed to be of the form given by the sum of a London–van der Waals attraction term and a repulsive contribution due to electrostatic repulsion between the particles. Extensive discussions of the forms of these interactions and of the origin and derivation of colloidal interaction forces are available in standard texts, such as Verwey and Overbeek [30], Kruyt [31], Vold and Vold [32], and Sonntag and Strenge [33]. The specific forms of the London–van der Waals interaction energy and of the electrostatic interaction used for the simulations described herein are given in Castillo, Rajagopalan, and Hirtzel [14] and so are not repeated here. The ranges of values of the physical and chemical parameters in the interaction energy expressions used in the computer experiments described herein are shown in Table 7.2; specific values are given in the individual figures where appropriate.

Figures 7.2 and 7.3 show radial distribution functions for colloidal fluids computed using Metropolis Monte Carlo simulations. The parameters of the dispersions are shown in each figure; all parameters have the same values in the two figures, with the exception of volume fraction ϕ. Figure 7.2 shows the radial distribution functions at a relatively low volume fraction, $\phi = 0.01$, and Fig. 7.3 shows the radial distribution functions at a moderate volume

TABLE 7.1. Expressions for thermodynamic properties in terms of the radial distribution function $g(r)$ and the pair potential of interaction $u(r)$.

Property	Expressions
Internal energy, E	$E = E_K + 2\pi N\rho \displaystyle\int_0^\infty g(r)u(r)r^2\,dr$
	$E_K = 3Nk_BT/2$
Pressure equation	$\dfrac{p}{\rho k_B T} = 1 - \left(\dfrac{2\pi\rho}{3k_B T}\right)\displaystyle\int_0^\infty g(r)\left[\dfrac{du(r)}{dr}\right]r^3\,dr$
Isothermal compressibility K_T	$K_T = -V^{-1}(\partial V/\partial p)_T$
	$= \rho^{-1}(\partial\rho/\partial p)_T$
High-frequency bulk modulus B_∞	$B_\infty = (2/3)\rho k_B T + p + \left(\dfrac{2\pi}{9}\right)\rho^2 \displaystyle\int_0^\infty \dfrac{d[r(du/dr)]}{dr}g(r)r^3\,dr$

TABLE 7.2. Physical and chemical parameters in the interaction energy expressions.

Parameter	Typical values[a]
Particle radius a_p	0.01–0.5 μm
Electrolyte concentration c	10^{-6}–10^{-2} M
Temperature of system T	280–360 K
Dielectric constant of the medium ε_r	70–90
Surface potential ψ_0	40–240 mV
Hamaker constant H	10^{-14}–10^{-12} erg
Valence of counterion z	1.0

[a] For a dispersion of polystyrene spheres in water.

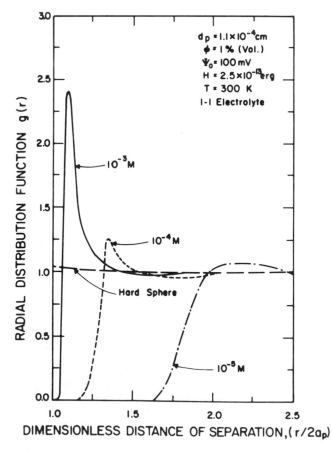

FIGURE 7.2. Radial distribution functions computed using Monte Carlo experiments at a volume fraction of 0.01. The parameters used are shown in the legend.

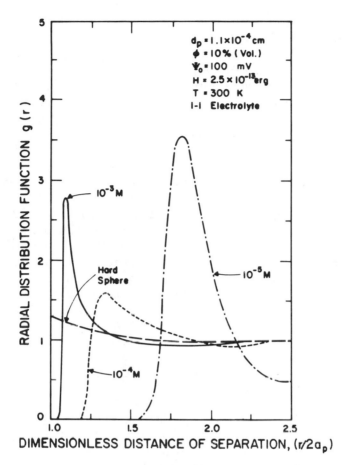

FIGURE 7.3. Radial distribution functions computed using Monte Carlo computer experiments at a volume fraction of 0.1. The parameters used are shown in the legend.

fraction, $\phi = 0.1$. In both figures, the hard-sphere $g(r)$ is also shown; these hard-sphere radial distribution functions were obtained from the Percus–Yevick analytical solution using a Verlet–Weis correction (see Castillo, Rajagopalan, and Hirtzel [14] for details). The hard-sphere case corresponds to that of absolutely no interactions, that is, no positional correlations are expected for the volume fractions considered here, and $g(r)$ is approximately unity for all distances of separation, r/a_p. For the case of the lower volume fraction, $\phi = 0.01$ (Fig. 7.2), the effect of increasing attraction can be seen as $g(r)$ changes for the different values of electrolyte concentration. At the lowest value of electrolyte concentration shown (10^{-5} M), the repulsion pushes the particles apart [compare this with the $g(r)$ shown for the hard-sphere dispersion]. As electrolyte concentration increases, the repulsion between the particles is screened more effectively, and the particles can get closer together.

At the highest value of electrolyte concentration shown (10^{-3} M), there is a high probability of finding a particle close to the surface of an arbitrary particle because of attraction. In such cases, pair interactions alone may suffice to explain the observed phenomena at sufficiently low volume fractions. On the other hand, at lower electrolyte concentrations, the dominance of repulsion and the large magnitudes of the screening length cause significant local or long-range ordering. This is illustrated in Fig. 7.3; as shown therein, significant long-range stucture develops for electrolyte concentrations of the order of $10^{-5}M$ or lower. At volume fractions larger than that shown in Fig. 7.3, these effects are even greater and significant long-range ordering can occur. Such effects cannot be accounted for by pair interactions only, and collective interactions must be considered.

As noted, the computer experiments yield exact results within the restrictions imposed by the usual assumptions invoked (e.g., pair-additivity of the potentials, an assumption that is also used in most of the analytical theories). Hence, the computer experiments can be used to test the accuracy of the analytical theories (i.e., the approximations made implicitly or explicitly in the analytical techniques). An illustration of this can be seen in Fig. 7.4, which compares radial distribution functions obtained from Monte Carlo simulations and from Weeks–Chandler–Andersen (WCA) perturbation theory. This figure also demonstrates the fact that perturbation theories [zeroth order or first order, depending on the strength of the perturbative part of the potential; see Section IV.A] are sufficiently accurate when the hard-core of the interaction determines the structure.

Other thermodynamic properties of the colloidal dispersions can be obtained, as described previously. Some examples of typical results are illustrated in Figs. 7.5 and 7.6. Osmotic pressure gradients [which are related to the structure factor at zero scattering angle, i.e., $S(0)$] are shown in Fig. 7.5 as functions of volume fraction ϕ for several values of electrolyte concentration. Results based on Monte Carlo simulations and two types of perturbation theories [i.e., second-order Barker–Henderson theory (BH PT) and first-order Weeks–Chandler–Andersen theory (WCA PT)] are compared. Note also that the collective diffusion coefficient D_c is related to the pressure gradient via the Stokes–Einstein relationship; the diffusion coefficient D_c is equal to the product of the gradient $(\partial p/\partial \rho)_T$ and the inverse of the friction coefficient. Finally, Fig. 7.6 shows the osmotic pressures (as a function of volume fraction) calculated from Monte Carlo computer experiments and from perturbation theories.

There are certain computational constraints associated with the Metropolis Monte Carlo experiments; these are related to the use of the minimum-image convention, periodic boundary conditions, and restrictions imposed by the existence of two-phase or multiphase coexistence regions or other situations in which Boltzmann sampling may be inadequate. For example, when the ordering over large interparticle distances cannot be ignored, the computer experiments become very expensive because of the need to use a

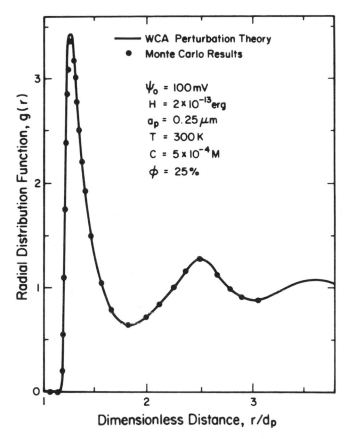

FIGURE 7.4. Comparison of the radial distribution function calculated from the perturbation theory with the results from a Monte Carlo simulation.

large number of particles in the simulations. This is a result of the fact that the distance over which the radial distribution function can be constructed is limited by the cutoff distance selected in the minimum-image convention (usually taken to be $L/2$, where L is the length of the simulation box; the number of particles used in the simulation, N, is equal to ρL^3 for a cubic simulation box). Details on these computational constraints are available in the references cited earlier.

C. Non-Boltzmann Computer Experiments

The Boltzmann sampling technique (i.e., Metropolis Monte Carlo algorithm) works well in those cases in which the property f of interest is not a sharply varying function of the configurational energy $U(\mathbf{r}^N)$, for example, osmotic pressure or internal energy [see Eq. (2)]. However, the Boltzmann sampling

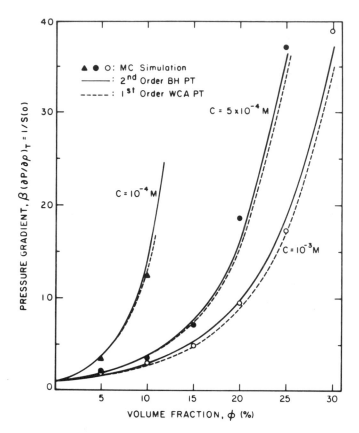

FIGURE 7.5. Osmotic pressure gradients calculated from perturbation theories compared with Monte Carlo results. The relevant parameters are shown in Fig. 7.6.

scheme is not efficient for estimating certain properties, such as the Helmholtz free energy, which are sensitive functions of $U(\mathbf{r}^N)$. Thus, there is a need for non-Boltzmann techniques in order to determine these properties. To see this, consider again Eq. (2) and suppose that the function f of interest is $f = \exp(\beta U)$. Then, from Eq. (2),

$$\langle f \rangle = \langle \exp(\beta U) \rangle \tag{11}$$

$$= \frac{\int \exp(\beta U) \exp(-\beta U) \, d\mathbf{r}^N}{\int \exp(-\beta U) \, d\mathbf{r}^N} \tag{12}$$

$$= V^N / Q^c, \tag{13}$$

where $d\mathbf{r}^N$ is the element of configurational space of volume V^N, and Q^c is the configurational portion of the partition function Q. Note that Q^c is related to the configurational Helmholtz free energy A^c by the relationship

$$A^c = -k_B T \ln Q^c. \tag{14}$$

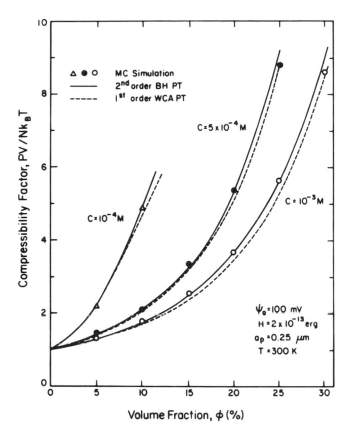

FIGURE 7.6. Comparison of osmotic pressures calculated using two types of perturbation theories and Monte Carlo experiments.

As can be seen from Eq. (11) through (13), the quantity Q^c is estimated as

$$Q^c = V^N/\langle\exp(\beta U)\rangle. \tag{15}$$

Since the exponential term $\exp(\beta U)$ increases very strongly with the configurational energy U, the high-energy configurations are quite important to the determination of the ensemble average $\langle\exp(\beta U)\rangle$ in Eq. (15) and, subsequently, are also very important to estimates of Q^c (and, hence, A^c). Recall, however, the fact that the Metropolis Monte Carlo algorithm focuses on low-energy regions (see previous discussion). Consequently, for estimation of properties such as the Helmholtz free energy, non-Metropolis (i.e., non-Boltzmann) computer sampling experiments are necessary.

The results of non-Boltzmann studies of colloidal dispersions are very few; Maleki, Hirtzel, and Rajagopalan [34] present results of the direct determination of the (Helmholtz) free energies of colloidal dispersions using a weighted, non-Boltzmann sampling technique known as umbrella sampling.

(The umbrella sampling technique, as developed for applications in statistical physics, is described in Valleau and Torrie [35] and Torrie and Valleau [36].) Maleki, Hirtzel, and Rajagopalan [34] compare the results of Boltzmann and non-Boltzmann computer experiments for the determination of Helmholtz free energies with free energies computed from perturbation theories. Their results also indicate the conditions under which Metropolis Monte Carlo experiments are adequate for the direct determination of Helmholtz free energies. In those situations in which Metropolis experiments are not adequate, results based on the non-Boltzmann, umbrella sampling algorithm are presented.

The basic idea underlying the direct determination of free energies can be described as follows. The Helmholtz free energy difference ΔA, between two systems, that is, a reference system (1) and a system of interest (2), is written as

$$\Delta A = A_2 - A_1 \tag{16}$$

$$= -k_B T \ln \int_{-\infty}^{+\infty} f_1(\Delta) \exp(-\Delta) \, d\Delta, \tag{17}$$

where Δ is the configurational energy difference (dimensionless) between systems 2 and 1, that is

$$\Delta = (U_2 - U_1)/k_B T, \tag{18}$$

and $f_1(\Delta) \, d\Delta$ is the probability of finding an energy difference of Δ in an interval $d\Delta$ around Δ. Thus, the density f_1 is the density that would be obtained in a conventional Boltzmann (Metropolis) sampling experiment. However, Boltzmann sampling is not always adequate to obtain reliable values of f_1 in the range of Δ important to the integral in Eq. (17). Alternatively, the umbrella sampling technique is based on weighting the sampling scheme so that both regions that contribute to the integral in Eq. (17) [i.e., the region where f_1 peaks and the region that contributes significantly to the integral in Eq. (17)] are covered approximately uniformly.

Several research needs, computational as well as physical, can be identified in the case of non-Boltzmann computer experiments. From the computational point of view, one drawback of the umbrella sampling technique is that the selection of weights used in the scheme is done via trial and error; no simple, nonarbitrary prescription for selection of these weights is, as yet, available. Second, there is a need to address the issue of identification and characterization of the conditions under which Boltzmann sampling can (or cannot) be used efficiently. Finally, from a physical point of view, there is a need to examine application of non-Boltzmann techniques (1) to the study of multiphase regions in electrostatic and steric systems [e.g., Eqs. (16) through (18) can be used with any two pair potentials and any two systems, one a reference system and the other a system of interest] and (2) to the testing of certain analytical theories and their approximate assumptions [e.g., perturbation and integral equation theories].

D. Transport Properties Derivable from Equilibrium Structure

The Monte Carlo method can be used to obtain dynamic properties of the dispersion provided the dynamic property can be expressed as an average of some time-independent quantities over an equilibrium ensemble. In general, if a property M can be expressed in terms of the radial distribution function $g(r)$ as

$$M = 2\pi N\rho \int_0^\infty Y(r)g(r)r^2 \, dr, \tag{19}$$

where $Y(r)$ is strictly a time-independent quantity, then the property M can be estimated as an equilibrium sample average of Y using a Monte Carlo experiment [see Eq. (8) and the discussion following it].

For example, the short-time diffusion coefficients can be expressed as equilibrium averages of the hydrodynamic mobilities. These mobilities depend only on the initial (static) structure of the dispersion and, hence, they can be estimated accurately using a Monte Carlo experiment. The short-time self-diffusion coefficient $D_{s,0}$ can be expressed in terms of the radial distribution function and the hydrodynamic mobility correction functions as

$$D_{s,0}/D_\infty = 1 + 2\pi N\rho \int_0^\infty Y(r)g(r)r^2 \, dr, \tag{20}$$

where

$$Y(r) = (2/3N)[A_{11}(r) + 2\,B_{11}(r) - 3], \tag{21}$$

and $A_{11}(r)$ and $B_{11}(r)$ are the hydrodynamic mobility correction functions. This procedure is known as direct Monte Carlo averaging [37,38]. The short-time collective-diffusion coefficients, especially at zero wave vectors, can also be estimated using a direct-averaging Monte Carlo procedure, although the relationships cannot be expressed as simply as the terms [e.g., $Y(r)$] in Eqs. (20) and (21). The short-time collective-diffusion coefficient, $D_c(0)$, at zero wave vector and without hydrodynamic interactions can be obtained from its relationships to the structure factor at scattering vector equal to zero, $S(0)$:

$$D_c(0)/D_\infty = 1/S(0). \tag{22}$$

The structure factor at zero angle, in turn, is related to the radial distribution function:

$$S(0) = 1 + 4\pi\rho \int_0^\infty [g(r) - 1]r^2 \, dr. \tag{23}$$

Thus, the coefficient $D_c(0)$ without hydrodynamic interactions can be obtained from a Monte Carlo experiment using direct averaging (for details, see Ref. 37). An alternative means for estimating $D_c(0)/D_\infty$ from a Monte Carlo experiment is also possible; this method is based on the relationship between $S(0)$ and the isothermal compressibility

$$S(0) = \rho k_B T K_T \tag{24}$$

and leads to a direct estimate of $[1/S(0)]$. Last, the determination of $D_c(0)$ but with hydrodynamic interactions deserves attention. This coefficient can also be obtained by direct-averaging Monte Carlo procedures, although the technique is not straightforward because of the complexity of the integrals and terms involved.

The procedures discussed here for the determination of the short-time collective-diffusion coefficients at zero wave vector can also be generalized to determine the collective-diffusion coefficients at nonzero wave vectors (this is possible provided that the wave vectors satisfy certain conditions [37]). For additional information on these calculations and on the equations involved, see Venkatesan [37] and the references cited therein. Snook and van Megen [26] have also calculated local self- and mutual-diffusion coefficients in colloidal dispersions using Monte Carlo experiments.

Finally, a comment on the limitations of Metropolis Monte Carlo experiments with respect to hydrodynamic interaction terms is appropriate. In employing Monte Carlo experiments, the hydrodynamic interactions have to be truncated to zero for distances greater than half the length of the simulation box. However, since hydrodynamic interactions are long-ranged, such truncation can lead to significant errors unless tail corrections, which account for the effects of truncation, are included in the calculations. Some methods to account for these tail corrections are described in Refs. 37 and 39.

III. Dynamic Structure and Phenomena

Since measurements of diffusion coefficients and rheological behavior offer yet another method of exploring the structure of supramolecular systems, there is considerable interest in simulating the dynamics of cooperative motions of strongly interacting supramolecular species. While, in principle, a large class of transport phenomena (e.g., ion transfer through membranes and diffusion-limited aggregation, in addition to rheological behavior) can be studied using computer experiments, the one that is perhaps the simplest is the diffusional relaxation (Brownian dynamics) of equilibrium structures from one to the other. The dynamics of equilibrium structural relaxation has a direct practical appeal as well, since the diffusion coefficients of supramolecular species can be measured directly and nonintrusively through radiation scattering techniques [40]. In addition, the extension of Brownian dynamics techniques (i.e., nonequilibrium experiments) to study structural transitions under externally imposed disturbances (e.g., shear field or electric field gradients) is also conceptually straightforward. Such experiments will be the focus of this section.

However, computational constraints (such as excessive storage and computational time requirements) have so far restricted the quality and the value of the results obtained from these experiments. Consequently, the following treatment will be confined primarily to computational outlines of the experiments and to the types of results one might expect to obtain from them.

A. Brownian Dynamics Experiments

The computational basis of the Brownian dynamics (BD) technique is analogous to the molecular dynamics (MD) method used to obtain time-dependent configurations of molecules in molecular fluids (see Alder and Wainwright [41], Rahman [42], and Barker and Henderson [43]). The major difference between the two is the random displacement of the particles in the BD simulation; this displacement enters the equations of motion through an appropriate stochastic force term and represents an average effect on the Brownian particles due to collisional momentum transfer between the particles and the molecules of the host fluid. (This basic idea has also been used in the simulation of the dynamics of single long-chain molecules suspended in molecular fluids; see Haile and Mansoori [21].) Thus, although many algorithmic details for a typical BD computer experiment (such as setting up the simulation box, calculating correlation functions, and determining time-dependent diffusion coefficients) can be borrowed from MD simulation methods, the actual procedure of obtaining time-dependent configurations is different from its MD counterpart and involves solving the underlying stochastic equations. In the following, a brief description of the solution procedure is provided. For convenience, simulation of hydrodynamically dilute systems is considered first. A list of time scales of interest in supramolecular dynamics is presented in Table 7.3.

For this case, the phase–space equations of motion are given by the following Langevin equations:

$$m_{pl}\frac{d\mathbf{u}_l}{d\tau} = -\zeta\mathbf{u}_l + \mathbf{F}_l + \mathbf{F}_{Bl}; \qquad l = 1, \ldots, N, \qquad (25)$$

where N is the number of particles used in the computer experiment and $\mathbf{F}_l$ and $\mathbf{F}_{Bl}$ are the net colloidal force and the Brownian force, respectively, on the lth particle. The particle mass is denoted by m_p and the velocity by $\mathbf{u}$. For times much greater than the viscous relaxation time (i.e., $\tau \gg \tau_B$), which are of primary interest here, the velocities of the particles would have already reached their equilibrium values governed by the Maxwell distribution, and, consequently, the acceleration terms on the left-hand side of Eq. (25) can be neglected. {The general method for simulating the motion of particles (in

TABLE 7.3. Some time scales of interest in supramolecular dynamics.

Time scale	Definition[a]	Significance
τ_B	$\tau_B \sim m_p/3\pi\mu d_p \sim 0.05\,(d_p^2/\nu)$ for neutrally buoyant spheres.	Relaxation of free Brownian motion[b]
τ_H	$\tau_H \sim l_H^2/\nu \sim a^2\,(d_p^2/\nu)$. $l_H = ad_p$ = distance over which hydrodynamic disturbances are of interest; $a \sim [(1/\phi)^{1/3} - 1]$ for hard-sphere dispersions.	Propagation time for hydrodynamic disturbances[b]
τ_I	$\tau_I \sim l_I^2/(2D_\infty) \sim (b^2/2)(\nu/D_\infty)(d_p^2/\nu)$; $l_I = bd_p$ = distance over which colloidal interactions extend; D_∞ = free Brownian diffusion coefficient of the particles; $b \sim [(1/\phi)^{1/3} - 1]$ for hard-sphere dispersions; $b \sim [(1/\phi)^{1/3} - 2N_{DL}^{-1} - 1]$ for charged dispersions with $N_{DL} = \kappa d_p$ (κ = inverse screening length).	Configurational relaxation time[b]
τ_{DL}	$\tau_{DL} \sim \kappa^{-2}/2D_i$; D_i = ionic diffusion coefficient.	Electrical double-layer relaxation time[c]
τ_{Ch}	$\tau_{Ch} \sim \sigma/i_0$; σ = charge density (on the particle); i_0 = exchange current density.	Time for charge adjustment on the particle[c]

[a] See *Nomeclature* for notations.

[b] Typical values of τ_{Br}, τ_H, and τ_I for $d_p \sim 1000$ Å, $\phi \sim 0.1$, and $\kappa^{-1} \sim d_p$ in water at 300 K: $\tau_{Br} \sim 10^{-9}$ s, $\tau_H \sim 10^{-8}$ s, and $\tau_I \sim 10^{-3}$ s.

[c] See Refs. 44 and 45. Typical values are $\tau_{DL} \sim 10^{-8}$ s and $\tau_{Ch} \sim 10^{-6}$–10^4 s.

hydrodynamically dilute dispersions) in phase space is given in Ermak and Buckholz [46].} Thus, one has

$$\zeta \mathbf{u}_l = \mathbf{F}_l + \mathbf{F}_{Bl}; \qquad l = 1, \ldots, N; \qquad \tau \gg \tau_B, \tag{26}$$

or

$$d\mathbf{r}_l/d\tau = b_\infty \mathbf{F}_l + b_\infty \mathbf{F}_{Bl}; \qquad l = 1, \ldots, N; \qquad \tau \gg \tau_B, \tag{27}$$

where $\mathbf{r}_l$ is the position of the lth particle, and b_∞ ($= \zeta^{-1}$) is the mobility coefficient. It is generally assumed that the properties of the fluctuating Brownian force $\mathbf{F}_{Bl}$ are

$$\langle \mathbf{F}_{Bl} \rangle = \mathbf{0}, \tag{28}$$

and

$$\langle \mathbf{F}_{Bl}(\tau) \mathbf{F}_{Bl}^T(\tau + \Delta\tau) \rangle = \mathbf{F}\, \delta(0), \tag{29}$$

where $\mathbf{F}$ is the strength of the Brownian force:

$$\mathbf{F} = 2k_B T \zeta \mathbf{I}. \tag{30}$$

The Brownian force vector $\mathbf{F}_{Bl}$ is therefore represented in terms of a probability density function with the mean value and autocorrelation kernel specified in Eqs. (28) and (29). However, to complete this representation, one needs higher order moments; these are usually taken to be zero in view of the common assumption that $\mathbf{F}_{Bl}$ is Gaussian (see Wang and Uhlenbeck [47]), that is,

$$\mathbf{F}_{Bl} \sim \hat{N}(\mathbf{0}, 2k_B T \zeta \mathbf{I}), \tag{31}$$

so that the first two moments of the probability density function of $\mathbf{F}_{Bl}$ are sufficient.

In view of Eq. (31), the equation for the positions of the particles given in Eq. (27) can be written as

$$d\mathbf{r}_l/d\tau = b_\infty \mathbf{F}_l + b_\infty \hat{N}(\mathbf{0}, 2k_B T \zeta \mathbf{I}); \qquad l = 1, \ldots, N; \qquad \tau \gg \tau_B. \tag{32}$$

Integrating Eq. (32) between the limits zero[1] and a small change in correlation time $\Delta\tau$, one has

$$\Delta\mathbf{r}_l = b_\infty \int_0^{\Delta\tau} \mathbf{F}_l \, d\tau + b_\infty \int_0^{\Delta\tau} \hat{N}(\mathbf{0}, 2k_B T \zeta \mathbf{I}) \, d\tau; \qquad l = 1, \ldots, N, \tag{33}$$

where

$$\Delta\mathbf{r}_l = \mathbf{r}_l(\tau + \Delta\tau) - \mathbf{r}_l(\tau). \tag{34}$$

---

[1] The zero limit refers to times much greater than the viscous relaxation time.

If we now choose the time step $\Delta\tau$ such that $\mathbf{F}_l$ does not change appreciably during that period, then the first term on the right-hand side of Eq. (33) becomes

$$b_\infty \int_0^{\Delta\tau} \mathbf{F}_l \, d\tau = b_\infty \mathbf{F}_l \, \Delta\tau. \tag{35}$$

For the second integral in Eq. (33), it can be shown[2] that for time periods much greater than the collision time between the particles and fluid molecules (i.e., $\Delta\tau \gg \tau_c$; $\tau_c \sim 10^{-20}$ s),

$$\int_0^{\Delta\tau} \hat{N}(0, 2k_\mathrm{B}T\zeta\mathbf{I}) \, d\tau = \hat{N}(0, 2k_\mathrm{B}T\zeta \, \Delta\tau \, \mathbf{I}). \tag{36}$$

Using the above results in Eq. (33), one obtains

$$\Delta\mathbf{r}_l = b_\infty \mathbf{F}_l \, \Delta\tau + b_\infty \hat{N}(0, 2k_\mathrm{B}T\zeta \, \Delta\tau \, \mathbf{I}); \qquad l = 1, \dots, N, \tag{37}$$

which can be also written as

$$\Delta\mathbf{r}_l = b_\infty \mathbf{F}_l \, \Delta\tau + \hat{N}(0, 2k_\mathrm{B}Tb_\infty \, \Delta\tau \, \mathbf{I}); \qquad l = 1, \dots, N, \tag{38}$$

where the relation $b_\infty = \zeta^{-1}$ has also been used.

The above result is the desired equation for describing the motion of the particles in coordinate space. For the purpose of using this equation as the basis for simulating the motion of particles (in hydrodynamically dilute systems), it is instructive to rewrite Eq. (38) in dimensionless form. To this end, we choose the dimensionless variables to be

$$\mathbf{R}_l = \mathbf{r}_l/d_\mathrm{p}, \tag{39}$$

$$\mathbf{c}_l = d_\mathrm{p}\mathbf{F}_l/(k_\mathrm{B}T), \tag{40}$$

and

$$\delta\tau = \Delta\tau/\tau_\mathrm{B}, \tag{41}$$

where d_p is the diameter of the particles, and $\delta\tau$ is the time step in units of the viscous relaxation time. In Eqs. (39)–(41), $\mathbf{R}_l$ and $\mathbf{c}_l$, respectively, are the dimensionless position vector and the dimensionless interaction force for the lth particle. Using Eqs. (39)–(41) one can show that the equation for the change in the position of the lth particle during the time period $\Delta\tau$ is given by

$$\Delta\mathbf{R}_l = L_B^2\mathbf{c}_l/2 + \hat{N}(0, L_B^2\mathbf{I}), \tag{42}$$

where

$$\Delta\mathbf{R}_l = \mathbf{R}_l(\tau + \Delta\tau) - \mathbf{R}_l(\tau), \tag{43}$$

and

$$L_B^2 = (2k_\mathrm{B}Tb_\infty\tau_\mathrm{B}/d_\mathrm{p}^2) \, \delta\tau. \tag{44}$$

---

[2] This is proven in many classical texts on Brownian motion; in particular, see Eqs. (146) and (147) in Chandrasekhar [48].

Equation (44) can also be written in terms of the single-particle diffusion coefficient D_∞ as

$$L_B^2 = 2D_\infty\,\Delta\tau/d_p^2, \tag{45}$$

where the Stokes–Einstein relation for D_∞ has been used. The quantity L_B^2 represents the mean-square displacement (in dimensionless form) of a particle due only to Brownian motion during the time period $\Delta\tau$. Finally, for computational convenience, Eq. (42) can be rewritten as

$$\Delta\mathbf{R}_l = L_B^2 \mathbf{c}_l/2 + L_B\hat{N}(\mathbf{0},\mathbf{I}). \tag{46}$$

This result is the basic equation for BD experiments in coordinate space for hydrodynamically dilute systems. For hydrodynamically dense dispersions, the appropriate equation is the set of coupled Langevin equations

$$m_{pl}(d\mathbf{u}_l/d\tau) = \mathbf{F}_l + \sum_{m=1}^{N} (\zeta_{lm}^{(N)}\cdot\mathbf{u}_m + \alpha_{lm}\cdot\mathbf{f}_m);\qquad l,m = 1,\ldots,N, \tag{47}$$

where $\zeta_{lm}^{(N)}$ is the pair-resistance tensor between the particles L and m in a system of N particles and $\mathbf{f}_m$ is related to the Brownian force on the mth particle. It can be shown (see Deutch and Oppenheim [49] and Stratonovich [50]) that the Brownian term $\mathbf{f}_m$ is Gaussian with

$$\langle\mathbf{f}_m(\tau)\rangle = 0, \tag{48}$$

and

$$\langle\mathbf{f}_i^a(\tau)\mathbf{f}_j^{\mathrm{T},b}(\tau+\Delta\tau)\rangle = 2k_B T\,\delta(\Delta\tau)\,\delta_{ij}\,\delta_{ab}\mathbf{I}, \tag{49}$$

where $\delta(\tau)$ is the dirac function, δ_{ij} and δ_{ab} are Kronecker δ's, $\mathbf{I}$ is the unit tensor, $\mathbf{f}^a$ is the ath component of $\mathbf{f}$, and $\mathbf{f}^{\mathrm{T},b}$ is the bth component of the transpose of $\mathbf{f}$. The coefficients α_{lm} in Eq. (47) are defined as

$$\sum_{k=1}^{N} \alpha_{lk}(\tau)\cdot\alpha_{km}(\tau) = \zeta_{lm}^{(N)}(\tau) \tag{50}$$

and specify the correlations (due to the hydrodynamic interactions) between the Brownian forces on the lth and mth particles.

Again, the above equation can be reduced to the following form for time steps $\Delta\tau$ much greater than the viscous relaxation time τ_B:

$$\Delta\mathbf{r}_l = \Delta\tau \sum_{m=1}^{N} [\mathbf{b}_{lm}^{(N)}\cdot\mathbf{F}_m + k_B T\,\nabla_{\mathbf{r}_m}\mathbf{b}_{lm}^{(N)}] + \hat{N}(0, 2k_B T\mathbf{b}_{lm}^{(N)}\,\Delta\tau);$$

$$l,m = 1,\ldots,N, \tag{51}$$

where the Gaussian term on the right-hand side of the above equation specifies the Brownian displacement during the period $\Delta\tau$. (It is also implicit that the colloidal force term $\mathbf{F}_l$ itself does not change appreciably during the period $\Delta\tau$.) Using the dimensionless variables given in Eq. (39) through (41), the above equation can be written (for the lth particle) as

$$\Delta\mathbf{R}_l = L_B^2 \left[\sum_{m=1}^{N} (\mathbf{H}_{lm}^{(N)} \cdot \mathbf{c}_m + \nabla_{\mathbf{R}_m} \cdot \mathbf{H}_{lm}^{(N)}) \right] + \hat{N}(0, L_B^2 \mathbf{H}_{lm}^{(N)});$$

$$m = 1, \ldots, N, \quad (52)$$

where

$$\mathbf{H}_{lm}^{(N)} = \mathbf{b}_{lm}^{(N)}/b_\infty \tag{53}$$

is the normalized mobility (or, diffusion) tensor between the particles l and m in a system of N particles. The other symbols in Eq. (52) have been defined earlier. Equation (52) is the desired displacement equation for describing the motion of particles in hydrodynamically interacting dispersions.

Thus, depending on the presence (or absence) of hydrodynamic interaction effects, Eq. (52) [or Eq. (46)] can be used as a basis for generating time-dependent configurations successively in a Brownian dynamics experiment.

Aside from the details of the relevant equations of motion, the basic simulation procedure is analogous to the molecular dynamics method and, hence, will not be discussed here. However, it is worth noting that the time step chosen in the simulation determines at least some of the details of the equation; this is summarized in Table 7.4. The coordinate–space simulation outlined previously, which is sufficient for the types of systems of interest in this chapter, is appropriate only when the time step $\Delta\tau$ is much larger than the Brownian motion relaxation time τ_B. (It is, of course, assumed that $\Delta\tau \ll \tau_I$, where τ_I is the configurational relaxation time; see Table 7.3.)

TABLE 7.4. Implication of time scales in computer experiments.

Magnitude of time step $\Delta\tau$	Implication
$\Delta\tau \ll \tau_B$	Phase-space simulation; Brownian motion approximation inadmissible for collisional momentum transfer
$\Delta\tau \sim \tau_B$	Phase-space simulation; continuum host fluid
$\Delta\tau \gg \tau_B$ (but $\ll \tau_I$)	Coordinate–space simulation; continuum host fluid
$\Delta\tau \ll \tau_H$	Hydrodynamic interaction negligible
$\Delta\tau \sim \tau_H$	Time-dependent hydrodynamics
$\Delta\tau \gg \tau_H$	Instantaneous hydrodynamic interaction

B. Nonequilibrium Experiments

A brief outline of the method of simulating a system under an externally imposed shear field follows.

The equation analogous to Eq. (25) for a collection of particles acted upon by an external field may be written as

$$\mathbf{m} \cdot \frac{d\mathbf{W}}{d\tau} = \mathbf{F}_h + \mathbf{F}_{nh}, \tag{54}$$

where $\mathbf{W}$ is a vector of dimension $6N$ containing the translational and angular velocities of all the N particles, $\mathbf{m}$ is a generalized mass/moment-of-inertia matrix of dimension $6N \times 6N$, $\mathbf{F}_h$ is the hydrodynamic force/torque vector of dimension $6N$, and $\mathbf{F}_{nh}$ is a vector of dimension $6N$ containing forces and torques of nonhydrodynamic origin. In the absence of significant Brownian motion, the hydrodynamic force/torque vector $\mathbf{F}_h$ has a linear relationship with the relative velocity vector $\mathbf{W}^r$ and the rate of strain tensor $\mathbf{E}$:

$$\mathbf{F}_h = \mathbf{R} \cdot \mathbf{W}^r + \mathbf{S} : \mathbf{E}, \tag{55}$$

where $\mathbf{R}$ is a grand resistance matrix (of dimension $6N \times 6N$) analogous to the collection of $\zeta^{(N)}$, used earlier, $\mathbf{W}^r$ are translational/angular velocities of the particles relative to the bulk translational–angular velocities of the particles (evaluated at the centers of the particles), and $\mathbf{S}$ is a shear resistance matrix of dimension $6N \times 3 \times 3$.

The grand matrices $\mathbf{R}$ and $\mathbf{S}$ possess many symmetry properties, and additional details on their structures may be obtained from Happel and Brenner [51] and Rallison [52] and references therein.

When the inertial effects can be neglected from Eq. (54), one can write, using Eq. (55),

$$\mathbf{W}^r = -\mathbf{R}^{-1} \cdot \{\mathbf{S} : \mathbf{E} + \mathbf{F}_{nh}\}. \tag{56}$$

Equation (56) forms the basis of nonequilibrium dynamics in supramolecular systems. The extension of this to Brownian particles becomes considerably more involved in view of the possible correlations between Brownian forces and hydrodynamic interactions [see, for example, the second term inside the summation sign in Eq. (47)] and will not be discussed further here since work on this topic is still in preliminary stages and is outside the scope of this chapter.

Aside from the differences in the equations of motion, nonequilibrium experiments differ from the earlier dynamics experiments in one major respect, and this difference lies in the need for changing the shape of the simulation box as a function of time (in the case of dispersions under steady shear).[3] Such changes in shapes are needed to maintain consistency in

---

[3] Other types of nonequilibrium experiments, in which the simulation box may remain in its original shape while other nonequilibrium conditions, such as a force gradient, are imposed on the dispersion, are not considered here.

applying periodic boundary conditions and are similar to the procedure used by Evans [53] in nonequilibrium molecular dynamics.

Nonequilibrium experiments are considerably more involved and time-consuming than the earlier experiments discussed in Section III.A, and consequently, only some preliminary attempts on lower dimensional systems are currently in progress. However, both classes of experiments do have some common features, particularly with regard to the precise manner in which hydrodynamic interactions must be accounted for in the computations. Some preliminary results on these topics are discussed in the following section.

C. Some Results and Discussion

As mentioned, incorporation of hydrodynamic interactions offers the major difficulty in the simulation of Brownian dynamics. This difficulty manifests itself in two ways. The first is the problem of specifying the complex many-body hydrodynamics theoretically. Even the simplest case of two-body interactions has been specified sufficiently accurately only in recent years (see [54] and related references cited in [1]). The second difficulty is the related problem of numerical manipulation of the large grand resistance matrices (see Sections III.A and III.B).

In the absence of hydrodynamic interactions, Brownian dynamics simulation is relatively simple (albeit time-consuming) and has been presented by a number of investigators [37,55-57]. Results of such simulations can be presented in terms of van Hove correlation functions, intermediate scattering functions, and diffusion coefficients, computed from mean-square displacements. The first two are equivalent. An example of the computed van Hove distinct correlation function has been shown in Fig. 7.7. The self-correlation function (not shown), which is Gaussian for noninteracting particles, deviates strongly from the free-diffusion case when either the particle concentrations or the colloidal interactions become significant. The Gaussian form is the optimum estimate if only the first two moments (the mean and the standard deviation of the displacements) are known. However, the general case of the self-correlation function (for interacting particles) can be expanded in terms of an infinite series (in Hermite polynomials) with the Gaussian form as the reference; this leads to marginal improvements for intermediate times (on the order of a few milliseconds). The Gaussian form becomes exact in the limits $t \rightarrow 0$ and $t \rightarrow \infty$.

The inclusion of hydrodynamic interactions in the above computations is a formidable task that has not yet been investigated satisfactorily. Usually, n-body interactions for $n > 2$ are assumed to be given approximately by the sum of pair interactions—an assumption that cannot be justified on physical grounds because of the long-range nature of the hydrodynamic interactions. More specifically, at large volume fractions of the dispersed species, pair interactions are strongly screened by the intervening particles, and the

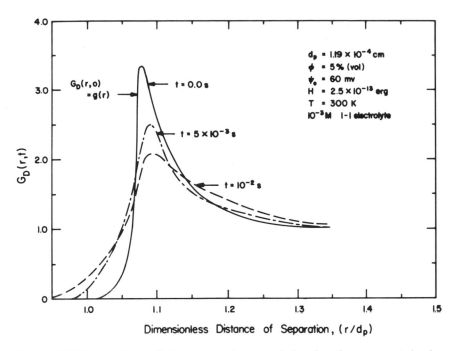

FIGURE 7.7. The van Hove distinct space-time correlation functions computed using Brownian dynamics.

nearest neighbors exert a much stronger influence than the particles farther away. In the absence of rigorous methods for including many-body hydrodynamic interactions, various approximations are currently being tried, although the accuracy of these is open to speculation. The computation of diffusion coefficients using Brownian dynamics has been hindered because of the above uncertainties and because of the extreme demands on the computational power required. Some general results, such as comparison of Brownian dynamics computations of mean-square displacements with the results based on Brownian motion in a harmonic potential (formed by the nearest neighbors), are presented elsewhere [1]. Typical examples of mean-square displacement as a function of time are also presented in Ref. 1.

Additional comments on dynamic experiments follow in Section IV.B.

IV. General Observations and Conclusions

This section presents a summary of major observations and conclusions that can be made based on the results of the computer experiments described in Sections II and III. This summary also includes some comments on the usefulness of a few analytical theories (e.g., integral equation theories and perturbation theories) in the case of dispersions of interest here. General observations for equilibrium structure and properties are presented first, followed by those on the dynamic aspects.

A. General Observations: Equilibrium Structure and Properties

In presenting these observations, it is convenient to divide the results according to the range of the pair potential since this range exerts an influence on the accuracy of the statistical mechanical approximations used in both analytical theories and computer experiments. Physically, dispersions at high electrolyte concentrations or with adsorbed, short-chain polymers have pair potentials with a range that is small relative to the particle diameters (i.e., $\kappa d_p \gg 1$). On the other hand, pair potentials in dispersions of charged macroions (at low electrolyte strengths and especially when the interaction is dominated by repulsion) have hard cores and soft, long-ranged tails (i.e., $\kappa d_p \lesssim 1$).

Short-Range Potentials ($\kappa d_p \gtrsim 6$)

In this case, since long-range cooperative effects are small unless the system is close to phase boundaries, a small number of particles (sometimes as low as 32) and minimum-image convention are sufficient in simulations.

Hard-sphere fluids, with an appropriate hard-sphere diameter (e.g., the Barker–Henderson diameter), serve as very good reference systems.

At moderate values of $\kappa d_p (\lesssim 100)$, the van der Waals attractive tail in the pair potential has a negligible effect on the structure. Usually, the zeroth-order radial distribution function from the Chandler–Weeks–Andersen perturbation theory is sufficient under such cases. For larger values of κd_p, including the first-order correction rectifies the deviations.

Integral equation theories have not been tested extensively using exact results from computer experiments. Preliminary results indicate that errors of up to 20–25 % may be expected in the location and the magnitude of the peak of the first coordination shell when the Percus–Yevick (PY) and hyper-netted chain (HNC) equations are used. Correspondingly, the thermo-dynamic properties based on these theories become substantially inaccurate (particularly at high volume fractions of the dispersed phase).

Long-Range Potentials ($\kappa d_p \lesssim 6$)

In this case, charged hard-sphere fluids or one-component plasmas may serve better as reference fluids than neutral hard spheres.

The HNC closure approximation and the mean spherical approximation (MSA) [58], or the rescaled MSA (RMSA) [59], seem to lead to better radial distribution functions than the PY approximation. The success of MSA and RMSA is, however, not surprising in view of the relation between the form of the closure relation used in Ref. 58 and 59 and the actual direct correlation function (see the article by Chen and Sheu [60] in this book).

The above observation is related to the recent attempts to circumvent a priori assumptions regarding the form of the interaction potential. Treating the dispersion as a highly asymmetric primitive model electrolyte, within the HNC scheme, leads to good estimates of the structure factor (see Ref. 60 and references therein).

Recent studies of order/disorder transitions in DLVO and Yukawa dispersions based on molecular dynamics and self-consistent phonon theory [61] appear to show that for long-range interactions a number of particles of about 1000 is needed in computer experiments for obtaining reliable results on freezing transitions and on the structure of crystalline phase. The results of this study also show no reentrant transitions, which had been reported previously [62,63].

The above observation indicates that the results for the case of long-range potentials are only tentative. Additional analytical and computational studies are needed.

Thermodynamic Properties

Values of the osmotic pressure and isothermal compressibility calculated from perturbation theories are in good agreement with results of the computer experiments. This is especially true when the repulsive part of the pair potential is steep and the secondary minimum is small. More work, however, is needed on the application of perturbation theories to dispersed systems to study the effects of softer repulsive potentials and longer ranged attractive potentials.

Perturbation theories and cell model theories are good statistical mechanical techniques for the disordered and ordered phases, respectively.

Applications of integral equation theories to estimate thermodynamic properties have not yet been examined extensively using computer experiments.

B. General Observations: Dynamic and Nonequilibrium Phenomena

Very little work has been done so far on dynamic simulations and nonequilibrium experiments (especially on the latter), and therefore the following comments are more qualitative than the ones presented in Section IV.A.

Brownian dynamics simulations that provide space–time correlation functions have been demonstrated to be possible in principle. These can also be used to generate intermediate scattering functions numerically.

The relation between the self-correlation function and the distinct correlation function in the case of colloidal dispersions is very similar to that observed in the case of classical fluids.

The major difficulty that must be overcome is the inclusion of indirect, hydrodynamic interactions in the computations. Pair-additivity of hydrodynamic interactions is unlikely to be of acceptable accuracy except at high dilution; this limitation is severe in the case of collective diffusion and sedimentation. A few theoretical and semi-empirical expressions for effective pair interaction are available, but have not been explored sufficiently.

The computational task is formidable, especially with the inclusion of hydrodynamic interactions. The long range of these interactions imposes severe restrictions on the periodic boundary conditions and box size used in the simulations, although the screening of the interactions by the intervening particles may ease these restrictions.

It is perhaps possible to study the range of many-body hydrodynamic interactions and to examine the available expressions for the mobility by computing diffusion coefficients and comparing these with experimental results.

V. Closing Remarks

Although considerable progress has been made in recent years with respect to various aspects of colloidal interactions and phenomena in dispersions, there are still many outstanding research problems that require attention. Some of these have been mentioned or alluded to in previous sections of this chapter. These research needs include

the systematic examination of integral equation theories,

studies of the structure and properties of steric and other supramolecular systems (e.g., polyelectrolytes),

examination of the structure and properties of interfaces (e.g., electrical double layers, interfacial structure of biocolloids at membrane surfaces, structure and properties of flocs),

examination of the dynamic behavior of dispersions (e.g., Brownian dynamics experiments using various forms of hydrodynamic coupling), nonequilibrium computer experiments,

understanding of the kinetics of flocculation processes, and

examination of the structure and properties of polydisperse systems (e.g., polydispersity of particle size, charge, or potential at the surface of the particles).

Acknowledgments. We would like to express our appreciation to the U. S. National Science Foundation and to the Donors of the Petroleum Research Fund administered by the American Chemical Society for partial support of some of the work reported here.

Nomenclature. This is a list of some of the more frequently used variables and abbreviations.

a_p	Radius of the particle
A	Helmholtz free energy
b_∞	Mobility coefficient of the particle
B_∞	High-frequency bulk modulus
c	Concentration of the electrolyte
d_p	Diameter of the particle
D_c	Collective diffusion coefficient
$D_{S,0}$	Self-diffusion coefficient at short times
D_∞	Single-particle free diffusion coefficient
E	Internal energy
E_k	Kinetic contribution to the total internal energy
$g(r)$	Radial distribution function
H	Hamaker constant
$\mathbf{I}$	Unit tensor
k_B	Boltzmann constant
K_T	Isothermal compressibility
L	Length of the simulation box
m_p	Mass of the particle
N	Number of particles in the computer experiment
$\hat{N}(x, y)$	Normal distribution with mean x and variance y
p	Pressure
Q	Partition function
r	Radial position
$S(\cdot)$	Structure factor
T	Absolute temperature
u	Pair potential
$\mathbf{u}$	Velocity vector of the particle
$U(\cdot)$	Configurational energy
V	Volume of the dispersion

Greek Letters

β	$(k_B T)^{-1}$
κ	Inverse screening length
ν	Kinematic viscosity of the host fluid
ρ	Number density of the particles in the dispersion
ϕ	Volume fraction of the particles in the dispersion

Abbreviations

BD	Brownian dynamics
BH	Barker–Henderson
HNC	Hypernetted chain
HS	Hard sphere
MC	Monte Carlo
MD	Molecular dynamics
MSA	Mean spherical approximation
PT	Perturbation theory
WCA	Weeks–Chandler–Andersen

References

1. C. S. Hirtzel and R. Rajagopalan, "*Colloidal Phenomena: Advanced Topics,*" Noyes Publications, Park Ridge, New Jersey, 1985.
2. S. A. Safran and N. A. Clark, eds., "*Physics of Complex and Supermolecular Fluids,*" Wiley, New York, 1987.
3. N. Ise and I. Sogami, eds., "*Ordering and Organisation in Ionic Solutions,*" World Scientific, Singapore, 1988.
4. B. Jönsson and H. Wennerström, *this book.*
5. P. Rossky, C. S. Murthy, and R. Bacquet, *this book.*
6. M. Wertheim, L. Blum, and D. Bratko, *this book.*
7. J. J. Wendoloski, S. J. Kimatian, C. E. Schutt, and F. R. Salemme, *Science* **243**, 636 (1989).
8. M. C. Woods, J. M. Haile, and J. P. O'Connell, *J. Phys. Chem.* **90**, 1875 (1986).
9. B. Jonsson, O. Edholm, and O. Teleman, *J. Chem. Phys.* **85**, 2259 (1986).
10. K. Watanabe, M. Ferrario, and M. L. Klein, *J. Phys. Chem.* **92**, 819 (1988).
11. E. Dickinson, in "*Specialist Periodical Reports, Colloid Science,*" Vol. *4*, pp. 150–179, D. H. Everett, ed., The Royal Soc. of Chem., London, 1983.
12. E. Dickinson, in "*Annual Reports of Progress of Chemistry,*" Vol. *80*, Sect. *C*, pp. 3–37, The Royal Soc. of Chem., London, 1983.
13. W. van Megen and I. Snook, *Adv. Colloid Interface Sci.* **21**, 119 (1984).
14. C. A. Castillo, R. Rajagopalan, and C. S. Hirtzel, *Rev. in Chem. Eng.* **2**, 237 (1984).
15. V. Degiorgio and M. Corti, eds., "*Physics of Amphiphiles: Micelles, Vesicles and Microemulsions,*" North-Holland, Amsterdam, 1985.
16. S.-H. Chen, *Ann. Rev. Phys. Chem.* **37**, 351 (1986).
17. J. C. Brown, P. N. Pusey, J. W. Goodwin, and R. H. Ottewill, *J. Phys. A* **8**, 664 (1975).
18. W. van Megen and I. Snook, *J. Chem. Phys.* **66**, 813 (1977).

19. K. Binder, ed., "*Monte Carlo Methods in Statistical Physics*," 2nd ed., Springer-Verlag, Berlin and New York, 1986.
20. K. Binder, ed., "*Applications of the Monte Carlo Method in Statistical Physics*," 2nd ed., Springer-Verlag, Berlin and New York, 1987.
21. J. M. Haile and G. A. Mansoori, eds., "*Molecular-Based Study of Fluids*," Adv. in Chem. Series, Vol. 204, American Chemical Society, Washington, D. C., 1983.
22. N. Metropolis, A. W. Rosenbluth, M. N. Rosenbluth, A. H. Teller, and E. Teller, *J. Chem. Phys.* **21**, 1087 (1953),
23. J. M. Hammersley and D. C. Handscomb, "*Monte Carlo Methods*," Chapman and Hall, London, 1964.
24. W. van Megen and I. Snook, *J. Colloid Interface Sci.* **57**, 40 (1976).
25. I. Snook and W. van Megen, *J. Colloid Interface Sci.* **57**, 47 (1976).
26. I. Snook and W. van Megen, *J. Colloid Interface Sci.* **100**, 194 (1984).
27. C. A. Castillo, "*A Monte Carlo Study of Order/Disorder Transitions in Colloidal Dispersions*," MS Thesis, Rensselaer Polytechnic Institute, Troy, New York, 1982.
28. C. A. Castillo, "*Equilibrium Structure of Interacting Colloidal Dispersions*," Ph. D. Dissertation, Rensselaer Polytechnic Institute, Troy, New York, 1984.
29. B. Svensson and B. Jönsson, *Mol. Phys.* **50**, 489 (1983).
30. E. J. W. Verwey and J. Th. G. Overbeek, "*Theory of the Stability of Lyophobic Colloids*," Elsevier, Amsterdam, 1948.
31. II. R. Kruyt, ed., "*Colloid Science, Volume 1: Irreversible Systems*," Elsevier, Amsterdam, 1952.
32. R. D. Vold and M. J. Vold, "*Colloid and Interface Chemistry*," Addison-Wesley, Reading, Massachusetts, 1983.
33. H. Sonntag and K. Strenge, "*Coagulation and Stability of Disperse Systems*," Halstead Press, New York, 1972.
34. S. Maleki, C. S. Hirtzel, and R. Rajagopalan, *Phys. Lett.* **97A**, 289 (1983).
35. J. P. Valleau and G. M. Torrie, in "*Statistical Mechanics*," Part A, pp. 169–194, B. J. Berne, ed., Plenum, New York, 1977.
36. G. M. Torrie and J. P. Valleau, *J. Comput. Phys.* **23**, 187 (1977).
37. M. Venkatesan, "*Diffusion in Interacting Colloidal Dispersions*," Ph. D. Dissertation, Rensselaer Polytechnic Institute, Troy, New York, 1985.
38. M. Venkatesan, C. S. Hirtzel, and R. Rajagopalan, *J. Chem. Phys.* **82**, 5685 (1985).
39. W. van Megen, I. Snook, and P. N. Pusey, *J. Chem. Phys.* **78**, 931 (1983).
40. R. Pecora, ed., "*Dynamic Light Scattering*," Plenum, New York, 1985.
41. B. J. Alder and T. E. Wainwright, *J. Chem. Phys.* **27**, 1208 (1957).
42. A. Rahman, *Phys. Rev. A.* **136**, 405 (1964).
43. J. A. Barker and D. Henderson, *Rev. Mod. Phys.* **48**, 587 (1976).
44. J. Th. G. Overbeek, *J. Colloid Interface Sci.* **58**, 408 (1977).
45. R. Rajagopalan and J. Kim, *J. Colloid Interface Sci.* **83**, 428 (1981).
46. D. L. Ermak and H. Buckholz, *J. Comput. Phys.* **35**, 169 (1980).
47. M. C. Wang and G. E. Uhlenbeck, *Rev. Mod. Phys.* **17**, 323 (1945).
48. S. Chandrasekhar, *Rev. Mod. Phys.* **15**, 1 (1943).
49. J. M. Deutch and I. Oppenheim, *J. Chem. Phys.* **54**, 3547 (1971).
50. R. Stratonovich, "*Topics in the Theory of Random Noise*," Gordon Breach, New York, 1963.
51. J. Happel and H. Brenner, "*Low Reynolds Number Hydrodynamics*," Martinus Nijhoff, The Hague, 1973.
52. J. M. Rallison, *J. Fluid Mech.* **88**, 529 (1978).

53. D. J. Evans, *Mol. Phys.* **37**, 1745 (1979).
54. G. K. Batchelor, *J. Fluid Mech.* **74**, 1 (1976).
55. K. J. Gaylor, I. Snook, W. van Megen, and R. O. Watts, *Chem. Phys.* **43**, 233 (1979).
56. K. J. Gaylor, I. Snook, W. van Megen, and R. O. Watts, *J. Phys. A* **13**, 2513 (1980).
57. S. Tutt, "*Brownian Dynamics Simulation of Colloidal Dynamics*," MS Thesis, Rensselaer Polytechnic Institute, Troy, New York, 1982.
58. J. B. Hayter and J. Penfold, *Mol. Phys.* **42**, 109 (1981).
59. J. P. Hansen and J. B. Hayter, *Mol. Phys.* **46**, 651 (1982).
60. S.-H. Chen and E. Y. Sheu, *this book*.
61. R. O. Rosenberg and D. Thirumalai, *Phys. Rev. A* **36**, 5690 (1987).
62. P. M. Chaikin, P. Pincus, S. Alexander, and D. Hone, *J. Colloid Interface Sci.* **89**, 555 (1982).
63. W. H. Shih and D. Stroud, *J. Chem. Phys.* **79**, 6254 (1983).

8
Analytical Results for the Scattering Intensity of Concentrated Dispersions of Polydispersed Hard-Sphere Colloids

A. VRIJ and C.G. DE KRUIF

The analyses of the experimental scattering data for colloidal dispersions often rely on the assumption that the dispersed particles are of uniform size. In practice, this is seldom the case, and at least a moderate level of polydispersity often exists. Analytical expressions for the scattering intensity of concentrated dispersions of polydispersed, hard-sphere colloidal particles are presented in this chapter. The basic theoretical concepts and scattering equations are outlined, and equations for scattering intensity at low and moderate scattering angles are presented. Some calculated results are presented for a log-normal distribution in hard-sphere diameters. The equations presented are simpler than others available in the literature, and their use in calculating the average structure factor is straightforward.

I. Introduction

Scattering techniques employing light, X-rays, or (cold) neutrons are being used more and more to study the interactions between colloidal particles in a liquid. In analyzing the data, it is customary to assume that these particles are identical in size, shape, and internal structure and interact identically with neighboring particles. In practice, however, this is never the case. Colloidal systems are in varying degrees *polydisperse*.

The oil-soluble silica spheres that we have synthesized during several years in our laboratory are good examples of a colloidal system [1]. The diameters of these spheres range from 30 to 1000 nm. The standard deviations are 10–20% for the diameters of the smallest spheres and less than 5% for the largest ones.

Another colloidal system is a microemulsion consisting of tiny water droplets (with a diameter of only a few nanometers) dispersed in oil. This may show standard deviations of up to about 30% [2]. More complicated situations are possible in which very broad distributions in diameter or even multimodal distributions occur.

In dilute colloidal dispersions where no interactions between the particles are perceptible, scattering techniques have been used for a long time to characterize particle populations. In nondilute systems, however, things are much more complicated. It would be very useful indeed to have indications of the influence that interactions between the particles have on the determination of sizes. At the same time, we need to know how polydispersity influences the analysis of interparticle interactions.

It is fortunate, however, that fairly accurate *theoretical* calculations can be made for one important case, namely, where there are so-called hard-sphere interactions between the particles; these interactions are characterized by a pair potential of the form

$$U_{ik}(r) = \infty \quad \text{if} \quad 0 < r < \tfrac{1}{2}(d_i + d_k),$$

$$U_{ik}(r) = 0 \quad \text{if} \quad r \geq \tfrac{1}{2}(d_i + d_k). \tag{1}$$

Here, r is the distance between the hard-sphere centers and d_i and d_k are the hard-sphere diameters associated with the particles i and k.

The thermodynamic and structural properties of this pair potential are described fairly accurately by the approximate Percus–Yevick (PY) theory [3,4]. In this theory, an assumption is made about the relation between the pair distribution function $g_{ik}(r)$ and the so-called direct correlation function $c_{ik}(r)$, known from liquid-state theory. But, after this approximation has been adopted, the matrix of interactions $c_{ik}(r)$ can be resolved exactly [3,4]. The scattering spectrum, that is, the intensity of the scattered radiation as a function of scattering angle, can even be expressed in closed form for any distribution of hard-sphere diameters, as was shown by us [5,6] and independently by Blum and Stell [7]. We also did calculations [8] based on Ref. 6 for some interesting cases, applying a Schulz distribution of hard-sphere sizes.

Recently, some simpler approaches have been suggested [9,10], but these have proved less adequate for the study of hard-sphere interactions [11].

In this chapter, we treat the polydisperse hard-sphere case again, but this time more briefly. The results are simpler to use, we think, than might appear from the complex derivations. Therefore, we explain in some detail the underlying theoretical concepts that may not be so well known by investigators who are interested mainly in the results. For this reason, we have explicitly included the final equation for the scattering intensity. We also include some results of computer simulations of hard-sphere mixtures that have recently become available. They do indeed confirm the accuracy of the PY theory for hard-sphere scattering.

II. Scattering Equations

Consider a colloidal system of particles in a solvent where the interactions between the particles are spherically symmetric. Let the number of particle species be p. Each particle contains a center of force from which the distance r

to other particles is measured. For each pair of species $i, k = 1, \ldots, p$, there is a radial distribution function $g_{ik}(r)$ that measures the probability of finding the center of a particle of species k at distance r from the center of a given particle of species i (and vice versa).

A. Intraparticle Interference

In our treatment, the spatial distribution of local scattering power inside a particle can be chosen arbitrarily. A schematic diagram is shown in Fig. 8.1. For most systems, a centro-symmetric distribution of scattering material will be an appropriate one; the center of the scattering distribution coincides with the center of force.

The scattering field of a particle j is then given by

$$B_j(\mathbf{K}) = f_j^{-1} \int d^3\mathbf{R}\,\zeta_j(\mathbf{R})e^{i\mathbf{K}\cdot\mathbf{R}}, \tag{2}$$

where $\zeta(\mathbf{R})$ is the local scattering amplitude density, and the origin of R is the center of force. The integration is performed over the particle volume. Here, $\mathbf{K}$ is called the scattering vector and has a magnitude $K - |\mathbf{K}| = (4\pi/\lambda)\sin(\theta/2)$, where λ is the wavelength of the radiation, and θ is the scattering angle. The (complex) quantity $B_j(\mathbf{K})$ is normalized to

$$f_j = \int d^3\mathbf{R}\,\zeta_j(\mathbf{R}), \tag{3}$$

so that, for $K \to 0$: $B_j \to 1$,

For spherically symmetric ζ_j, Eqs. (2) and (3) become

$$B_j(K) = f_j^{-1} \int 4\pi R^2\, dR\, \zeta_j(R)\frac{\sin KR}{KR}\, dR, \tag{4}$$

and

$$f_j = \int 4\pi R^2\, dR\, \zeta_j(R), \tag{5}$$

where B_j is now real.

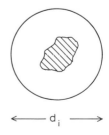

FIGURE 8.1. Particle with hard-sphere diameter of interaction d_i and scattering material depicted by the hatched area. The scattering density is described by the function $\xi_i(\mathbf{r})$, where $\mathbf{r}$ is measured from the hard-sphere center. Only spherically averaged values of the function $\xi_i(\mathbf{r})$ are needed.

In the case of neutron scattering, ζ_j is proportional to the local scattering length density inside the particle minus that of the solvent. For light scattering, ζ_j is proportional to the difference between the refractive index of the solvent and the local refractive index of the particle. For X-rays, ζ_j is the difference between the local electron density and the electron density of the solvent. Two operations are performed with B_j; these will appear in the equation for the scattering intensity of the system, to be formulated later. (1) First, the average scattering *intensity* of a single particle is obtained by squaring the absolute value of the scattering amplitude and taking the angular average (denoted by $\langle \cdots \rangle_a$). Thus,

$$P_j(K) = \langle |B_i(K)|^2 \rangle_a. \tag{6}$$

Here, $P_j(K)$ is called the particle scattering factor. (2) Second, the average scattering *amplitude* of a single particle is obtained by rotating the particle around the center of force over all angles:

$$B_j(K) = \langle B_j(K) \rangle_a. \tag{7}$$

Notice that if the angular average in Eq. (6) is taken then the choice of the center of rotation is immaterial and the averages in Eqs. (6) and (7) are both real (i.e., not complex) quantities depending only on the scalar $K = |\mathbf{K}|$.

For spherically symmetric ζ_j, the rotation of the particle has no effect on the scattering. The angular brackets are omitted from Eq. (6); thus,

$$P_j(K) = B_j^2(K). \tag{8}$$

B. Interparticle Interference

The intensity of the scattered radiation per unit volume of the particle dispersion is given by

$$I(K) = \sum_{i=1}^{p} \rho_i f_i^2 [\langle |B_i(K)|^2 \rangle_a - \langle B_i(K) \rangle_a^2]$$

$$+ \sum_{i,k=1}^{p} f_i f_k \langle B_i(K) \rangle_a \langle B_k(K) \rangle_a (\rho_i \rho_k)^{1/2} [\delta_{ik} + \tilde{H}_{ik}(K)]. \tag{9}$$

Here, ρ_i is the number density of particles of species i, δ_{ik} is the Kronecker delta ($\delta_{ik} = 0$ if $i \neq k$ and $\delta_{ik} = 1$ if $i = k$) and $\tilde{H}_{ik}(K)$ is given by

$$\tilde{H}_{ik}(K) = (\rho_i \rho_k)^{1/2} \int_0^\infty 4\pi r^2 h_{ik}(r) \frac{\sin Kr}{Kr} dr. \tag{10}$$

In the integral, $h_{ik}(r) = g_{ik}(r) - 1$. The function $h_{ik}(r) = h_{ki}(r)$ is called the total correlation function. It is, of course, the presence of the (matrix of)—usually unknown—functions $\tilde{H}_{ik}(K)$ that makes most of Eq. (9) intractable as a basis for making calculations.

Equation (9) consists of two parts. The first part contains a single sum over species and contributes to the scattering only when the distribution is not spherically symmetrically positioned around the center of force. This contribution is sometimes called incoherent scattering, because it does not contain information about interparticle interference. The scattering is proportional to the particle concentration. In our further discussion, we can drop this term without the general validity of the equation being affected: the term can always be added afterward when required. Using Eq. (7), we are then left with the equation

$$I(K) = \sum_{i,k=1}^{p} f_i f_k B_i(K) B_k(K) (\rho_i \rho_k)^{1/2} [\delta_{ik} + \tilde{H}_{ik}(K)]. \tag{11}$$

For further details about the development of the scattering equations see Refs. 5, 6, 8, and 9.

C. Very Dilute Systems

For a very dilute system, the interparticle interferences $\tilde{H}_{ik}(K)$ vanish, and one obtains

$$I_0(K) = \sum_{i=1}^{p} \rho_i f_i^2 B_i^2(K) = \sum_{i=1}^{p} \rho_i f_i^2 P_i(K), \tag{12}$$

with

$$P_i(K) = B_i^2(K). \tag{13}$$

D. (Quasi) Monodisperse Systems

For systems containing only a single-particle species, Eq. (11) reduces to

$$I(K) = \rho f^2 B^2(K) S(K), \tag{14}$$

with $S(K)$, the structure factor, given by

$$S(K) = 1 + \tilde{H}(K) = 1 + 4\pi\rho \int_0^\infty r^2 h(r) \frac{\sin Kr}{Kr} \, dr, \tag{15}$$

and at small concentration,

$$I_0(K) = \rho f^2 P(K). \tag{16}$$

A somewhat similar equation is applicable when particles have different scattering properties but (nevertheless) the same interactions. This situation occurs in neutron scattering experiments where $f_i B_i(K)$ can sometimes be varied by the substitution of different isotopes. But, it may also arise in light scattering when the refractive index of the particles shows a (slight) variation. When the refractive index of the particle and that of the solvent differ only slightly (solvent matching), the above effect is appreciable. In this case, Eq. (11) reduces to

$$I(K) = \rho \sum_{i=1}^{p} x_i \left\{ f_i B_i(K) - \left[\sum_{i=1}^{p} f_i B_i(K) x_i \right] \right\}^2$$

$$+ \rho \left[\sum_{i=1}^{p} f_i B_i(K) x_i \right]^2 [1 + \tilde{H}(K)], \tag{17}$$

with $\rho = \sum_{i=1}^{p} \rho_i$ and $x_i = \rho_i/\rho$.

The second part of Eq. (17) contains the same information as Eq. (14), the only difference being that fB is replaced by its average value: $\sum_i x_i f_i B_i$. The first term in Eq. (17) is extra. It contains no information about the interparticle interference and is therefore also called incoherent. In this respect, it is analogous to the first term in Eq. (9).

The degree to which the first term in Eq. (17) contributes to the scattering intensity depends on the variation in the scattering amplitudes around the average value. In the special case where the average scattering amplitude of the particle system is zero, only the incoherent contribution will remain.

E. Average Structure Factor

We now define an average particle factor $\bar{P}(K)$ and structure factor $\bar{S}(K)$ by introducing simple operational definitions that are directly accessible experimentally:

$$\bar{P}(K) = \left(\sum_i \rho_i f_i^2 \right)^{-1} \left[\sum_i \rho_i f_i^2 B_i^2(K) \right], \tag{18}$$

$$\overline{f^2} = \left(\sum_i \rho_i \right)^{-1} \left(\sum_i \rho_i f_i^2 \right), \tag{19}$$

and

$$\rho = \sum_i \rho_i. \tag{20}$$

Thus, according to Eq. (12),

$$I_0(K) = \rho_0 \overline{f^2} \bar{P}(K), \tag{21}$$

where $I_0(K)$ is the intensity of scattering of a (very) dilute system with total particle number density ρ_0. We further define $\bar{S}(K)$ by

$$I(K) = \rho \overline{f^2} \bar{P}(K) \bar{S}(K), \tag{22}$$

so that

$$\bar{S}(K) = \frac{I(K)/\rho}{I_0(K)/\rho_0}, \tag{23}$$

where $I(K)$ and $I_0(K)$ are given by Eqs. (11) and (12).

III. Hard-Sphere Interactions

To evaluate the interparticle interference functions $\tilde{H}_{ik}(K)$, the total correlation functions $h_{ik}(r) = g_{ik}(r) - 1$ are needed. Even for the (simple) hard-sphere pair potential, these functions are not exactly known, except when the particle number densities are very low.

A. Dilute Systems

When the number densities are so low that only interactions between particle pairs are of importance the $g_{ik}(r)$ are very simple, that is,

$$g_{ik}(r) = e^{-U_{ik}/kT}. \tag{24}$$

Thus, for hard spheres,

$$g_{ik}(r) = \begin{cases} 0 & \text{for} & 0 < r < \frac{1}{2}(d_i + d_k) \\ 1 & \text{for} & r > \frac{1}{2}(d_i + d_k) \end{cases} \tag{25}$$

It is then relatively simple to find the resulting scattering equation, using Eqs. (10) and (11). The intensity (times $\pi/6$) can be written as

$$(\pi/6)I(K) = \langle f^2 B^2 \rangle - 2\langle d^3 f B\Phi \rangle\langle f B \cos X \rangle - 6\langle d^2 f B\Psi \rangle\langle d f B\Psi \rangle, \tag{26}$$

where, for convenience, we have used the angular brackets $\langle \ \rangle$ (without subscript a) to express a summation over species:

$$\langle Y \rangle \equiv (\pi/6) \sum_{i=1}^{p} \rho_i Y_i, \tag{27}$$

where Y_i is any function of the hard-sphere diameter d_i; thus, e.g.

$$\langle f B \cos X \rangle = (\pi/6) \sum_{i=1}^{p} \rho_i f_i B_i \cos X_i \tag{28}$$

(see Appendix). Further,

$$X_k = \frac{1}{2}Kd_k, \tag{29}$$

$$\Psi_k(K) = \frac{\sin X_k}{X_k}, \tag{30}$$

and

$$\Phi_k(K) = \frac{3}{X_k^3}(\sin X_k - X_k \cos X_k), \tag{31}$$

with $\Psi_k(0) = \Phi_k(0) = 1$.

In particular, we define

$$\xi_v \equiv \langle d^v \rangle = (\pi/6) \sum_{k=1}^{p} \rho_k d_k^v. \tag{32}$$

Note that ξ_3 is the total volume fraction of hard spheres.

The first term in Eq. (26) is proportional to ρ and gives the contribution of the single particles to the total scattering intensity. The second and third terms are proportional to ρ^2 and represent the interparticle interference. Equation (26) becomes inaccurate when ξ_3 is larger than about 0.04.

B. Higher Number Densities of Hard Spheres

For higher number densities, approximate theories must be invoked to obtain $\tilde{H}_{ik}(K)$. For hard-sphere mixtures, the (approximate) PY integral equation is rather successful. It permits an exact solution for the case of the hard-sphere pair potential [3,4]. Baxter's solution even enables us to obtain an equation for the scattering intensity in *closed* form [5–7]. Liquid-state theory uses not only the (matrix) elements $\tilde{H}_{ik}(K)$ of the total correlation functions $h_{ik}(r)$, but also the $\tilde{C}_{ik}(K)$ of the so-called direct correlation functions $c_{ik}(r)$, which usually have a simpler mathematical structure [3]. Baxter introduces a new auxiliary function $q_{ik}(r)$ having even simpler properties to formulate matrix elements $\tilde{Q}_{ik}(K)$.

It is then possible to show that the $\tilde{H}_{ik}(K)$ can be obtained [5] with

$$\delta_{ik} + \tilde{H}_{ik}(K) = [\Delta(K)]^{-1} \sum_{j=1}^{p} |\tilde{\mathbf{Q}}(-K)|^{jk} |\tilde{\mathbf{Q}}(K)|^{ji}, \qquad (33)$$

where $\Delta(K) = |\tilde{\mathbf{Q}}(K)|$ is the determinant of the matrix $\tilde{\mathbf{Q}}(K)$.

Since the evaluation of Eq. (33) is complicated, we refer for details to the original papers. The route followed by Blum and Stell [7] is different.

The scattering intensity can be described by

$$(\pi/6)I(K) = \langle f^2 B^2 \rangle + \langle d^3 f B \Psi \rangle (L_2 + L_2^*)$$
$$+ 3 \langle d^2 f B \Psi \rangle (L_3 + L_3^*) + \langle A A^* \rangle. \qquad (34)$$

The (complex) functions L_2, L_3, and A are given in the Appendix. Here, L_2^* is the complex conjugate of L_2, etc.

Equation (34) is exact up to and including terms of the order of ρ^3. Furthermore, when Eq. (34) is developed to the order ρ^2, it is easy to prove that the equation becomes identical to Eq. (26), as it should.

C. Limiting Case: $K \to 0$

For small scattering angles ($K \to 0$), Eq. (34) leads to

$$I(K = 0) = \sum_{k=1}^{p} \rho_k [f_k - d_k^3 \langle f \rangle + \frac{3d_k^2}{1 + 2\xi_3} (\xi_4 \langle f \rangle - \langle df \rangle)]^2. \qquad (35)$$

For instance, when putting $f_i = d_i^3$ (solid scattering sphere), we obtain

$$(\pi/6)I(K = 0) = (1 - \xi_3)^2 \left[\xi_6 - \frac{6\xi_4 \xi_5}{1 + 2\xi_3} + \frac{9\xi_4^3}{(1 + 2\xi_3)^2} \right], \qquad (36)$$

whereas for particles with an infinitely thin scattering shell with diameter d_i and $f_i = d_i^2$, we find

$$(\pi/6)I(K = 0) = \xi_4 - 2\xi_2\xi_5 + \xi_2^2\xi_6 + \frac{9\xi_4}{(1 + 2\xi_3)^2}(\xi_3 - \xi_2\xi_4)^2$$

$$- \frac{6}{1 + 2\xi_3}(\xi_3 - \xi_2\xi_4)(\xi_4 - \xi_2\xi_5). \tag{37}$$

For the monodisperse case ($\xi_v = d^{v-3}\xi_3$), both Eqs. (36) and (37) lead to

$$I(K = 0) \sim \xi_3 \frac{(1 - \xi_3)^4}{(1 + 2\xi_3)^2}. \tag{38}$$

This is discussed further in Refs. 5 and 13.

D. Derivative of Chemical Potential

The expression $\delta_{ik} + \tilde{H}_{ik}(K = 0)$ is equal to the thermodynamic quantity $kT(\partial\rho_i/\partial\mu_k)_\mu$. It can be formulated in closed form for hard spheres in the PY approximation. We refer to the original papers for the complete equations [12,13].

IV. Results

Some results will be shown for a *log-normal* distribution of hard-sphere diameters:

$$\psi(d) = \frac{1}{\beta(2\pi)^{1/2}d}\left\{\exp - \left[\frac{\ln(d/d_0)}{\beta\sqrt{2}}\right]^2\right\}. \tag{39}$$

Here, $\psi(d) \, \Delta d$ is proportional to the number of particles with diameters between d and $d + \Delta d$. The (relative) standard deviation σ_d, is given by

$$\sigma_d^2 = [\overline{d^2} - \bar{d}^2]/\bar{d}^2 = \exp(\beta^2) - 1. \tag{40}$$

For $\beta \ll 1$, $\beta \simeq \sigma_d$. For the distribution [Eq. (39)], the ξ_v are given by

$$\xi_v = \phi d_0^{v-3} \exp[\tfrac{1}{2}\beta^2(v^2 - 9)], \tag{41}$$

where we introduced the symbol $\phi = \xi_3$ for the overall volume fraction. The distribution function for $\beta = 0.1$ and 0.3 ($\sigma_d = 0.100$ and 0.307, respectively) is plotted in Fig. 8.2(a).

For the intraparticle interference, we take two models, the homogeneous sphere with radius $d_i/2$,

$$B_i(K) = 3X_i^{-3}[\sin X_i - X_i \cos X_i], \tag{42}$$

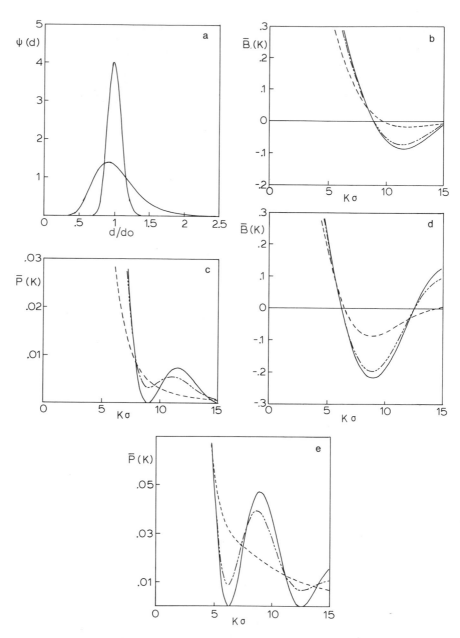

FIGURE 8.2. (a) Log-normal distribution of hard-sphered diameters d. High peak, $\beta = 0.1$ (standard deviation $\sigma_d = 0.100$). Low peak, $\beta = 0.3$ (standard deviation $\sigma_d = 0.307$). (b) Average intraparticle amplitude function $\bar{B}(K)$ for homogeneous spheres. $\beta = 0$ ($\sigma_d = 0$), ———; $\beta = 0.1$ ($\sigma_d = 0.100$), —·—·—; $\beta = 0.3$ ($\sigma_d = 0.307$), ————. (c) Average particle scattering factor $\bar{P}(K)$ for homogeneous spheres. Details as in (b). (d) Average intraparticle amplitude function $\bar{B}(K)$ for thin shells. Details as in (b). (e) Average particle scattering factor $\bar{P}(K)$ for thin shells. Details as in (b).

with $X_i = Kd_i/2$ and $f_i = d_i^3$, and the infinitely thin shell with radius $d_i/2$,

$$B_i(K) = X_i^{-1} \sin X_i, \tag{43}$$

with $f_i = d_i^2$.

$\bar{B}(K)$ and $\bar{P}(K)$ are calculated according to the equations

$$\bar{B}(K) = \sum_{k=1}^{p} \rho_k f_k B_k(K) \bigg/ \sum_{k=1}^{p} \rho_k f_k, \tag{44}$$

and

$$\bar{P}(K) = \sum_{k=1}^{p} \rho_k f_k^2 B_k^2(K) \bigg/ \sum_{k=1}^{p} \rho_k f_k^2. \tag{45}$$

The structure factor $\bar{S}(K)$ for polydisperse hard spheres in the PY approximation is calculated from

$$\bar{S}(K) = \sum_{k=1}^{p} \rho_k |f_k B_k + A_k|^2 \bigg/ \sum_{k=1}^{p} \rho_k f_k^2 B_k^2, \tag{46}$$

where A_k is given in the appendix. The ρ_k are obtained by taking p equidistant samples from $\psi(d)$, The number p is chosen so high ($p \sim 70$) that the results have reached their limiting values within 4 decimal places. Plots of $\bar{B}(K)$ and $\bar{P}(K)$ are given in Figs. 8.2(b) through (e).

A. Zero Scattering Angle

With the help of Eqs. (36), (37), and (41), it is simple to calculate the effect of polydispersity on the scattering intensity at (very) small scattering angles. For example, the value of $\bar{S}(K = 0)$ is larger for the polydisperse solid sphere scatterer than it is for the monodisperse case, at $\xi_3 = \phi = 0.3$, about 60%. For the polydisperse thin shell scatterer, $\bar{S}(K = 0)$ is slightly *smaller* than it is for the monodisperse case, at $\xi_3 = \phi = 0.3$, about 15%. For further details, we refer to Ref. 5.

B. Finite Scattering Angle

$\bar{S}(K)$ is plotted as a function of $K\sigma$ (with $\sigma \equiv (\overline{d^3})^{1/3}$) in Figs. 8.3 and 8.4 for a homogeneous scattering sphere and in Figs. 8.5 and 8.6 for an infinitely thin shell. In Fig. 8.3, the $S(K)$ of a one-component hard sphere is given as well.

The results clearly show the dependence of $\bar{S}(K)$ on the $B_i(K)$ chosen, because the scattering amplitude of a sphere is very different from that of a

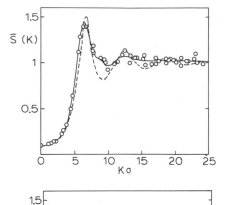

FIGURE 8.3. The average structure of a polydisperse system of hard spheres having an overall volume fraction of $\xi_3 = \phi = 0.3$. $B_i(K)$ is the scattering amplitude of a solid sphere with radius $d_i/2$. The hard-sphere diameter distribution is log-normal with $\beta = 0.1$ and thus a standard deviation of $\sigma_d = 0.100$. For the monodisperse case, the dashed line is $\sigma_d = 0$.

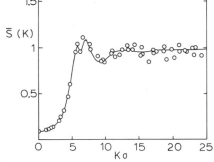

FIGURE 8.4. Symbols are as in Fig. 8.3, but for $B_i(K)$ the scattering amplitude is that of an infinitely thin spherical shell with radius $d_i/2$.

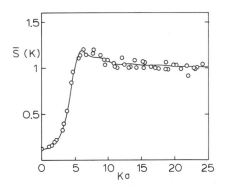

FIGURE 8.5. Symbols as in Fig. 8.3, but for $\beta = 0.3$ and (thus) $\sigma_d = 0.307$.

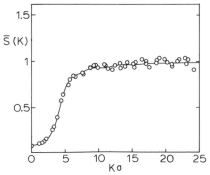

FIGURE 8.6. Symbols as in Fig. 8.4, but for $\beta = 0.3$ and (thus) $\sigma_d = 0.307$.

shell. (This dependence would not, of course, exist in a one-component particle system.)

Figs. 8.3 and 8.4 reveal some special aspects of this dependence. In Fig. 8.3, one sees a slight irregularity near $K\sigma = 9$, and in Fig. 8.4, near $K\sigma = 6$, where the maximum in $\bar{S}(K)$ shows a dent, which is real.

These irregularities can be seen to originate in the zero points of the functions $B_i(K)$. For a homogeneous scattering sphere, $\Phi(X)$ has a zero point at $X = 4.5$, thus at $Kd \sim 9$. For a thin scattering shell, the zero point is situated at $X = \pi$, thus at $Kd \sim 6$. See also the Figs. 8.2(b) and (c).

At higher degrees of polydispersity, many of the details in $\bar{B}(K)$, $\bar{P}(K)$, and $\bar{S}(K)$ disappear, in particular, the oscillations. The main maximum in $\bar{S}(K)$ is also considerably depressed, and it disappears altogether for the spherical shell. In the last case, the contributions of $\tilde{H}_{ik}(K)$ and $B_i(K)$ interfere strongly with each other.

The circles shown in Figs. 8.3–8.6 are results of Monte Carlo simulations by Frenkel et al. [14]. The simulations were carried out on a system of 108 polydisperse hard spheres, chosen at random from a log-normal distribution. The structure factors were obtained from the scattering amplitudes, which were computed with a fast Fourier routine.

The Monte Carlo calculations clearly show that the PY approximation gives very satisfactory results for these ϕ. The results are, in fact, within the range of statistical variation of the simulations and are sufficiently accurate for comparisons to be made with experiment. A more detailed discussion of the simulations is beyond the scope of this paper and is given elsewhere [14].

The equations show that the calculations needed to obtain $\bar{S}(K)$ are not very involved, although they are extensive. They contain single summations (or integrals) over the population of hard-sphere diameters. The equations, however, are also valid for any (small) number of components, that is, down to $p = 1$ or 2. Our equations are, in fact, simpler (we think) than the equation which Ashcroft and Langreth [15] derived for a two-particle system ($p = 2$) and Hoshino [17] for a three-particle system ($p = 3$) and which are based on the Lebowitz [3] and Hiroike [18] formalism. Nevertheless, identical results were obtained [16].

After this manuscript was written, two papers appeared [19] in which the average structure factor of a Schulz polydisperse ($p = \infty$) system of hard spheres was obtained in closed form. The particle scattering functions are, however, restricted to point scatters and to homogeneous spheres with radius equal to half of the hard sphere diameter, which is rather restrictive.

Recently a paper appeared [20] in which calculations are performed on polydisperse hard spheres with adhesion.

Acknowledgment. We thank Dr. D. Frenkel for making the Monte Carlo calculations.

Appendix[1]

The solution obtained for $I(K)$ in Ref. 6 can be written as

$$(\pi/6)I(K) = (\pi/6) \sum_{k=1}^{p} \rho_k |f_k B_k + A_k|^2$$
$$= \langle (fB + A)(fB + A^*) \rangle = \langle f^2 B^2 \rangle + \langle d^3 f B\Phi \rangle (L_2 + L_2^*)$$
$$+ 3 \langle d^2 f B\Psi \rangle (L_3 + L_3^*) + \langle AA^* \rangle. \tag{A1}$$

Here, L_2, L_3, and A are complex quantities. Their complex conjugates L_2^*, L_3^*, and A^* are obtained by replacing i, the imaginary unit, by $-i$. Further,

$$L_2 = T_2/T_1; \qquad L_3 = T_3/T_1, \tag{A2}$$

$$T_1 = F_{11}F_{22} - F_{12}F_{21}, \tag{A3}$$

$$T_2 = F_{21}\langle df Be^{iX} \rangle - F_{22}\langle f Be^{iX} \rangle, \tag{A4}$$

$$T_3 = F_{12}\langle f Be^{iX} \rangle - F_{11}\langle df Be^{iX} \rangle, \tag{A5}$$

$$F_{11} = 1 - \xi_3 + \langle d^3 \Phi e^{iX} \rangle, \tag{A6}$$

$$F_{12} = \langle d^4 \Phi e^{iX} \rangle, \tag{A7}$$

$$F_{22} = 1 - \xi_3 + 3\langle d^3 \Psi e^{iX} \rangle, \tag{A8}$$

$$F_{21} = \tfrac{1}{2}(1 - \xi_3)iK - 3\xi_2 + 3\langle d^2 \Psi e^{iX} \rangle, \tag{A9}$$

$$\Psi_k = (\sin X_k)/X_k, \tag{A10}$$

$$\Phi_k = (3/X_k^3)(\sin X_k - X_k \cos X_k), \tag{A11}$$

$$A_k = d_k^3 \Phi_k L_2 + 3d_k^2 \Psi_k L_3, \tag{A12}$$

$$X_k = \tfrac{1}{2}Kd_k. \tag{A13}$$

The brackets indicate averages over diameters, that is,

$$\langle y \rangle \equiv (\pi/6) \sum_{k=1}^{p} \rho_k y(d_k), \tag{A14}$$

where $y(d)$ is any function of d. Examples of these averages are

$$\langle f^2 B^2 \rangle = (\pi/6) \sum_{k=1}^{p} \rho_k f_k^2 B_k^2, \tag{A15}$$

$$\langle f Bd^2 \Psi \rangle = (\pi/6) \sum_{k=1}^{p} \rho_k f_k B_k d_k^2 \Psi_k, \tag{A16}$$

$$\langle df Be^{iX} \rangle = (\pi/6) \sum_{k=1}^{p} \rho_k d_k f_k B_k e^{iX_k}, \tag{A17}$$

$$\langle df Be^{iX} \rangle^* = (\pi/6) \sum_{k=1}^{p} \rho_k d_k f_k B_k e^{-iX_k}, \tag{A18}$$

---

[1] A listing of the computer program can be obtained from the authors.

and particularly,

$$\xi_v = (\pi/6) \sum_{k=1}^{p} d_k^v \rho_k = \langle d^v \rangle. \tag{A19}$$

Thus, ξ_3 is the overall volume fraction of spheres. Furthermore,

$$B_k(K) = f_k^{-1} \int_0^\infty 4\pi r^2 \zeta_k(r) \frac{\sin Kr}{Kr} dr, \tag{A20}$$

and

$$f_k = \int_0^\infty 4\pi r^2 \zeta_k(r) \, dr, \tag{A21}$$

where $\zeta_k(r)$ is the (spherically symmetric) distribution of scattering amplitude in particle k as a function of the distance from the center of the hard sphere k.

References

1. A.K. van Helden, J.W. Jansen, and A. Vrij, *J. Colloid Interface Sci.* **81**, 354 (1981).
2. D.J. Cebula, R.H. Ottewill, J. Ralston, and P.N. Pusey, *J. Chem. Soc. Faraday Trans.* **77**, 2585 (1981).
3. J.L. Lebowitz, *Phys. Rev. Sect.* **A133**, 895 (1964).
4. R.J. Baxter, *J. Chem. Phys.* **52**, 4559 (1970).
5. A. Vrij, *J. Chem. Phys.* **69**, 1742 (1978).
6. A. Vrij, *J. Chem. Phys.* **71**, 3267 (1979).
7. L. Blum and G. Stell. *J. Chem. Phys.* **71**, 42 (1979); with Erratum in *J. Chem. Phys.* **72**, 2212 (1980).
8. P. van Beurten and A. Vrij, *J. Chem. Phys.* **74**, 2744 (1981).
9. M. Kotlarchyk and S.-H. Chen, *J. Chem. Phys.* **79**, 2461 (1983).
10. J.B. Hayter, *J. Chem. Soc. Faraday Discussions* **76**, 7 (1983).
11. See, for example, *J. Chem. Soc. Faraday Discussions* **76**, 93, 95, 97 (1983).
12. A. Vrij, *J. Chem. Phys.* **72**, 3735 (1980).
13. A. Vrij, *J. Colloid Interface Sci.* **90**, 110 (1982).
14. D. Frenkel, R.J. Vos, C.G. de Kruif, and A. Vrij, *J. Chem. Phys.* **84**, 4625 (1986).
15. N.W. Ashcroft and D.C. Langreth, *Phys. Rev.* **156**, 685 (1967).
16. J.B. van Tricht, private communication.
17. K. Hoshino, *J. Phys. F: Met. Phys.* **13**, 1981 (1983).
18. K. Hiroika, *J. Phys. Soc. Japan* **27**, 1415 (1969).
19. W.L. Griffith, R. Triolo, and A.L. Compere, *Phys. Rev. A.* **33**, 2197 (1986); **35**, 2200 (1987).
20. C. Robertus, W.H. Philipse, J.G.H. Joosten, and Y.K. Levine, *J. Chem. Phys.* **90**, 4482 (1989).

Part Two   Statistical Thermodynamics of Phase Behavior and Critical Phenomena

9
Theory of Structure and Phase Transitions in Globular Microemulsions

S.A. SAFRAN

A phenomenological theory for the structure and phase behavior of a dilute, three-component microemulsion is reviewed. The equilibrium structure of a system of noninteracting globules is calculated from the elasticity of the surfactant layer. Spherical, cylindrical, and lamellar structures are predicted as a function of concentration. The effects of thermal fluctuations on these structures are considered. For spheres, the main effect is a polydispersity of sphere sizes and shapes; for cylinders, thermal fluctuations result in worm-like structures that can behave like polymers in solution. Interactions between microemulsion globules are discussed for spherical systems. Evidence for attractive interactions is reviewed. We show how these attractions modify the scattering, phase diagrams, and transport properties of spherical microemulsions. The model exhibits a critical point prior to two-phase coexistence, a first-order phase separation (emulsification failure), and a three-phase equilibrium. In the single-phase region, clustering due to the attractive interactions tends to dramatically lower the percolation threshold (assuming the particles are conducting), while the dynamic rearrangements of the equilibrium clusters of globules modify the critical exponents for the onset of the percolation transitions.

I. Introduction

Microemulsions [1] are multicomponent fluids that are characterized by equilibrium globular, domain, or network-like structures on length scales of ~ 100 Å. There has been some controversy [2] over whether a new term is needed to describe these systems—exemplified by water, oil, and surfactant mixtures or whether they are best classified as particular examples of multicomponent or micellar solutions. However, these systems are novel and of unique interest precisely because two of the components (e.g., the oil and the water) tend to form domains with the third component (e.g., the surfactant) at the interfaces between these domains—and not because they form random, multicomponent microscopic mixtures. Indeed, microemulsions are a prime example of systems that form domains in equilibrium; most systems with domain structures coarsen while approaching an equilibrium state of complete phase separation [3,4]. It is the spontaneous solubilization of water in oil or oil in water via the surfactant that distinquishes microemulsions from (nonequilibrium) emulsions. The system is self-organizing and determines its microstructure by minimizing its free energy. The challenging experimental problem is to characterize the structure and properties of the domains and interfaces found in microemulsions. The primary theoretical focus is to predict the number, size, and shape of the interfaces formed by the system as a function of the concentrations.

Most of the theoretical treatments of microemulsions can be divided into two groups. The first [5,6,7,8] consists of quasimicroscopic studies of dilute systems of spherical globules with an emphasis on the dependence of the optimal radius of the sphere on parameters such as polar-head chemistry or salinity. The second group [9,10,11,12] focuses upon the characterization of the structures in the concentrated regime, where the microemulsion is thought to consist of bicontinuous [13] regions of both water and oil. These theories have used phenomenological models of the interfacial energy and have focused on the phase behavior of the system as opposed to its detailed structure.

However, recent work [14,15,16,17,18,19,20] by this author (and collaborators L. Turkevich and P. Pincus) has shown that even in the dilute regime, where the microemulsions are globular, both the structure and phase behavior are considerably richer than had been expected. A phenomenological model of the surfactant interface is used to predict the physical trends in the structure of these globules and their phase behavior; these models are at best qualitative in predicting chemical trends. However, a wide variety of possible structures were examined in terms of a small number of parameters. Experimental studies of microemulsion structure are most simply interpreted in the dilute limit [21,22,23,24]. These studies are in good agreement with many of the theoretical predictions of the structure [16,25], polydispersity [14,15,23], phase diagrams [18], and transport properties [20,26,27,28,29,30].

In this Chapter, the results of the theory for the dilute limit are reviewed (for details see the papers referenced previously). A theoretical model that predicts the structure of the domains and interfaces in three-component microemulsions in the dilute limit, that is, for small volume fractions of water in oil or oil in water is presented. In this limit, the system is assumed to form globular structures whose positions are correlated because of interglobular interactions. The building blocks of the microemulsion, the individual droplets, are discussed in Section II. The globules of water in oil or oil in water with surfactant at the interfaces have a structure that is a result of the minimization of the free energy of the interface. Assuming that the surfactant molecules are incompressible on the water–oil interface, the interfacial energy is due to the orientational interactions between the surfactant molecules; this results in a splay or bending energy. Depending on the concentrations, spherical, lamellar, or cylindrical structures are favored. The effects of thermal fluctuations on these structures are then calculated. In the spherical phase, these fluctuations result in an equilibrium polydispersity, as well as some small shape fluctuations of the globules. The effects of fluctuations on the lamellar phases have been discussed in Ref. 9. In the cylindrical phase, these fluctuations cause a wandering of the cylindrical axis. The structural and rheological behavior is similar to those of polymers. In contrast to polymers, however, these flexible, cylindrical microemulsions have a molecular weight or polymerization index that is determined by equilibrium considerations. Both the persistence length and the polymerization index are controlled by varying the concentrations—they are not determined by the preparation kinetics as in most polymers.

The macroscopic properties of microemulsions are not determined by the structure of the building blocks alone. The organization of the globules in solution is important in determining long wavelength properties, such as phase behavior and transport. This organization is a function of the interactions between globules and is discussed in Section III for the case of spherical globules. Experimental evidence for attractive interactions in some microemulsions is reviewed [22,23,24,31]. The dependence of the interactions on the globule size can be analyzed from geometrical considerations; this is in contrast to the situation in atomic gases or fluids, where the details of the quantum chemistry determine these quantities. Finally, it is shown how these interactions result in phase separation and liquid or gas-like critical behavior in microemulsions. A generic phase diagram for a system of spherical microemulsions is presented. Even though the system is dilute, the phase behavior is quite rich. The existence of two two-phase equilibria, as well as a region of three-phase coexistence, is due to the fact that the globules have internal degrees of freedom that can lead to additional instabilities.

Section IV presents some recent theoretical work on the effects of interactions and dynamics on percolation transitions in microemulsions. Recent conductivity measurements on water in oil systems have shown that interacting spherical microemulsions undergo a nonconducting to conducting transition at very small ($\sim 10\%$) volume fractions of dispersed phase [27-29]. This section summarizes the results of Monte Carlo simulations that show how clustering due to attractive interactions plays an important role in dramatically lowering the percolation threshold of colloidal systems. The microemulsions are particularly interesting, since they are a prime example of an interacting colloidal system that shows a percolation threshold; most other materials that show percolation transitions consist of quenched mixtures of conducting and nonconducting materials. In addition, the dynamical nature of the microemulsion droplets is shown to change the functional form of the conductivity near the percolation transition. Section V concludes the chapter with a brief look at some outstanding problems. More current references can be found on p. 181.

II. Bending Energy and Microemulsion Structures

This section reviews the theory of the curvature energy of the surfactant interface. The relative energies of microemulsion globules with different shapes are compared. Thermal fluctuations about the state of minimum curvature energy are considered, and their effects on the structure are predicted.

A. Curvature Energy

To predict the structure of a simple, three-component microemulsion, an expression for the energy of any arbitrary structure is needed. Since the focus of this work is on microemulsions as examples of an equilibrium interfacial system, it is assumed that all the surfactant is at the water–oil interface. This assumption is based on the low solubility of water in oil and of the surfactant in either pure water or pure oil. Since the range of surfactant concentrations is typically $> 1\%$ for microemulsions, while the critical micelle concentration (CMC) of equivalent two-component systems is well below 1%, the assumption of negligible surfactant in either the bulk water or oil regions is justified.

The energetics of the system is then governed by the energy of the surfactant layer at the water–oil interface. This energy has a contribution from the compressibility of the surfactants, as well as from their orientational interactions. In this phenomenological model, all the interactions are *effective* ones, taking into account the environment of the surfactant in the oil/water regions. Experimental studies in several systems have indicated that the area per surfactant is fairly constant as a function of water, oil, and surfactant concentrations. It is therefore assumed that the surfactant layer at the interface is incompressible. This is equivalent to the well-known Schulman criterion [9,32] for a saturated interface. Recent theoretical analysis has shown that this approximation breaks down only for globules whose size is comparable to a molecular length [11,33].

Since the total surface area of all the microemulsion globules is assumed to be constant, the surface energy terms are irrelevant. The energy is then related to the curvature of the interface. In this section, the microemulsion is pictured as a collection of noninteracting globules, and the total energy is the sum of the curvature or bending energy of the individual droplets. One of the reasons that microemulsions are more amenable to a general treatment compared with micelles is that the characteristic globule size is (much) greater than a typical molecular length. Details of the molecular packing become less crucial; the bending energy can then be written as an expansion in the radius of curvature since it is large compared to molecular dimensions. Keeping terms up to second order in the curvatures, we write the bending energy of a single globule as

$$F_b = \frac{K}{2}\int dS \left(\frac{1}{R_1} + \frac{1}{R_2} - \frac{2}{\rho_0}\right)^2 + \frac{\bar{K}}{2}\int dS \left(\frac{1}{R_1} - \frac{1}{R_2}\right)^2. \tag{1}$$

In Eq. (1), the integral is over the globule surface S. The natural radius of curvature ρ_0 describes the tendency of the interface to bend towards either the water ($\rho_0 > 0$) or the oil ($\rho_0 < 0$) side of the interface. The first term of Eq. (1) is identical to the splay energy of a single layer of a smectic liquid crystal [34] with the splay constant K typically on the order of 0.1 eV, although the addition of cosurfactant can reduce this value. Smectic molecules have no preferred top or bottom, so that bulk smectics have $\rho_0^{-1} = 0$. For microemulsions, the heads and tails are, of course, distinguishable, and in general, $\rho_0^{-1} \neq 0$. However, the tendency of the interface to bend (and hence the value of ρ_0) can be changed by varying the temperature and salinity which change the relative packings of the heads and tails on the interface. The term with the characteristic energy $\bar{K}$ is the so-called saddle-splay term, which vanishes in bulk liquid crystals [35]. For microemulsions where the number of globules is not fixed, this term contributes to the total energy of all the globules [36, 37]. As written here, the saddle-splay term favors structures with equal radii of curvature (for $\bar{K} > 0$), for example, spheres and lamellae instead of cylinders or ellipsoids. If the surfactant molecules were rotationally anisotropic (even in the presence of thermal rotation), this simple form for F_b would have to be further generalized.

Although the expression for the bending energy can be derived from an expansion in the curvatures, an identical form for F_b has been derived for a more microscopic model [17]. This model has also been used to study the elasticity of lipid bilayers and vesicles [35] and uses a harmonic approximation for the head–head and tail–tail interactions. In the approximation that the distance between the centers of mass of the surfactants is independent of curvature, Eq. (1) is derived with explicit expressions for K, $\bar{K}$, and ρ_0 in terms of the microscopic lengths and interaction strengths [17].

The total curvature energy of the (noninteracting) system is the sum of F_b over all the globules. However, it is important to note that the number of globules is not fixed, but is self-determined by the system, which chooses the size, shape, and number of globules that minimize its free energy. In the present model, the curvature energy is minimized by a system of monodispersed spheres with radius ρ_0. However, the system cannot attain this state for arbitrary concentrations. The total surface area of all the globules is fixed by the incompressibility conditions. In addition, the total volume of all the globules is fixed by the incompressibility of the internal water or oil. Therefore, the system cannot—for an arbitrary concentration of water, oil, and surfactant—make a monodisperse set of globules with radius ρ_0. It is this competition between the tendency to make globules with the optimal radius of curvature (ρ_0) and the necessity to use up all the water, oil, and surfactant molecules that leads to a variety of structures in the system.

With these constraints, we have calculated the energies of a system of n identical globules per unit volume for various shapes. These constraints incorporate the conservation of surfactant with volume fraction v_s and internal water or oil. This conservation also applies to the total volume fraction of globules, ϕ. In our calculations, we have assumed that the globule size is much larger than the typical surfactant length δ.

The expressions for the bending energy for spheres, cylinders, lamellae, and spheroids are given in Refs. 16 and 17. By comparing their energies, we can construct a stability plot that is shown in Fig. 9.1. Defining, $\rho = 3\delta\phi/v_s$ (it is the radius of the spheres in the spherical phase), we see that the structure is determined primarily by the ratio ρ/ρ_0. At $\rho = \rho_0$, the energy is at its global minimum—the globules have their optimal radius of curvature—and the structure is spherical. If the volume fractions of internal phase are increased, the globules fail to grow any further and reject the excess water or oil. This has been termed an emulsification failure instability. In the simple model, it results in a phase of spherical globules with $\rho = \rho_0$ coexisting with a bulk phase of excess water or oil (whichever is internal to the globule). More realistic situations are probably better described by an equilibrium of globules with water or oil containing a small amount of surfactant, most probably in a lamellar phase [38]. Note that Fig. 9.1, which considers only

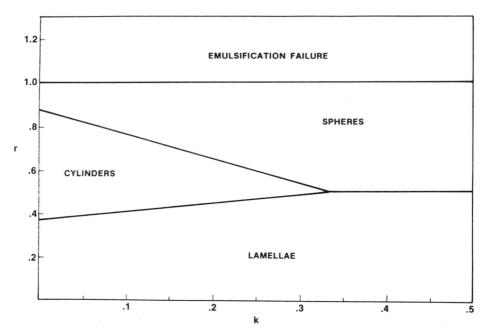

FIGURE 9.1. The mean-field shape stability of spherical, cylindrical, and lamellar microemulsion phases, as a function of the ratios $k = \bar{K}/K$ and $r = \rho/\rho_0$. K and $\bar{K}$ are the splay and saddle-splay elastic constants, and ρ_0 is the natural radius of curvature. The length scale ρ is related to the ratio of internal phase to surfactant; $\rho = 3\delta\phi/v_s$, where δ is a typical surfactant length, ϕ is the volume fraction of dispersed phase, and v_s is the volume fraction of surfactant. This figure is obtained by comparing the curvature energies only. For the effects of entropy of dispersion, see the text.

free energy crossings, is not a true phase diagram; in general, there are two-phase coexistence regions at all the phase boundaries.

As the ratio ρ/ρ_0 is decreased, other shapes can better accommodate the conservation constraints and keep the average radius of curvature close to ρ_0. Thus, Fig. 9.1 shows that lamellar structures are preferred for $\rho/\rho_0 < \frac{1}{2}$ for large values of $k \equiv \bar{K}/K$. If the anisotropic splay energy is small, a cylindrical structure (which has a finite saddle-splay energy cost) is of lower energy than either the spheres or lamellae. In the absence of thermal fluctuations, these cylinders are infinitely long and rigid.

The phase diagram presented in Fig. 9.1 was obtained by minimization of the total bending energy only. For a dilute system (ϕ small), at finite temperatures, the entropy of dispersion (the translational entropy of the centers of mass) must be added to the curvature energy. In the absence of interactions, the noninteracting free energy is given by

$$F_0 = nF_b + nT(\log nv_0 - 1), \tag{2}$$

where n is the number density of drops, v_0 is a molecular volume, and T is the temperature measured in units of Boltzmann's constant. Since the number of globules is not fixed, both the curvature energy and the entropy of dispersion must be considered in minimizing the free energy—subject to the conservation constraints. The inclusion of the entropy results in a stabilization of the smaller structures over the larger ones, for example, the domain of stability of the spherical phase is increased compared with the lamellar phase. The value of $\rho = \rho_{SL}$ at which this transition occurs is obtained by equating F_0 of Eq. (2) for spheres $[F_b \sim (1 - \rho/\rho_0)^2]$ and lamella ($F_b \sim 1/\rho_0^2$). Thus, it is found that

$$\frac{\rho_{SL}}{\rho_0} = \frac{1}{2}\left[1 + \frac{T}{8\pi K}(\ln nv_0 - 1)\right].$$

Spheres are thus stable to even smaller values of ρ/ρ_0. Similarly, emulsification failure (coexistence with a bulk phase of the component internal to the microemulsion) occurs at a value of $\rho = \rho_{EF}$, which depends on T and ϕ. This relationship is determined by the equality of the chemical potential of the internal component in both the microemulsion and the bulk phase. If, for example, the chemical potential in the bulk phase is set to zero (the bending energy is the only relevant energy), then chemical equilibrium requires $\partial F_0/\partial \phi = 0$. Remembering that ρ also depends on ϕ, it is found that

$$\frac{\rho_{EF}}{\rho_0} = 1 + \frac{T}{8\pi K}\ln nv_0.$$

B. Fluctuation Effects

However, in addition to the effects of temperature via the translational entropy of the globules, there exist thermal fluctuations of the internal degrees of freedom of the globules as well. The entropy of the interface results in globules whose size and shape depart from the ideal shapes discussed previously. The existence of these internal degrees of freedom, that is, the ability of the globule to change its internal structure, differentiates micro-emulsions from systems such as solid colloids. The present model of the curvature energy for a general globule shape [Eq. (1)] allows the quantitative calculation of the effects of thermal fluctuations on the size and shape of the globules.

The probability that an arbitrary deformation of the globules will occur in thermal equilibrium is proportional to the Boltzmann factor $e^{-\Delta F/T}$, where ΔF is the energy cost of the deformation. This energy cost is calculated from Eq. (1). The calculation is careful in including the conservation constraints. The expression for ΔF for an ensemble of spherical globules was derived in Refs. 14 and 15 for $k = \bar{K}/K = 0$. Here, we extend those results for finite values of k (note that our definition of the splay constant differs from that of Refs. 14 and 15 by a factor of 2). The change in free energy is given by

$$\Delta F = K \sum_{ilm} |a^i_{lm}|^2 \, \Delta F_l, \tag{3a}$$

where

$$\Delta F_l = (\log nv_0 - 1)\tau(1 - b_l) + \tilde{k}(3 + b_l^2 - 4b_l) + 4\frac{\rho}{\rho_0}(b_l - 1)$$

$$+ 6k(b_l - 1), \tag{3b}$$

where $\tilde{k} = 2(k + 1)$, i sums over globules, and $b_l = l(l + 1)/2$. If the globule is defined in spherical coordinates by $r = R(\theta, \phi)$, then the mean square deviation of R from the value for monodisperse spheres is given by

$$\left\langle \frac{R^2}{\rho^2} - 1 \right\rangle = \sum \langle |a_{lm}|^2 \rangle.$$

Here, a^i_{lm} is the amplitude of the deformation expanded in spherical harmonics and normalized to the radius of the sphere in the monodisperse case (ρ defined previously). The energy of the deformation depends on the index of the spherical harmonic; $l = 0$ deformations correspond to fluctuations in the sizes of the globules in the ensemble, while $l = 2$ fluctuations are the lowest modes for shape changes. The entropy of dispersion contribution is given by the term proportional to $\tau \equiv 3T/4\pi K$. By the equipartition theorem, the amplitude of the fluctuation in thermal equilibrium is given by $\langle |a^i_{lm}|^2 \rangle = (2\pi/3)\,\tau/\Delta F_l$.

In the limit of small τ, the polydispersity diverges at $\rho/\rho_0 = \frac{3}{2}$. However, since emulsification failure occurs when $\rho/\rho_0 \approx 1$, this divergence cannot be observed in equilibrated systems as long as $k > 0$. Similarly, the $l = 2$ shape fluctuations diverge when $\rho/\rho_0 = -3k/2$. Again, this divergence may not be observable, since the spherical phase may be unstable to the cylindrical or lamellar phases at a larger value of ρ/ρ_0. However, these fluctuations are nonnegligible in the region where the spherical phase is stable; a typical magnitude is a root mean square (rms) value of the polydispersity of 25 %, in good agreement with experiment [23]. In addition, Eq. (3) provides a quantitative prediction of the dependence of the fluctuation amplitude on the ratio of surfactant to water or oil (whichever is the internal phase) via the behavior of Eq. (3) as a function of ρ/ρ_0. This trend remains to be explored experimentally.

Finally, it should be noted that these fluctuations are not dynamical fluctuations of the individual globules (these have been treated in Ref. 39), but rather fluctuations from the ideal monodisperse sphere in the ensemble of droplets. In the ensemble, in thermal equilibrium, any one droplet is not exactly a sphere of radius ρ; it is this deviation that is reflected in the present calculations. One mechanism whereby such ensemble fluctuations occur is via collisions in which the globules may exchange surfactant and dispersed phase molecules. It should also be noted that the calculation of the fluctuation amplitude is quantitatively (but not qualitatively) different from a similar calculation for vesicles by Helfrich [37]. These differences are due to the assumption of constant globule number by Helfrich. While this may be a good approximation for vesicles that cannot readily exchange either surfactant or internal phase, it is not applicable to microemulsions as previously discussed.

While the effects of thermal fluctuations on an ensemble of monodisperse spheres is relatively minor, this is not the case for lamellar or cylindrical microemulsions. The thermal undulations of lamellae have been analyzed theoretically by de Gennes and Taupin [9, 40]. The main conclusion is that the lamellae are flat on length scales smaller than a persistence length $\xi_l \sim e^{K/T}$; for larger length scales, the interfaces are wrinkled, with disordered domains of the characteristic size ξ_l. It is suggested that this size is related to the scale size in a bicontinuous microemulsion, where K/T is small. Experimentally [40,41,42], the situation for lamellar microemulsions is somewhat complex with some systems showing well-defined stacking of oil and water sandwiched by surfactant and others showing a more random, but not completely disordered, structure.

For quasi-one-dimensional cylindrical microemulsions, the effects of thermal fluctuations are even greater than for lamellae. The cylinders are only rigid (rod-like) on length scales $\xi_c \sim (K/T)b$, where the cylinder radius $b = 2\rho/3$ with $\rho \sim \phi/v_s$ as previously defined. The persistence length ξ_c is calculated in Ref. 16 using the procedure outlined above. For length scales shorter than ξ_c, the cylinders are rigid, while for length scales greater than ξ_c, the cylinder axis wanders randomly in space.

The random wandering of the cylindrical axis suggests a polymer-like description of these long, flexible microemulsions. The number N of persistence lengths per chain is calculated with a Flory–Huggins expression for the entropy [43]. The end-cap energy is positive compared with the cylindrical section since cylinders have lower bending energies in this region of the phase diagram. This end-cap energy competes with the entropy of dispersion, which is maximized when the chains are as short as possible. The relevant free energy is

$$F_N \sim T\frac{\phi}{N}\log\frac{\phi}{N} + K'\frac{\phi}{N},$$ (4)

where it is shown in Ref. 17 that K' (the energy difference between the end-cap region and the cylindrical region) is given by

$$K' = \pi K(14 - 10\bar{K}/K - 32\rho/\rho_0)/3.$$ (5)

(Note that this corrects the value of K' in Ref. 16 by a factor of $\frac{9}{2}$). A crude estimate of N is obtained by minimizing Eq. (4) with respect to N. A more sophisticated treatment takes into account the fact that the system is polydisperse [44]. The crude estimate yields

$$N = \phi e^{K'/T+1},$$ (6a)

while the calculation that includes polydispersity estimates yields

$$N = 2\phi^{1/2}(e^{K'/T+1})^{1/2}.$$ (6b)

The important point to note is that the polydispersity is self-consistently determined by the system; it is not a result of the kinetics of preparation as for simple, molecular polymers. In addition, the polymerization index is a strong function of ρ/ρ_0, which enters in the end-cap energy. By varying the surfactant to internal phase ratio (or by varying ρ_0), N can vary from 1 (short, rigid rods) to values $N \gg 1$. For $N \sim 1$, the rods should align in a nematic phase as the volume fraction ϕ is increased [45,46].

For systems with $N \gg 1$, polymer-like behavior should be observed. The polymers, though, are of the order of 100 Å in diameter, instead of several angstroms! Indeed, giant flexible micelles have recently been studied [25,47]. Their osmotic pressure and cooperative diffusion constant have been measured as a function of ϕ and have been found to obey polymer scaling laws [25]. In addition, their viscosity shows a stronger dependence on ϕ compared with polymers, with $\eta \sim \phi^\beta$ with values of $3 \lesssim \sim \beta \lesssim 7$. A careful study of reptation dynamics [48] is needed to understand the scaling of the viscosity in self-assembling, polymer-like systems.

Future studies of such effects in *microemulsion* systems will be interesting since the scaling laws also involve the persistence length, which can be varied as a function of ρ/ρ_0. In addition, N is a strong function of ρ/ρ_0. Although, for simple polymers in a theta solvent, the radius of gyration, R_G, is simply related to the persistence length and the degree of polymerization, polymer-like microemulsions may have a complicated dependence of R_G on ϕ and v_s, as can be seen from Eqs. (5) and (6). However, in the limit of constant persistence length (i.e., $\xi_c \sim \phi/v_s$ constant), Eq. (6) predicts the dependence of R_G on the volume fraction ϕ in the dilute limit, where $R_G \sim N^{1/2}\xi_c$, and in the self-avoiding limit, where $R_G \sim N^{3/5}\xi_c$.

Thus, the structure of microemulsions, even in the dilute limit of noninteracting globules, is rich and varied. The competition between the curvature energy and the conservation constraints results in transitions between globules of different shapes. Thermal fluctuations are responsible for the polydispersity of the spherical phase and for possible polymer-like behavior of the cylindrical phase.

III. Interactions and Phase Separation

This section focuses on the effects of interactions between globules and upon their organization in solution. In particular, interactions, fluctuations, and the phase behavior of spherical microemulsion globules are analyzed.

While an analysis of the curvature energy of microemulsions is useful for determining globule size and shape, the macroscopic behavior of microemulsion systems, for example, phase equilibrium, is determined by the global structure of the globules in solution. The correlations between globule positions that lead to this organization are determined by the interactions between globules. The effects of interactions are most simply treated, both theoretically and experimentally, for spherical droplets. To treat the interactions, the internal degrees of freedom of the individual globules are assumed to be such that the globules always remain spherical, although they may change their size as the conservation laws dictate. The globule radius ρ and the density of globules n (number per unit volume) are determined for monodispersed spheres from the constraints of surfactant and internal phase conservations:

$$\phi = 4\pi\rho^3 n/3 \quad \text{and} \quad v_s = 4\pi\rho^2 \delta n,$$

where δ is a typical surfactant length. The radius ρ is then proportional to ϕ/v_s. From Eq. (1), the bending energy for a sphere is proportional to $(1 - \rho/\rho_0)^2$; for $\rho \gtrsim \rho_0$, the emulsification failure instability is expected.

The hard-core interactions of the spheres are accounted for by the entropy of mixing for hard spheres [49]:

$$S_m = -n_0[\phi(\log \phi - 1) + 4\phi^2 + 5\phi^3 + \cdots], \tag{7}$$

where $n_0^{-1} = 4\pi\rho^3/3$. The attractive interaction between globules is written phenomenologically as

$$F_i = -\tfrac{1}{2}n_0 TA[\rho,T]\phi^2, \tag{8}$$

where the dimensionless virial coefficient A is related to the microscopic interglobular interaction $U(r, \rho)$ by the relation

$$A[\rho, T] = n_0 \int d\mathbf{r} [e^{-U(r, \rho)/T} - 1]. \tag{9}$$

Because of the dependence of the virial coefficient on the drop size, $\rho = 3\delta\phi/v_s$, the strength of the interactions varies as the ratio of ϕ/v_s is changed. The microscopic origin of the attractive interaction is discussed later. Here, it is noted that (for short-range interactions) the number of contacts per drop between two small drops is smaller than for large drops—consider the limiting case of two planes—where there are many more contacts. Thus, $U(r, \rho)$ is expected to increase as the concentrations are varied such as to make ρ increase.

On a thermodynamic level, the competition between the attractive interactions and the entropy of mixing permits a collective liquid/gas instability wherein the system separates from a single phase of spheres into two coexisting phases of spheres immersed in the same continuous phase. Such phase separations are observed in a wide variety of water in oil microemulsions. The phase boundary is calculated by looking for two phases of globules where the chemical potential of both the internal phase and the surfactant are the same and where the total amounts of surfactant, oil, and water are conserved. In the limit $K \gg T$, the internal degrees of freedom act as constraints. Two globular phases in equilibrium must then have the same radius to have the same chemical potential since the bending energy is dominant and cannot be balanced by the temperature or the interactions. In this limit, the density of globules in the two coexisting phases is determined by the entropy and interaction terms alone, and the usual common-tangent construction yields the phase boundary. Recent neutron scattering work in aerosol-OT surfactant (AOT), water, and oil microemulsions has shown that these systems are indeed in the large bending energy limit; the radius of the globules in the two coexisting phases is measured to be the same within 3% [23].

The generic phase diagram for attractively interacting spherical microemulsion globules is shown in Fig. 9.2 as a function of the interaction strength (or virial coefficient) and volume fraction of dispersed phase ϕ at fixed temperature. When the attractions are small, the system consists of a single phase of globules with radius $\rho = 3\delta\phi/v_s$. The phase separation to two coexisting phases of globules (with the same radius if $K \gg T$) occurs as the radius ρ is increased, that is, by varying the water/surfactant ratio. The radius attains a critical value ρ_c, such that the total free energy ($F_{tot} = nF_b - TS_m + F_i$) has both its second and third derivatives with respect to ϕ equal to zero. The attractions are then strong enough so that the microemulsion separates into coexisting liquid (high number density of globules) and gas (low number density of globules) phases. This occurs when ρ is such that $A(\rho_c, T) \approx 21$ and $\phi_c \approx 13\%$, within a mean-field calculation.

Of course, the radius ρ can never exceed the natural bending radius ρ_0 (for $K \gg T$) at which the emulsification failure instability intervenes. Thus, if $\rho_0 > \rho_c$, the liquid–gas collective instability occurs and is then followed by emulsification failure. However, since emulsification failure happens after (i.e., at larger values of ρ) the liquid–gas transition, the excess phase coexists

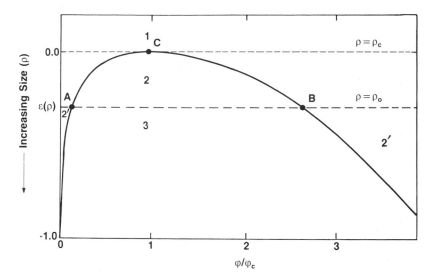

FIGURE 9.2. Generic phase diagram for attractively interacting spherical microemulsions as a function of $\varepsilon = [A(\rho_c, T) - A(\rho, T)]$ and volume fraction ϕ. Here, A is the virial coefficient of the attraction, ρ is the globule size, and T is the temperature; ρ_c is the critical value of the globule size at which liquid–gas-type phase separation occurs ($\phi_c \approx 13\%$), and the critical point is marked C. The region marked 1 is a single phase of spherical globules; 2 is a two-phase region with two phases of equal-size spheres, but with unequal densities; 2′ refers to the coexistence of the microemulsion with excess internal phase (emulsification failure); 3 denotes a three-phase region with low-density and high-density microemulsion phases coexisting with excess internal phase. The three-phase region consists of excess internal phase in equilibrium with globular phases whose concentrations are those at points A (low-density microemulsion) and B (high-density microemulsion). This diagram is drawn for the case $\rho_c < \rho_0$, so that the liquid–gas critical point can be obtained before emulsification failure. In the case $\rho_c > \rho_0$, there is only the single phase and the emulsification failure instability, 2′.

with two microemulsion phases, leading to a three-phase equilibrium. For example, for water in oil systems, one has a dense phase of water in oil globules coexisting with a dilute phase of droplets (almost a pure oil phase) and with a phase that is (mostly) excess water. These instabilities are shown in Fig. 9.2, where the dashed lines indicate the emulsification failure instability from either the single phase (1) to the excess internal phase (2′) (outside the liquid–gas coexistence region) or two-phase (2) to three-phase (3) (inside the liquid–gas coexistence region). If, however, $\rho_c > \rho_0$, the spheres can never grow larger than ρ_0 before emulsification failure takes place. Their interactions never become strong enough to drive the collective phase instability; in this case, only the emulsification failure instability is observed. Thus, the entire topology of the phase diagram in the dilute limit is determined by the competition between the two characteristic lengths ρ_0 (a property of a single

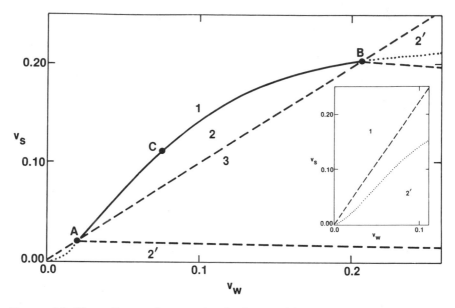

FIGURE 9.3. Phase diagram for water in oil microemulsion at fixed temperature from the parameters described in Ref. 18. The notation is the same as in Fig. 9.2. In addition, v_w is the volume fraction of water, and v_s is the volume fraction of surfactant. The dotted line is the continuation of the liquid–gas coexistence curve below the emulsification failure. The inset shows the phase diagram for the case $\rho_c > \rho_0$, where the liquid–gas phase separation is lost (dotted line) below the emulsification failure instability.

globule) and ρ_c (which depends on the interaction strength). Figure 9.3 illustrates this competition with two possible phase diagrams for dilute water in oil microemulsions as a function of the water and surfactant volume fractions. In the main part of the figure, $\rho_c < \rho_0$, so that the critical behavior and three-phase equilibrium occur, while in the inset, $\rho_c > \rho_0$, so that only the emulsification failure can occur.

The attractive interactions between globules are also responsible for more microscopic effects than the phase equilibrium alone. Attractions lead to clustering, which can be measured by scattering probes. Monte Carlo simulations described in Ref. 24 give a quantitative measure of this clustering. For a system with hard-sphere particles that interact with a square-well potential $V(r) = \infty$ if $r < 1$, $V(r) = -\varepsilon$ if $1 < r \leq (1 + \lambda)$, and $V(r) = 0$ for $r \geq (1 + \lambda)$, the effects of clustering are evident in Fig. 9.4, where the interparticle structure factor $S(Q)$ is shown. Here, r is the separation between the spheres measured in units of the sphere diameter, ε is the depth of the attractive well, and λ is the range of the attraction. Comparison of these calculations with experimental measurements of $S(Q)$ show that for AOT–water–oil systems, the attraction is extremely short-ranged, with a range of

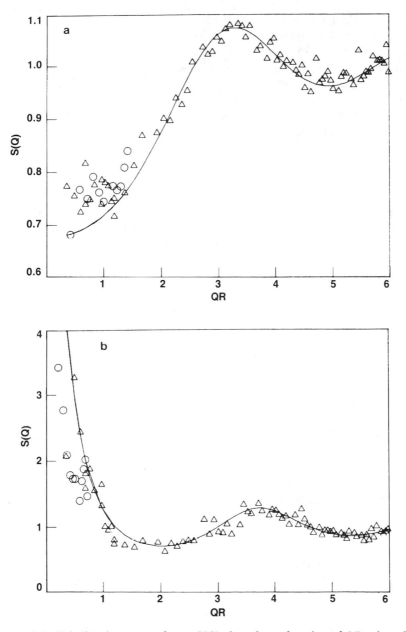

FIGURE 9.4. Calculated structure factor $S(Q)$ plotted as a function of QR, where R is the radius of the globule, and Q is the scattering vector. (a) Results for micelles: $R = 15$ Å; volume fraction, $\phi = 0.07$; well depth, $\varepsilon = 0.92$; and range, $\lambda = 0.1$. (b) Results for microemulsions: $R = 60$ Å, $\phi = 0.07$, $\varepsilon = 3.83$, and $\lambda = 0.02$. Both results use an interaction range of 3 Å; λ is scaled to the particle diameter. The solid line is a modified mean spherical approximation. The triangles are Monte Carlo results for 108 particles, and the circles are the results for 500 particles.

about 3 Å ($\lambda \approx 0.03$ for 100 Å-diameter globules). In addition, it was found that the interaction strength ε was approximately linear in the globule size. Figure 9.4. shows this effect; the larger size droplets show much more scattering at small Q, indicating the existence of larger clusters due to the larger interaction strength.

The results which are also consistent with recent analysis of both neutron [24] and light scattering [31] measurements, strongly suggest that the origin of the attractive interactions in these systems is not the long-range part of the van der Waals attractions of the water cores [21]. $S(Q)$ for a micellar system with $R = 15$ Å also showed the effects of some attraction, so the water cores are not crucial. Instead, the analysis suggests that a short-ranged attraction due to the effective interactions of the surfactant tails (as modified by the presence of oil) may be the origin of the observed interaction [50]. The linearity of ε with ρ is consistent with this hypothesis. For short-ranged interactions, the number of contacts between spheres (which is proportional to the interaction strength ε) is given by their overlap. The geometrical overlap of two spheres that interpenetrate a distance $\lambda \ll \rho$ is simply proportional to $\lambda^2 \rho$, consistent with the neutron scattering fits [24].

More recent scattering studies of the temperature dependence of ε and λ have suggested that λ increases as the critical point is approached [51]. However, mean-field fits (of which the mean spherical model is but one) can yield erroneous values of the interaction parameters near the critical point. The long-wavelength fluctuations, which are not correctly accounted for in these theories, can mimic very long-ranged attractions. Thus, the most reliable information concerning the microscopic interaction parameters is best obtained away from the critical point, although the temperature dependence of the interaction parameters is naturally of great interest.

IV. Percolation in Microemulsions

The previous section discussed the consequences of attractive interactions on the static properties of microemulsions, such as the phase diagram and equilibrium density fluctuations. In this section, we examine the effects of interactions on the conductivity of microemulsions with an analysis of percolation in interacting colloidal systems.

The concept of a percolation transition is often used in interpreting the conductivity of disordered systems. In systems with completely random distributions of particles, the percolation transition signifies the first emergence of an infinite cluster at some critical value of the volume fraction ϕ_p. If the particles are conducting and the matrix (or background fluid) is insulating, the system will show no conductivity for $\phi < \phi_p$. There will be a continuous transition at ϕ_p to a conducting state. For microemulsion systems, this transition is measurable through the ionic conductivity. However, a detailed treatment of the microscopic mechanism of the charge transfer is beyond the scope of this work.

Recent measurements of the conductivity of microemulsions have suggested the existence of a percolation threshold as a function of volume fraction, temperature, and globule size [27-29,52]. The transition, which occurs as either temperature or size is varied, shows percolation for very small volume fractions ($\sim 10\%$). Furthermore, the transition occurs near the temperature or size at which liquid-gas-phase separation [type (2) as discussed in Section III] takes place. Some studies have attributed this transition to the onset of a bicontinuous phase [53]. However, recent scattering experiments in AOT-water-oil microemulsions have shown that these systems maintain the integrity of the spherical globules up to rather high volume fractions [51]. In addition, there is no change in the particle size and shape as the critical point for phase separation is approached (or for that matter even in the two-phase regime). Thus, instead of indicating a transition from a spherical to a bicontinuous structure, the conductivity transition is interpreted to occur at the percolation threshold of rigid particles. The low value of the threshold is due to interactions between globules [28,29,51].

Recent Monte Carlo simulations of the percolation behavior of interacting colloidal particles by the author (and collaborators G. Grest, I. Webman, and A. Bug) [20,26,30] have shown that the percolation transition for spherical particles with a short-ranged attraction is a strong function of their interactions. As a critical point (or coexistence curve) is approached, the clustering of the particles that occurs tends to increase the connectivity of the system; the percolation threshold is decreased. Although this behavior is not universal—there are situations where the interactions can increase the percolation threshold—the lowering of ϕ_p by attractive interactions is a general feature of systems with short-range hopping of the conducting species. In addition to the interactions, the dynamical nature of the Brownian motion of the microemulsion globules has subtle effects on the dependence of the conductivity of the system for $\phi < \phi_p$ as discussed later.

The first misconception that was elucidated by these simulations was the notion of a universal value of ϕ_p for spheres. Although there are systems with universal values of ϕ_p (e.g., a binary mixture of conducting and insulating closely packed spheres [54] where $\phi_p \approx 0.17$), microemulsions are a unique experimental realization of the percolation [55,56] of interacting spheres with a hard-core in a continuum background. The threshold depends on the distance over which two adjacent globules can transfer charge. To define connectedness in this continuum system, it is necessary to introduce a shell of width $\delta/2$ (in units of the hard-core diameter) around each hard core. When two of these δ shells overlap, the corresponding spheres are said to be connected and hence in the same cluster. The physical origin of δ is the finite range of hopping of charge between two globules or a finite range of deformation of two adjacent spheres that may transfer charge via a local exchange of water and/or surfactant.

The Monte Carlo simulation is described in detail in Refs. 20, 26, and 30. The interparticle potential is the square-well interaction described in Section III. Standard Metropolis [57] algorithms for systems of 108, 500, and 2048 particles were studied in both two and three dimensions. In the noninteracting case ($\varepsilon = 0$), the percolation probability is a function of both the shell size δ and the volume fraction ϕ, as shown in Fig. 9.5, where the shell dependence of the effective volume fraction for percolation, $\bar{\phi}_p = \phi_p (1 + \delta)^3$ is shown. If δ is small, the system percolates only at high volume fractions in the absence of interactions. In the limit $\delta \to 0$ (and $\varepsilon = 0$), $\phi_p \approx 0.65$, the value for random close packing. At large values of δ, the results approach the known value of $\phi_p \approx 0.35$ for overlapping spheres. Away from the coexistence curve (i.e., when the attractive interactions are small), the transition is observed [29] at high values of ϕ. This is consistent with a small value of δ for these systems. (For a discussion of the equivalent problem in two dimensions, see Ref. 30).

As the attractive interaction strength is increased ($\varepsilon > 0$), ϕ_p drops dramatically (see Fig. 9.5). The lowering of the percolation threshold is associated with the formation of correlated clusters in the interacting system. These clusters are both anisotropic and fractal-like and their percolation threshold is expected to be lower than that of random spheres. For small values of the interaction range λ and the conductivity shell δ, ϕ_p typically drops from values near random close packing ($\phi_p \approx 0.65$) to values of ϕ on

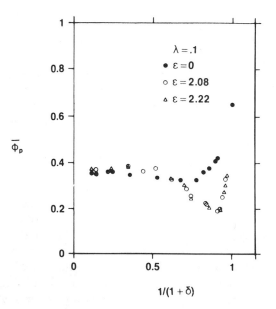

FIGURE 9.5. Monte Carlo results for the critical volume fraction for percolation ϕ_p as a function of the conductivity shell size (δ) for both the noninteracting ($\varepsilon = 0$) and attractively interacting systems. The effective volume fraction $\bar{\phi}_p = \phi_p(1 + \delta)^3$ is plotted vertically, while $1/(1 + \delta)$ is plotted horizontally.

the order of 10% as ε is increased. The largest effects occur near the critical point; the minimum in Fig. 9.5 occurs very close [18] to the critical volume fraction, where concentration fluctuations are the largest. The observed percolation transitions in microemulsions are therefore sensitive indicators of the effects of concentration fluctuations on the transport behavior. Again, the lowering of ϕ_p with increasing interaction is nonuniversal; it occurs only for small values of the shell δ. For large values of δ, the interactions can actually increase the percolation threshold; moderate attractions cause the clusters to become more compact, and ϕ_p is increased. The competition between these two effects of attractive interactions on percolation is discussed in more detail in Ref. 30.

In addition to being the prime physical example of percolation in an interacting equilibrium system, microemulsions are also a model system for the study of dynamical effects on percolation [52,58,59]. In the usual static models of systems of conducting regions embedded in an insulating matrix, the conductivity vanishes below ϕ_p. However, for microemulsions, the clusters continuously rearrange because of Brownian motion, resulting in a finite conductivity for $\phi < \phi_p$. Unlike the static case, the charge carriers are not trapped in the finite clusters. A charge on a water globule can propagate either by hopping to a neighboring globule or via the diffusion of the host globule. If the typical hopping time of the carriers between globules is much shorter than some characteristic time related to the motion of the globules, a steep increase in the overall carrier diffusion is expected as the $\phi \rightarrow \phi_p$. This is due to a transition from a regime of transport dominated by globular motion and cluster rearrangement to a regime of transport dominated by the motion of charge carriers on a large, connected cluster of globules.

The effects of globule dynamics on the conductivity in microemulsions were discussed by Lagues [58], who suggested that for $\phi < \phi_p$ the conductivity σ increases as $(\phi_p - \phi)^{-\tilde{s}}$, where $\tilde{s}$ differs from its static value. (The static percolation system with finite σ below ϕ_p is one where the charge carriers have finite conductivity in the matrix, i.e., the system is a mixture of good and bad conductors.) In our work [26], a scaling argument, different from that of Lagues, is presented, that suggests $\tilde{s} = 2v - \beta = 1.3$ in three dimensions for percolation in a system where the charge carriers hop on slowly moving colloidal particles. The value of $\tilde{s}$ differs from the corresponding exponent $s \approx 0.7$ for the static case. This value of $\tilde{s}$ is in agreement with the Monte Carlo simulations, as well as with the experimental results on microemulsions [26,52,58]. In all these cases, the conductivity transition is smeared around the value of ϕ_p appropriate to the static percolation problem. Thus, the previous discussion of the shift of ϕ_p with interaction strength (as modified by either temperature or globule size) is completely applicable as far as the value of the threshold, while the dynamical effects of globule motion and cluster rearrangement are responsible for the change in the exponent.

The scaling argument for the dynamical case has also been extended to predict the frequency dependence of the conductivity at $\phi = \phi_p$. For small values of the frequency $\omega < 1/T_R$ (where T_R is the cluster rearrangement time), the conductivity is frequency independent. This behavior is followed by a power law regime with $\sigma(\phi = \phi_p) \sim \omega^{\chi}$ with $\chi = \mu/(\tilde{s} + \mu)$, where μ is the conductivity exponent for $\phi > \phi_p$. Again, this relation is in agreement with recent measurements [26,52].

V. Discussion

This paper has reviewed a phenomenological theory for the structure and phase transitions in three-component microemulsions in the dilute limit, where a globular model applies. The elasticity of the surfactant layer has been shown to give rise to a curvature energy that distinguishes microemulsions with different shapes. The effects of thermal fluctuations are to introduce small, but measurable, intrinsic equilibrium polydispersity (in both size and shape) in the case of spherical globules. Cylindrical globules are predicted to have a behavior resembling those of polymers.

The effects of interactions were analyzed for spherical globules. Even in this simple case, the collective behavior of the system is subtle because of these interactions. Attractive interactions are responsible for density fluctuations in the one-phase regime. Experimental measurement of these fluctuations via scattering spectroscopy yields information on the microscopic interactions; the attractions are shown to be strong and of short-range compared to the globule size. The attractive interactions modify the phase behavior of the system. A two-phase equilibrium resembling that of a critical liquid–gas system is possible. Furthermore, three-phase equilibrium—a dense microemulsion coexisting with two phases, one of which is mostly oil and one of which is mostly water—is also possible even for spherical globules. Finally, the interactions are responsible for the observed lowering of the percolation threshold as the coexistence curve is approached by either changing temperature or globule size (both of which affect the strength of the attractions). The dynamical nature of the microemulsion globules results in finite ionic conductivity even below the percolation threshold with an exponent that differs from its static value.

Future studies of these systems can build on the experimental and theoretical framework presented here. The topic of most current interest is the transition from the globular states discussed here to the bicontinuous phases that are necessary to explain the continuous transitions from oil in water to water in oil microemulsions. Do they proceed via the close packing of spherical objects, the deformation of lamellae, or a polymer-like cylindrical phase? What are the effects of interactions on the lamellar and cylindrical structures?

In addition to being a necessary first step for the understanding of the transition to bicontinuity, the present work has indicated that the globular phases of microemulsions are of interest in their own right. They exhibit properties similar to but different from atomic systems. An area that is ripe for further study is the detailed nature of equilibrium polymerization in cylindrical microemulsions, including their phase behavior, transport [60], and rheology. These quantities should differ from the corresponding properties of molecular polymers, since the cylindrical microemulsions have their degree of polymerization (or molecular weight) determined by equilibrium conditions and not by the kinetics of preparation. The degree of polymerization, N, is predicted to be a sensitive function of the concentrations. Another area of both experimental and theoretical interest is the percolation transition in interacting dynamical systems. Now that the phenomenological description of the percolation threshold and exponents has been clarified, perhaps the more difficult problems of the microscopic mechanism of the interglobular interactions (and their temperature dependence) and conductivity mechanisms can be addressed.

Acknowledgments. The author is grateful for fruitful theoretical collaborations with A. Bug, G. Grest, P. Pincus, L. Turkevich, and I. Webman as well as stimulating interactions with experimental colleagues S. Bhatacharaya. W. Dozier, J. S. Huang, M. W. Kim, C. R. Safinya, and J. Stokes.

References

1. For a general survey, see "*Surfactants in Solution.*" K. Mittal and B. Lindman, eds., Plenum, New York, 1984.
2. B. Lindman and P. Stiles, Ref. 1, p. 1651.
3. S. M. Allen and J. W. Cahn, *Acata Mettal.* **27**, 1085 (1979).
4. I.M. Lifshitz, *Sov. Phys. JETP* **15**, 939 (1982).
5. E. Ruckenstein and J.C. Chi, *JCS Faraday Trans. II* **71**, 1690 (1975) and Ref. 1, p. 1551.
6. C.A. Miller and P. Neogi, *AIChE J.* **26**, 212 (1980); J. Jeng and C.A. Miller, Ref. 1, p. 1829.
7. C. Huh, *J. Coll. and Int. Sci.* **97**, 201 (1984) and **71**, 408 (1979).
8. M. Robbins, in "*Micellization, Solubilization, and Microemulsions,*" Vol. 2, p. 713, K.L. Mittal, ed., Plenum, New York, 1977.
9. P.G. de Gennes and C. Taupin, *J. Phys. Chem.* **86**, 2294 (1982).
10. J. Jouffroy, P. Levinson, and P.G. de Gennes, *J. Phys. (Paris)* **43**, 1241 (1982).
11. B. Widom, *J. Chem. Phys.* **81**, 1030 (1984).
12. Y. Talmon and S. Prager, *J. Chem. Phys.* **69**, 2984 (1978) and **76**, 1535 (1982).
13. L.E. Scriven, in Ref. 8, p. 877.
14. S.A. Safran, *J. Chem. Phys.* **78**, 2073 (1981).
15. S.A. Safran, Ref. 1, p. 1781.
16. S.A. Safran, L.A. Turkevich, and P.A. Pincus, *J. de Phys. Lett.* **45**, L69 (1984).
17. L.A. Turkevich, S.A. Safran, and P.A. Pincus, "*Surfactants in Solution,*" p. 1177, K. Mittal, and P. Bothorel, eds., Plenum, New York, 1986.
18. S.A. Safran and L.A. Turkevich, *Phys. Rev. Lett.* **50**, 1930 (1983).
19. S.A. Safran, L.A. Turkevich, and J.S. Huang, in Ref. 17, p. 1167.

20. S.A. Safran, I. Webman, G.S. Grest, *Phys. Rev.* **A32**, 506 (1985).
21. A.J. Calje, W. Agterof, and A. Vrij, in Ref. 8, p. 779.
22. S. Brunetti, D. Roux, A.M. Bellocq, G. Forche, and P. Bothorel, *J. Chem. Phys.* **83**, 1028 (1983); *J. Coll. Int. Sci.* **88**, 302 (1982). R. Ober and C. Taupin, *J. Phys. Chem.* **84**, 2418 (1980).
23. M. Kotlarchyk, S.-H. Chen, J.S. Huang, and M.W. Kim, *Phys. Rev.* **A29**, 2054 (1984): B.H. Robinson, in Ref. 8.
24. J.S. Huang, S.A. Safran, M.W. Kim, G.S. Grest, M. Kotlarchyk, and N. Quirke, *Phys. Rev. Lett.* **53**, 592 (1984).
25. S.J. Candau, E. Hirsch, R. Zana, *J. Coll. and Int. Sci.* **105**, 521 (1985).
26. G.S. Grest, I. Webman, S.A. Safran, and A.L.R. Bug, *Phys. Rev.* **A33**, 2842 (1986).
27. H.F. Eicke, R. Kubick, R. Hasse, and I. Zschokke, Ref. 1, p. 1533.
28. A.M. Cazabat, D. Chatenay, F. Guering, D. Langevin, I. Meunier, O. Sorba, J. Lang, R. Zana, and M. Pailette, in Ref. 1, p. 1737; A.M. Cazabat, D. Langevin, J. Meunier, and A. Pouchelon, *J. Phys. Lett.* **43**, L89 (1982).
29. M.W. Kim and J.S. Huang, unpublished.
30. A.L.R. Bug, S.A. Safran, G.S. Grest, and I. Webman, *Phys. Rev. Lett.* **55**, 1896 (1985).
31. M.W. Kim, W.D. Dozier, and R. Klein, *J. Chem. Phys.* **84**, 5919 (1986) and **87**, 1455 (1987).
32. J.H. Schulman and J.B. Montagne, *Ann. N.Y. Acad. Sci.* **92**, 366 (1961).
33. C. Borzi, *J. Chem. Phys.* **82**, 3817 (1985).
34. P.G. de Gennes, "*Physics of Liquid Crystals*," Clarendon Press, Oxford, 1974.
35. A.G. Petrov, and A. Derzzhanski, *J. Phys. Coll.* **37**, C3-155 (1976): A.G. Petrov, M.D. Mitov, and A. Derzhanski, *Phys. Lett.* **65A**, 374 (1978).
36. W. Helfrich, *Z. Naturforsch* **38**, 6693 (1973).
37. W. Helfrich, *J. de Phys.* **47**, 321 (1986).
38. J.S. Huang and M.W. Kim, *Phys. Rev. Lett.* **47**, 1462 (1981). T. Assih, F. Larch, and P. Delord, Ref. 1, p. 1821; P. Delord, and F.C. Larche, *J. Coll. Int. Sci.* **98**, 277 (1984).
39. S. Ljunggren, and J.C. Eriksson, *J. Chem. Soc. Faraday Trans. II* **80**, 489 (1984).
40. J.M. di Meglio, M. Dvolaitzky, R. Ober, and C. Taupin, *J. Phys. Lett.* **44**, L229 (1983).
41. E.W. Kaler, K.E. Bennett, H.T. Davis, and L.E. Scriven, *J. Chem. Phys.* **79**, 5673 and 5685 (1983).
42. M. Kotlarchyk, S.-H. Chen, J.S. Huang, and M.W. Kim, *Phys. Rev. Lett.* **53**, 941 (1984).
43. P. Flory, "*Principles of Polymer Chemistry*," Cornell University Press, Ithaca, New York, 1971.
44. J.N. Israelachvili, D.J. Mitchell, and B.W. Ninham, *J. Chem. Soc. Faraday Trans. II* **72**, 1525 (1976).
45. J. Apell, G. Porte, and Y. Poggi, *J. Coll. Int. Sci.* **87**, 492 (1982); W. Gelbart, A. Ben-Shaul, W. McMullen, and A. Masters, *J. Chem. Phys.* **88**, 8861 (1984).
46. N. Kumar, J.D. Litster, and C. Rosenblatt, *Phys. Rev. Lett.* **50**, 1672 (1983).
47. G. Porte, *J. Phys. Chem.* **87**, 3541 (1983); G. Porte, J. Mairgnan, J. Appell, Y. Poggi, and G. Maret, in Ref. 17; J. Appell and G. Porte, *J. Phys. Letts.* **44**, L689 (1989).
48. S.F. Edwards, *Proc. Phys. Soc.* **92**, 9 (1967).
49. R.K. Pathria, "*Statistical Mechanics*," p. 268, Pergammon, New York, 1972.

50. D. Roux, A.M. Bellocq, and P. Bothorel, in Ref. 1, p. 1843.
51. S.-H. Chen, *Ann. Rev. Phys. Chem.* **37**, 351 (1986).
52. S. Bhattacharaya, J. Stokes, M.W. Kim, and J.S. Huang, *Phys. Rev. Lett.* **55**, 1884 (1985).
53. D. Chatenay, W. Urbach, A.M. Cazabat, and D. Langevin, *Phys. Rev. Lett.* **20**, 2253 (1985).
54. H. Scher and R. Zallen, *J. Chem. Phys.* **53**, 3759 (1970): R. Blanc and E. Guyon, *"Percolation Structures and Processes—Annals of the Israel Physical Society,"* Vol. 5, p. 229, G. Deutcher, R. Zallen, and J. Adler, eds., Israel Physical Society, Jerusalem, 1983.
55. A. Coniglio, U. De Angelis, A. Forlani, *J. Phys. A* **10**, 1123 (1977).
56. Y. Chiew and E. Glandt, *J. Phys. A* **16**, 2599 (1983).
57. K. Binder, ed., *"Monte Carlo Methods in Statistical Physics,"* Springer-Verlag, Berlin and New York, 1979.
58. M. Lagues, *J. Phys. Lett.* **40**, L331 (1979).
59. R. Kutner and K.W. Kehr, *Phil. Mag.* **A48**, 199 (1983).
60. S.J. Chen, D.F. Evans, and B.W. Ninham, *J. Phys. Chem.* **88**, 1631 (1984).

The topics explored here have been more fully studied in the following references:

1. *General references*:
   (a) *"Surfactants in Solution,"* K. Mittal and P. Bothorel, eds. (Plenum, N.Y., 1986) (b) *"Physics of Complex and Supermolecular Fluids,"* S.A. Safran and N.A. Clark, eds. (Wiley, N.Y., 1987) (c) *"Physics of Amphiphilic Layers,"* J. Meunier, D. Langevin, and N. Boccara, eds. (Springer-Verlag, N.Y., 1987).
2. The elastic theory described here for globules has been extended to bicontinuous systems:
   S.A. Safran, D. Roux, M. Cates, and D. Andelman, *Phys. Rev. Lett.* **57**, 491 (1986), and in *Surfactants in Solution: Modern Aspects*, K. Mittal, ed. (Plenum, N.Y., in press); D. Andelman, M. Cates, D. Roux, and S.A. Safran, *J. Chem. Phys.* **87**, 7229 (1987); D. Andelman, S.A. Safran, D. Roux, and M. Cates, *Langmuir* **4**, 802 (1988); L. Golubovic and T.C. Lubensky, (*Phys. Rev.*, in press).
3. The experimental results for the AOT system have been summarized in:
   S.-H. Chen, T.L. Lin, and J.S. Huang, in *"Physics of Complex and Supermolecular Fluids,"* S.A. Safran and N.A. Clark, eds. (Wiley, N.Y., 1987), p. 285.
4. Dynamical fluctuations of spherical droplets are discussed in:
   S.T. Milner and S.A. Safran, *Phys. Rev.* **A36**, 4371 (1987) (theory); J.S. Huang, S.T. Milner, B. Farago, and D. Richter, *Phys. Rev. Lett.* **59**, 2600 (1987) (experiment).
5. Polymer-like micelles have been extensively studied:
   (a) Experimental studies: S.J. Candau, E. Hirsch, and R. Zana, *J. Phys.* **45**, 1263 (1984); *J. Colloid Int. Sci.* **105**, 521 (1985); in *"Physics of Complex and Supermolecular Fluids,"* S.A. Safran and N. Clark, eds. (Wiley, N.Y., 1987), p. 569; S.J. Candau, E. Hirsch, R. Zana, and M. Adam, *J. Coll. Int. Sci.* **122**,

430 (1988); S.J. Candau, E. Hirsch, R. Zana, and M. Delsanti, *Langmuir* (in press); R. Messager, A. Ott, D. Chatenay, W. Urbach, and D. Langevin, *Phys. Rev. Lett.* **60**, 1410 (1988).
(b) Theory of Dynamics: M.E. Cates, *Macromolecules* **20**, 2289 (1987); *Europhys. Lett.* **4**, 497 (1987); *J. de Phys.* **49**, 1593 (1988).

10
Theory of Thermodynamic Properties and Phase Separation of Self-Associating Micellar Solutions

D. BLANKSCHTEIN, G.M. THURSTON, M.R. FISCH, and G.B. BENEDEK

The need to develop a new theoretical framework to study the thermodynamic properties and the phenomena of phase separation of micellar solutions is discussed. Accordingly, we have recently developed a thermodynamic theory that is capable of describing self-consistently the single-phase equilibrium properties and the phase separation of two-component amphiphile-water micellar solutions. In this chapter, we review the results of our theory and the comparison of these results with experimental findings in micellar solutions.

I. Introduction

When amphiphilic molecules are placed in an aqueous environment, they can segregate their hydrophobic regions from water by self-associating into a variety of aggregate structures known as micelles [1]. The formed micellar aggregates often exhibit a broad distribution of sizes. Micelles are noncovalently bonded macromolecular aggregates that are continually and reversibly exchanging amphiphiles with one another and with the amphiphilic monomers in the solution [2]. As a result, the individual micelles do not maintain a distinct, unchanging identity. Instead, a description of the state of the micellar solution requires a specification of the distribution of micellar sizes that exists on average in the solution and is governed by the thermodynamic principle of multiple chemical equilibrium [3]. It is essential to recognize, in distinction to other multicomponent mixtures, that the micellar size distribution is sensitively dependent upon solution conditions, such as amphiphile concentration, temperature, pressure, and salt concentration. In this respect, micellar solutions are fundamentally different from previously considered multicomponent mixtures in which the polydispersity is a fixed and unchanging feature determined solely by the initial concentrations of the nonassociating solute molecules.

At amphiphile volume fractions below 20%, micellar solutions often exist as homogeneous isotropic liquid phases [4]. Phase separation and critical phenomena can be induced in this concentration range by changing the temperature, pressure, salt concentration, and other solution conditions [5]. In many such phase separations, a single isotropic micellar phase separates into two isotropic phases, both of which contain micelles and water, but differ in total amphiphile concentration [4,5]. In particular, phase separations of this type can be induced by lowering the temperature, as in solutions of the zwitterionic amphiphile dioctanoyl phosphatidylcholine (C_8-lecithin) and water [6,7], and by raising the temperature, as in solutions of the nonionic amphiphile n-dodecyl hexaoxyethylene glycol monoether ($C_{12}E_6$) and water [8,5,9]. These cases lead to coexistence curves that exhibit upper and lower consolute (critical) points, respectively. In addition, micellar solutions can exhibit both types of consolute points as the temperature is varied monotonically over a finite range, leading to closed loop coexistence curves. An example of the latter behavior is found in solutions of n-decyl pentaoxyethylene glycol monoether ($C_{10}E_5$) and water [10]. Typically, consolute points in these phase transitions occur at very dilute amphiphile concentrations, for example, in C_8-lecithin at about 2% volume fraction. The coexistence curves usually show a pronounced asymmetry between the dilute and concentrated branches (See Fig. 10.1). Although polymers and other macromolecules in solution can exhibit similar behavior [11], micellar solutions are fundamentally different, as previously emphasized.

In order to formulate a theory of the equilibrium properties of micellar solutions including the phase separation phenomena, it is necessary to incorporate the unique characteristics of micellar aggregates and their size distribution, which distinguish micellar solutions from previously considered multicomponent mixtures. Previous theoretical investigations of phase separation in micellar solutions [4,12–15] did not include these unique features, which we have previously described.

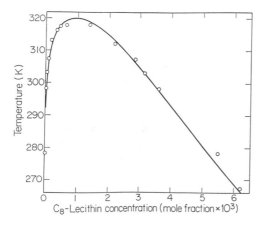

FIGURE 10.1. Theoretical coexistence curve for the phase separation in a micellar solution of C_8-lecithin and water. The open circles are experimental points.

The present theory [7,16,17] represents an effort to include in a single, unified theoretical framework the effect of both intermicellar interactions and multiple chemical equilibrium on the micellar size distribution and on the equilibrium thermodynamic properties in both the single-phase and two-phase regions of micellar solutions. As usual, the fundamental mechanism leading to phase separation involves a competition between the entropy of the solution, which favors random mixing of micelles and water molecules, and net attractive micelle–micelle and water–water interactions, which favor demixing into micelle-poor and micelle-rich phases. In addition, as the system is brought toward phase separation, the micellar size distribution can change according to the principle of multiple chemical equilibrium, and this is the essential new physical feature that we have included in our theory. We have adopted a thermodynamic approach in which we introduce a phenomenological Gibbs free energy that incorporates the essential physical ingredients of the micellar solution. All the equilibrium properties of the solution, including phase separation, can be calculated from this free energy. As a result, we have put the theory of micellar solutions in the same category as existing mean-field type theories of simple binary mixtures or polymer solutions.

In this chapter, we review the theoretical framework and its application to micellar solutions of C_8-lecithin and water, which exhibit upper consolute points. We also briefly describe the application of the theory to micellar solutions of $C_{12}E_6$ and water, which exhibit lower consolute points. Further details may be found in Refs. 7, 16 and 17.

II. Theoretical and Experimental Results

Consider a solution of N_s solute molecules (amphiphiles) and N_w solvent molecules (water) in thermodynamic equilibrium at temperature T and pressure p. The self-association of the amphiphiles produces a distribution $\{N_n\}$ of micellar sizes, where N_n is the number of micelles having n amphiphiles. We provisionally consider micelles of different sizes as distinct chemical species and model [7] the Gibbs free energy G of the solution as consisting of three additive parts, G_f, G_m, and G_{int}. These parts are chosen so as to provide a heuristically appealing identification of the factors that are responsible for micellar formation and growth on the one hand and for phase separation on the other.

It should be kept in mind at the outset that the procedure presented here may be carried out on Gibbs free energy models that differ from the present one. For example, other analytic forms may be adopted for G_m and G_{int}, or one might proceed by modeling the partition function instead of the Gibbs free energy.

Our model for G_f has the form

$$G_f = N_w \mu_w^0 + \sum_n N_n \tilde{\mu}_n^0, \tag{1a}$$

where $\mu_w^0(T, p)$ is the free energy change of the solution when a water molecule is added to pure water, and $\tilde{\mu}_n^0(T, p)$ reflects the free energy change of the solution when a single aggregate of size n is placed at a given position in pure water. G_f summarizes the many complex physical factors that are responsible for the formation of micelles. These contributions are evaluated for a dilute reference solution that lacks intermicellar interactions, as is reflected in the definitions of μ_w^0 and $\tilde{\mu}_n^0$. These factors include hydrophobic, hydrogen bonding, electrostatic, steric, and Van der Waals interactions, as well as other subtle considerations that are the topics of active experimental and theoretical research [4,18–21].

The chemical potentials $\{\tilde{\mu}_n^0(T, p)\}$ and $\mu_w^0(T, p)$ depend on n, T, and p, as well as on other quantities, such as electrolyte concentration, which are not explicitly included in this presentation. All of these dependencies contribute to the response of the entire micellar size distribution to changes in the thermodynamic state of the solution. Fortunately, it is possible to proceed quite far in an examination of the thermodynamics of micellar solutions without adopting specific models for the detailed dependence of the sequence of chemical potentials $\{\tilde{\mu}_n^0(T, p)\}$ and of $\mu_w^0(T, p)$ on n, T, and p, while at the same time accounting, in principle, for these dependencies.

The model for G_m has the form

$$G_m = k_B T\left[N_w \ln(X_w) + \sum_n N_n \ln(X_n) \right],$$ (1b)

where $X_w = N_w/(N_w + N_s)$, $X_n = N_n/(N_w + N_s)$, and k_B is the Boltzmann constant. Here, $-G_m/T$ models the entropy of mixing of the formed aggregates, the monomeric amphiphiles, and the solvent. This entropy of mixing reflects the number of geometric configurations that describes the possible positions of the micelles, the monomeric amphiphiles, and the water molecules in the solution as a function of the relative proportions of each of these constituents. These relative proportions are represented by the mole fractions X_w and $\{X_n\}$.

A fundamental calculation of the entropy of mixing is available only for a very few simple models [22,11]. There is no known first principles calculation of the mixing entropy, which includes the effect of size, shape, flexibility, and polydispersity of micellar solutions. In the absence of such information, we have adopted in Eq. (1b) an ad hoc expression to represent the entropy of mixing. The present model for G_m has a form that is similar to that for ideal solutions, but its justification must be a posteriori. It is important to emphasize that we have found Eq. (1b) suitable to describe micelles that grow in dilute solution [7,16,17]. Different forms for G_m may be needed to describe other experimental micellar solutions: in particular, these forms may have to reflect the size, shape, and flexibility of the micellar aggregates.

Our model for G_{int} reflects the interactions among the formed micellar aggregates, the monomeric amphiphiles, and the water molecules. Like the free energy of mixing, G_{int} is very difficult to calculate from first principles.

Our choice of G_{int} results from a simple mean-field type approximation, in which interactions are averaged uniformly over spatial and orientational configurations, thus neglecting correlations. Since the aim of the present theory is to describe isotropic micellar phases, which lack both positional and orientational long-range order, we believe that this approximation preserves the essential physical ingredients of the interactions.

G_{int} has the form

$$G_{int} = -(\tfrac{1}{2})CN_s\phi, \tag{1c}$$

where the parameter $C(T, p)$ represents an effective interaction free energy, mediated by the solvent, between pairs of amphiphiles on different micelles, and ϕ is the total volume fraction of amphiphile. Denoting the volume of an individual amphiphile by Ω_1, that of a water molecule by Ω_w, and their ratio by $\gamma = \Omega_1/\Omega_w$, the volume fraction of an amphiphile is given by $\phi = \gamma N_s/(N_w + \gamma N_s)$. Physically, Eq. (1c) reflects the fact that in the actual solution a certain fraction ϕ of the environment of each micelle consists of amphiphilic material found in other micelles, rather than water. Our choice for G_{int} reflects a simple mean-field approach, whose justification can also be properly made a posteriori. It is instructive to recognize that our choice for G_{int} has precisely the quadratic form widely used in the description of polymer solutions and binary mixtures [22,11].

We next review the thermodynamic consequences of this model [7,16,17]. The self-association equilibrium and the phase separation are both governed by the chemical potentials of the water, μ_w, and each n-mer, μ_n. These are calculated from G and are given by

$$\mu_w = \mu_w^0 + k_BT[\ln(1 - X) + X - M_0] + (C/2)\gamma X^2/[1 + (\gamma - 1)X]^2, \tag{2a}$$

and

$$\mu_n = \mu_n^0 + k_BT[\ln(X_n) + n(X - 1 - M_0)] + (C/2)n\{(1 - X)^2/[1 + (\gamma - 1)X]^2 - 1\}, \tag{2b}$$

where $\mu_n^0(T, p) = \tilde{\mu}_n^0(T, p) + k_BT$, $X = N_s/(N_w + N_s)$ is the total amphiphile mole fraction, and $X_n = N_n/(N_w + N_s)$ is the mole fraction of n-mers. M_0 is the zeroth moment of the distribution of micellar sizes, where the kth moment of this distribution is given by $M_k = \sum_m m^k X_m$.

Until this point, we have treated micellar aggregates of different sizes as independent chemical species in the solution. However, as emphasized in the introduction, the aggregates continually and reversibly exchange amphiphiles with one another. These reversible material exchanges can be conveniently described using the thermodynamic principle that governs multiple chemical equilibrium. This implies that the chemical potential per amphiphile must be the same in all the aggregates: $(\mu_n/n) = \mu_1$, independently of n. Using Eq. (2b) in this condition yields the distribution of micellar sizes

$$X_n = (X_1)^n \exp[-\beta(\mu_n^0 - n\mu_1^0)], \tag{3}$$

where $\beta = 1/k_B T$. Equation (3), combined with the conservation of the total number of amphiphilic monomers in solution, $X = \sum_n nX_n$, determines the micellar size distribution as a function of total amphiphile concentration X, provided that the sequence of chemical potentials $\{\mu_n^0\}$ is known as a function of T, p and other solution conditions.

A particularly important consequence of our choice of G_{int} is that intermicellar interactions do not affect the micellar size distribution. This can be immediately seen from Eq. (1c), since G_{int} depends only on the total amount of amphiphile N_s and not on the specific way that amphiphiles are distributed. Nevertheless, since the micellar size distribution $\{X_n\}$ depends explicitly on X, it follows that after phase separation the dilute and concentrated phases, having mole fractions Y and Z, respectively, will have different micellar size distributions $\{X_n(Y, T, p)\}$ and $\{X_n(Z, T, p)\}$. Clearly, other models of G_{int} may lead to distributions that will depend explicitly on intermicellar interactions [23]. In that case, certain aggregate structures and sizes may be preferred over the others and stabilized by these interactions [24]. These interesting possibilities are the subject of current theoretical research.

Another important consequence of the form of Eq. (3) is that, in conjunction with the definition of the moments M_k of the distribution of micellar sizes, it implies [7,25] that all these moments are related to the second moment M_2 through $M_{k+1} = M_2(dM_k/dX)$. This conclusion is valid regardless of the form of the $\{\mu_n^0\}$. Note that $M_1 = X$, and

$$M_0(X, T, p) = \int_0^X dX'[X'/M_2(X')].$$

As a result, all of the influence of the micellar size distribution on equilibrium properties of the solution, such as the osmotic pressure $\pi = (\mu_w^0 - \mu_w)/\Omega_w$, the osmotic compressibility $(\partial\pi/\partial X)_{T,p}^{-1}$, the coexistence curve, and the spinodal line, is determined by the dependence of M_2 on X, T, p, and other solution conditions [7,16,17].

This remarkable simplification becomes even more significant in view of the fact that a direct experimental determination of $M_2(X, T, p)$ can be obtained from measurements that yield the weight-average association number of the micelles

$$\langle n \rangle_w(X, T, p) = \left[\sum_n n(nN_n)\right]\bigg/\left[\sum_n (nN_n)\right] = M_2(X, T, p)/X.$$

Indeed, at constant T and p, the observed dependence of $\langle n \rangle_w$ on X can serve as a useful indicator of the applicability of our theory to specific experimental systems.

To compare the theoretical predictions with experiments in a real micellar solution, it is first necessary to evaluate $M_2(X, T, p)$ for that system. To do that we choose to model the sequence of chemical potentials $\{\mu_n^0\}$ as a function of n, T, p, etc. The sequence $\{\mu_n^0\}$ is then used in Eq. (3), which along with the constraint imposed by the conservation of amphiphiles, $X = \sum_n nX_n$, determines the distribution $\{X_n(X, T, p)\}$ of micellar sizes. Finally, $\{X_n\}$ is used to calculate $M_2(X, T, p)$, which can then be used in the relevant equations to determine the thermodynamic properties of the solution.

We have carried out this program for micellar solutions of C_8-lecithin and water, which exhibit an upper consolute point [7,17]. Experimental information suggests [6,7] that at constant T and p $\langle n \rangle_w$ is proportional to $X^{1/2}$. In the context of our free energy, such an $X^{1/2}$ behavior results from a very simple model of one-dimensional micellar growth [6,26–29], in which the number of monomers, n_0, in the micellar periphery is independent of the number of monomers $(n - n_0)$ in the micellar interior. Micelles can grow by incorporation of monomers into the micellar interior. Chemical potentials per amphiphile, μ_p^0 and μ_i^0, are associated with the periphery and the interior, respectively, yielding $\mu_n^0 = n_0\mu_p^0 + (n - n_0)\mu_i^0$, for $n \geq n_0$ [29,7]. In the experimental concentration range, where $\langle n \rangle_w \gg n_0$, a single parameter $K(T, p) = \exp[\beta n_0(\mu_p^0 - \mu_i^0)]$, which reflects the tendency of an amphiphile to favor either the interior or the periphery of a micelle, completely characterizes [29,7] the micellar size distribution through $M_2(X, T, p) = 2K^{1/2}X^{3/2}$. In particular, this model predicts that $\langle n \rangle_w = 2(KX)^{1/2}$.

Using this expression for M_2, we obtain [7] the following equations for the coexisting amphiphile mole fractions $Y(T, p)$ and $Z(T, p)$:

$$K(T, p) = \exp[\beta\Delta\mu(T, p)] = \frac{[6/(3\gamma - 2)]^2}{[YZ(Y^{1/2} + Z^{1/2})^6]}, \tag{4a}$$

$$\beta\gamma C(T, p) = 1 + [(3\gamma - 2)/3][2(Y^{1/2} + Z^{1/2})^2 - 3(YZ)^{1/2}], \tag{4b}$$

where $\Delta\mu(T, p) = n_0[\mu_p^0(T, p) - \mu_i^0(T, p)]$. Similarly, the osmotic compressibility is given by

$$(\partial\pi/\partial X)_{T,p}^{-1} = -\Omega_w^{-1}(\partial\mu_w/\partial X)_{T,p}^{-1}$$
$$= \beta\Omega_w\{X/(1 - X) + 1/[2(KX)^{1/2}] - \beta C\gamma X/[1 + (\gamma - 1)X]^3\}^{-1}. \tag{5}$$

Using this model of micellar growth, the theory has two independent parameters. The first, $\Delta\mu(T, p)$, reflects the growth properties of the individual noninteracting micelles, and the second, $C(T, p)$, reflects the effective attractive micellar interactions leading to phase separation. In addition, the theory contains the ratio γ, which describes the relative effective volumes of the amphiphile and water molecules. This quantity can be estimated experimentally. Hence, at constant pressure, a knowledge of $\Delta\mu(T)$ and $C(T)$ is sufficient to invert Eqs. (4a) and (4b) and derive a theoretical coexistence curve, $Y(T)$ and $Z(T)$. In the C_8-lecithin and water system, we found [7] that by assuming that $\Delta\mu$ and C are temperature independent, and determining their values at the experimentally observed critical concentration X_c and critical temperature T_c, we were able to obtain excellent agreement with the experimentally measured coexistence curve, as shown in Fig. 10.1. Independently, another estimate of the parameter $\Delta\mu$ was obtained by performing quasielastic light scattering measurements in the single-phase region of Fig. 10.1, at temperatures and concentrations at which intermicellar interactions could be neglected. The deduced value of $\Delta\mu$ was in good agreement with the value deduced independently from the coexistence curve. This was evidence that the theory could self-consistently describe both the phase separation and the single-phase region in that micellar systems. Details may be found in Refs. 7 and 17.

More recently, we have extended the theory to describe micellar solutions that phase separate and exhibit lower consolute points [16]. In particular, for the $C_{12}E_6$ and water system, we have shown that a strictly linear dependence of $\Delta\mu$ and C on temperature is consistent with (1) single-phase measurements of $\langle n \rangle_w$ versus X and of the osmotic compressibility along the critical isochore and (2) measurements of the two overall amphiphile concentrations as a function of temperature on the coexistence curve and the location of the critical point. Details may be found in Refs. 16 and 17.

It is clear that our analysis can also be implemented to describe amphiphile–water micellar solutions that exhibit closed-loop coexistence curves. Having a model for $M_2(X, T, p)$, an appropriate temperature dependence of M_2 and C can lead to such a closed-loop curve. This is a subject of current theoretical investigation.

III. Summary and Discussion

In this Chapter, we have indicated the need to develop a new theoretical framework to study the equilibrium properties, including the phase separation, of two-component amphiphile–water micellar solutions. Our theory incorporates the existence of multiple chemical equilibrium, leading to a distribution of micellar sizes that depends on amphiphile concentration, temperature, pressure, and other solution conditions, and it incorporates intermicellar interactions. The theory is capable of describing in a unified manner a variety of experimentally observable phenomena in both the single-phase and two-phase regions. We have found that with our choice of free energy the presence of multiple chemical equilibrium implies that all the moments of the distribution of micellar sizes can be expressed solely in terms of the second moment, M_2, of that distribution. This observation holds regardless of the free energies of formation of the individual micelles in the distribution. As a result, all of the influence of the micellar size distribution on equilibrium properties of the solution, such as the coexistence curve, the spinodal line, and the osmotic compressibility, is determined solely by the dependence of M_2 on amphiphile concentration, temperature, pressure, and other solution conditions.

So far, the applicability of our analysis has been tested by comparing our results with experimental findings in two real micellar solutions. In the micellar solution of C_8-lecithin and water, which exhibits an upper consolute point, the theory very accurately reproduced the experimental coexistence curve, as shown in Fig. 10.1. Furthermore, the theoretical predictions regarding the single-phase micellar size distribution that were deduced from the coexistence curve were in good agreement with independent experimental characterizations of that distribution.

In the micellar system of $C_{12}E_6$ and water, which exhibits a lower consolute point, we showed that the theory not only self-consistently describes the experimental coexistence curve and the single-phase micellar size distribution in this system, but also provides an accurate prediction for the experimentally measured osmotic compressibility along the critical isochore.

The success of the theory in describing these experiments suggests that the link between the phenomena of phase separation and self-association, via low-order moments of the distribution of micellar sizes, may be valid for more complex self-associating colloidal systems.

The theoretical analysis presented here permits us to describe the physically important features of (1) the interactions between micellar aggregates, (2) the entropy of mixing of the micellar solution, and (3) the micellar multiple chemical equilibrium, with the aid of only two phenomenological parameters, $\Delta\mu$ and C. The parameter $\Delta\mu$ describes the free energy advantage associated with micellar growth, and the parameter C describes the magnitude of the effective attractive intermicellar free energy. The theory in its present form enables us to provide an analytic representation of the following equilibrium properties of the micellar solution: the shape and location of the coexistence curve and spinodal line, the osmotic compressibility, and the concentration and temperature dependence of the micellar size distribution. This representation has proven to be an accurate description of the experimental data in two distinct experimental systems, one showing an upper consolute point and the other a lower consolute point. In the latter case, the theory is in excellent agreement with data taken by four different groups [8,30–32]. We believe that this advance provides researchers with a mean-field description of micellar systems that places this field on the same theoretical footing as has previously existed for binary mixtures, polymer solutions, and the gas–liquid phase transition.

Acknowledgments. This research was supported in part by the National Science Foundation under Grant No. DMR81–19295.

References

1. For an introduction to the field of self-associating colloidal systems, see "*Micellization, Solubilization and Microemulsions,*" Vols. 1 and 2, K.L. Mittal, ed., Plenum, New York, 1977.
2. C. Tanford, "*The Hydrophobic Effect,*" Wiley, New York, 1980, and *J. Phys. Chem.* **78**, 2469 (1974).
3. E. Ruckenstein and R. Nagarajan, *J. Phys. Chem.* **79**, 2622 (1975).
4. For comprehensive experimental and theoretical surveys of the field of micellar systems, see (a) "*Surfactants in Solution,*" Vols. 1, 2 and 3, K.L. Mittal and B. Lindman, eds., Plenum, New York, 1984; (b) "*Proceedings of the International School of Physics Enrico Fermi—Physics of Amphiphiles: Micelles, Vesicles and Microemulsions,*" V. Degiorgio and M. Corti, eds., North-Holland Physics Publishing, Amsterdam, 1985.
5. V. Degiorgio, R. Piazza, M. Corti, C. Minero, *J. Phys. Chem.* **82**, 1025 (1985).
6. R.J.M. Tausk, C. Oudshoorn, and J.Th.G. Overbeek, *Biophysical Chem.* **2**, 53 (1974).
7. D. Blankschtein, G.M. Thurston, and G.B. Benedek, *Phys. Rev. Lett.* **54**, 955 (1985).
8. R.R. Balmbra, J.S. Clunie, J.M. Corkill, and J.F. Goodman, *Trans. Faraday Soc.* **58**, 1661 (1962).
9. R. Strey and A. Pakusch, "*Surfactants in Solution,*" Vol. 4, p. 465, K.L. Mittal and P. Bothorel, eds., Plenum, New York, 1985.
10. J.C. Lang and R.D. Morgan, *J. Chem. Phys.* **73**, 5849 (1980).
11. P.J. Flory, "*Principles of Polymer Chemistry,*" Chapters 12 and 13, Cornell University Press, Ithaca, New York, 1953.
12. R. Kjellander, *J. Chem. Soc., Faraday Trans. 2,* **78**, 2025 (1982).
13. L. Reatto and M. Tau, *Chem. Phys. Lett.* **108**, 292 (1984) and Ref. 4(b), p. 448.
14. J.C. Lang, in Ref. 4(a), Vol. 1, p. 35.
15. C.A. Leng, *J. Chem. Soc., Faraday Trans. 2,* **81**, 145 (1985) and Ref. 4(b), p. 469.
16. G.M. Thurston, D. Blankschtein, M.R. Fisch, and G.B. Benedek, *J. Chem. Phys.* **84**, 4558 (1986).
17. D. Blankschtein, G.M. Thurston, and G.B. Benedek, *J. Chem. Phys.* **85**, 7268 (1986).
18. K.A. Dill and P.J. Flory, *Proc. Natl. Acad. Sci. USA* **77**, 3115 (1980); **78**, 676 (1981).
19. A. Ben-Shaul, I. Szleifer, and W.M. Gelbart, *Proc. Natl. Acad. Sci. USA* **81**, 4601 (1984).
20. D.W.R. Gruen, *J. Phys. Chem.* **89**, 146 (1985); **89**, 153 (1985).
21. B. Owenson and L.R. Pratt, *J. Phys. Chem.* **88**, 2905 (1984).
22. For a review, see, for example, E.A. Guggenheim, "*Mixtures: The Theory of the Equilibrium Properties of Some Simple Classes of Mixtures, Solutions and Alloys,*" Clarendon Press, Oxford, 1952.
23. A. Ben-Shaul and W.M. Gelbart, *J. Phys. Chem.* **86**, 316 (1982).
24. W.M. Gelbart, A. Ben-Shaul, W.E. McMullen, and A. Masters, *J. Phys. Chem.* **88**, 861 (1984).
25. See also J.M. Corkill, J.F. Goodman, T. Walker, and J. Wyer, *Proc. Roy. Soc. A* **312**, 243 (1969).
26. J.M. Corkill, J.F. Goodman, and T. Walker, *Trans. Faraday Soc.* **63**, 759 (1967).

27. P. Mukerjee, *J. Phys. Chem.* **76**, 565 (1972).
28. J.N. Israelachvili, D.J. Mitchell, and B.W. Ninham, *J. Chem. Soc., Faraday Trans.* 2, **72**, 1525 (1976).
29. P.J. Missel, N.A. Mazer, G.B. Benedek, C.Y. Young, and M.C. Carey, *J. Phys. Chem.* **84**, 1044 (1980).
30. M. Corti, C. Minero, and V. Degiorgio, *J. Phys. Chem.* **88**, 309 (1984).
31. J.M. Corkill, J.F. Goodman, and R.H. Ottewill, *Trans. Faraday Soc.* **57**, 1627 (1961).
32. D. Attwood, P.H. Elworthy, and S.B. Kane, *J. Phys. Chem.* **74**, 3529 (1970).

11
Film Flexibility of Amphiphilic Layers and Structure of Middle-Phase Microemulsions

C. Taupin, L. Auvray, and J.-M. di Meglio

Two applications of recent theories about microemulsions are presented. In the first part, the theoretical ideas are applied to birefringent microemulsions, leading to the first determination of the rigidity constant of the interfacial surfactant film and the demonstration of the role of the cosurfactant as decreasing this rigidity. In the second part, the structure of Winsor microemulsions is investigated by small angle X-ray and neutron scattering, which leads to the demonstration of the existence of random bicontinuous microemulsions.

I. Introduction

Much activity has been devoted to the intensive study of lyotropic systems and, more specifically, of microemulsions [1] because of their practical applications to enhanced oil recovery, the cosmetic industry, pharmacology, etc. [2]. From a more fundamental point of view, these systems are very interesting because of their fascinating properties: their critical behavior [3,4], percolation transition [5], ultralow interfacial tensions [6], and extraordinarily rich phase diagrams [7,8]. The usual definition of microemulsions states that they are isotropic, thermodynamically stable dispersions of water in oil (or oil in water) whose characteristic size is of the order of 100 Å and thus are transparent to visible light. These oil–water systems are stabilized by the addition of surfactants that form an interfacial film separating the oil and water. Very often one needs to add a cosurfactant (usually a short-tail alcohol) in order to form stable microemulsions.

The very first microemulsion model is attributable to Schulman [9] and he only took into account the interfacial tension between oil and water and the surfactant free energy. In particular, this model was at a very local scale (i.e., at the scale of the surfactant molecule) and neglected entropy of the film (in the case of a randomly bent interface), electrostatic energies between surfactants, interactions between interfacial films, and curvature energies. The essential difficulty for completing a microemulsion theory is that all these effects are small, compete with each other, are not independent from each other, and depend on the system. However, a particular emphasis has been given to the role of the curvature energy [1,10,11]. The curvature energy per unit area is written as [1]

$$E_c = \frac{1}{2} K \left(\frac{1}{R} - \frac{1}{R_0} \right)^2,$$

(1)

where K is the rigidity constant of the film, R is the radius of curvature, and R_0 is the spontaneous radius of curvature, that is, the radius of curvature that the film would adopt in the absence of any interaction [14].

For highly flexible films, one can define a persistence length [1] as

$$\xi_K = a \exp\left(\frac{2\pi K}{kT} \right),$$

(2)

where a is a molecular length (typically the minimum radius of curvature, that is, the length of the alkyl tail of the surfactant molecule). Thus, a very small decrease of the rigidity would lead to a drastic decrease of the persistence length. Schulman has supposed that one role of the cosurfactant could be to decrease the rigidity of the interfacial film; this would explain the transition that is observed from the lamellar birefringent phase to the isotropic state.

II. Birefringent Microemulsions and Interfacial Flexibility

We have investigated lamellar systems that are very close to isotropic microemulsions in the phase diagram [12]. These systems present lamellar textures when observed by microscope between crossed polarizers, and they are birefringent. But, their structure is easily destroyed through a gentle stirring, revealing an extremely fragile interfacial film. They can be swollen by oil and thus have reticular distances of up to several hundred angstroms, as observed by small-angle X-ray scattering. These particular phases are present in some other microemulsion systems [8,13].

We expect that, in their domain of existence in the phase diagram, the lamellar phases have a zero spontaneous curvature; this assertion is confirmed by the fact that the phases remain lamellar even for very large oil/water ratios (swelling ratios).

Our systems were prepared from commercially available compounds. The surfactant was sodium dodecyl sulfate (SDS) (Serlabo), the cosurfactant was 1-pentanol (Serlabo), the oil was cyclohexane (Merck), and the water was distilled three times before use. The water/surfactant ratio was 2.5 (by weight) in all our samples. The phases were obtained through careful titration by the cosurfactant pentanol. There is about one molecule of cosurfactant per surfactant in the interfacial film for the lamellar phase to be compared to two for the isotropic droplet microemulsion. We have performed two sets of experiments: one by varying the swelling ratio and staying at the linear border of appearance of the lamellar phase (to ensure that the chemical composition of the interfacial film remains constant) and the other by varying the amount of cosurfactant at a fixed swelling ratio. We do not exclude the possibility that the cosurfactant may affect the spontaneous curvature (as simple steric models tend to show), but we do not think that this is important in our case as we deal with lamellar phases. We benefit from the fact that our system is so close to the isotropic phase that we may imagine a scenario for the transition to this isotropic phase without involving the influence of other parameters.

These lamellar phases are easily oriented in parallel wall glass containers (path length 100 μm) by heating them to approximately 75°C. The anchoring is of the homeotropic type, and oily streaks are visible, but could almost be completely eliminated through the heating process. The fact that we can orient the samples is very advantageous when using methods sensitive to orientation, such as the spin labeling technique [15]; we will be able to test the orientation of the lamellae.

Spin labeling consists of studying the electronic paramagnetic resonance of a nitroxyde radical probe. This probe is grafted on the alkyl chain of a surfactant molecule [16] whose alkyl chain length is identical to that of the tensioactive molecules of the interfacial film (SDS); see following scheme.

$$CH_3-(CH_2)_{11}-C-(CH_2)_3-N^+ \diagup \diagdown O, \; CH_3SO_4^-$$

We have used 1 probe molecule per 1000 surfactant molecules, so that the interfacial film was not perturbed; indeed, no modification of the phase diagram of the labeled samples was observed.

The spectra of these peculiar lamellar systems revealed an unusual feature [17]: the spectra obtained with oriented samples could not be simply deduced from the powder spectra, that is, the spectra obtained in a geometry where all the orientations have the same occurrence of probabilities. This is because of the fact that the lamellae exhibit long wave length undulations. The amplitude of these undulations increases with the swelling ratio of the samples and tends to a limit. In the following, we will describe this phenomenon through the theory developed by Helfrich and de Gennes.

The main idea is that the interfacial film is undulated because of its high flexibility and that the amplitude of the undulations is limited by the interactions between adjacent lamellae. The free energy per unit area is thus written as

$$E = \frac{1}{2} \frac{K}{R^2} + \frac{1}{2} U'' z^2 \qquad \text{with} \qquad U'' = \frac{\partial^2 W}{\partial z^2}, \tag{3}$$

where W is the interlamellae potential, and z is the distance of the lamella from an ideally planar reference lamella. We have distinguished two types of interactions:

1. attractive interactions (Van der Waals) and
2. repulsive interactions (Helfrich) [18].

These latter are predominant and are steric repulsions due to the fluctuations of the highly flexible lamellae. We thus have

$$U'' = 5.04 \frac{(kT)^2}{K d^4}, \tag{4}$$

where d represents the average distance between lamellae that can be computed from the relative amounts of oil and water and from the area per polar head of the surfactant (determined by X-ray scattering). The theory developed by de Gennes and Taupin allows one to estimate the average θ^2:

$$\theta^2 = \frac{kT}{\pi K} \ln\left(\frac{\xi u}{a}\right) \qquad \text{with} \qquad \xi u = \left(\frac{K}{U''}\right)^{1/4} \tag{5}$$

We first undertake a temperature investigation of a sample with a swelling ratio of four [19]. This sample presents a lamellar texture as observed through an optical microscope up to 70°C. Figure 11.1 represents the square angular spread of the normal to the lamellae with respect to the temperature. From the slope, one can deduce an estimation of the rigidity constant K by assuming that it is fairly independent of the temperature. We find $K = 6.4 \times 10^{-15}$ erg.

This behavior ($\theta^2 \propto T$) is not very surprising and just describes thermal fluctuations. Another more interesting prediction of the de Gennes model is that

$$\theta^2 = \frac{kT}{\pi K} \ln d + \text{constant}, \tag{6}$$

where d is the distance between lamellae. This is shown in Fig. 11.2. This leads to another determination of the rigidity constant [19] that is independent of the constants of the model (Hamaker constant, minimum radius of curvature a). We find $K = 4 \times 10^{-14}$ erg. The fact that this prediction is well verified corroborates that the chemical composition of the interfacial film is the same in all of the samples studied. The rigidity of the interfacial film is 10^2 to 10^3 times less than in lecithin systems ($K_{lec} = 2 \times 10^{-12}$ erg) [20,21]. We attribute this difference to the fact that in lecithin systems there is no cosurfactant.

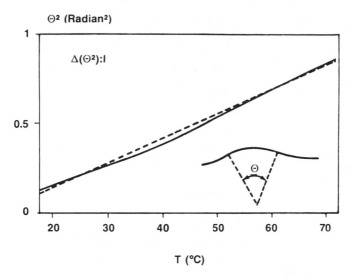

FIGURE 11.1. Amplitude of the undulations as a function of the temperature for an oil/water ratio of 4 (full line). The dashed line corresponds to $\theta^2 \propto 10^{-2}T$.

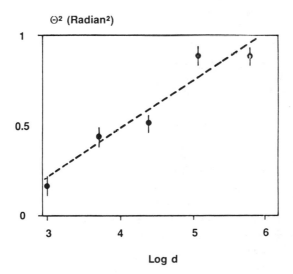

FIGURE 11.2. Amplitude of the undulations as a function of the reticular distance computed from the swelling ratio and the mean area per polar head obtained through X-ray experiments.

These undulations have also been observed by us in nonionic binary lamellar systems [22], where a transition of flexibility has been revealed.

Our aim as we began this study was, of course, to investigate the influence of the cosurfactant alcohol on the rigidity of the interface. We have thus performed spin labeling experiments on a sample with a swelling ratio of four and have varied the amount of cosurfactant starting from the minimum needed to get the phase. All of the samples studied present characteristic lamellar structures. The amplitude of the undulations increases with the cosurfactant, and as the added amount is relatively small with respect to the volume of the sample, we expect that the reticular distance remains the same. Thus, the observed increase of the amplitude of the undulations corresponds to a decrease of the rigidity constant (Fig. 11.4).

The classical spin labeling technique is not useful to investigate dynamics longer than 10^{-6} s. The fact that we find $\theta^2 \propto T$ indicates that the undulations we see are the thermal fluctuations of the lamellae. In order to investigate the dynamics of these fluctuations, we have performed a quasi-elastic light scattering experiment, this technique being sensitive to very long dynamics ($>10^{-4}$ s) for the apparatus we set up [23].

We wanted to observe undulations of the lamellae out of an ideally flat reference plane so that the scattering vector $\mathbf{q}$ is parallel to the lamellae; this condition is not so easy to fulfill and may explain some of the difficulties that we had in obtaining a signal. We used the same sample containers used in the spin labeling study (2 mm wide, around 5 cm long, and with a 100-μm path), and the samples were oriented the same way. The incident beam was from an ionized krypton laser (wave length of 5309 Å), and we have analyzed the scattered intensity for an angle of diffusion lying between 9 and 60°. This corresponds to q^{-1} between 845 and 5345 Å; thus, we expect to see mainly collective modes (120 Å $< d_{ret} <$ 360 Å).

From a careful analysis of the theoretical predictions available for this scattering geometry [24], it appears that the pure undulation mode is the most probable. Its dispersion relation is

$$\omega = i\frac{K}{\eta d}q^2, \tag{7}$$

where η is the viscosity of the interlayer solvent, d is the reticular distance, and q is the module of the scattering vector.

In all the experiments, the dynamic/static signal ratio is small and decreases strongly with the swelling ratio of the phases. We attribute that to the presence of textural defects in the samples, mainly induced by their manipulation to set them in the spectrometer. These defects are more likely present for the samples with a large swelling ratio. These defects lead to a large static scattering.

We have analyzed the autocorrelation of the scattered intensity and found that the correlation time τ ($= \omega^{-1}$) is proportional to q^{-2} according to the theoretical predictions. The results are plotted in Fig. 11.3 for three different

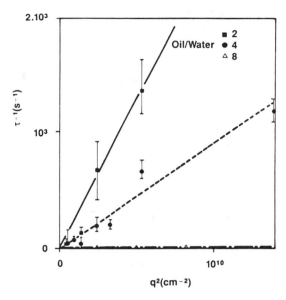

FIGURE 11.3. Relaxation times of the pure undulation hydrodynamical mode as a function of the square of the scattering vector for three different swelling ratios.

swelling ratios (2, 4, and 8). From the slopes, one may deduce the rigidity of the interfacial film. The results are reported in this table.

Oil/water	$K/\eta d$ (cgs)	d (Å)	K (erg)
2	2.6×10^{-7}	120	$(3.1 \pm 0.5) \times 10^{-15}$
4	9.2×10^{-8}	200	$(1.9 \pm 0.3) \times 10^{-15}$
8	3×10^{-9}	360	10^{-16}

We find that the values of K are roughly the same for samples with swelling ratios 2 and 4, and this is in good agreement with the spin labeling experiments. By contrast, the sample with ratio 8 has a rigidity lower by one order of magnitude; but, for this very swollen sample, the very high fluidity should lead to the presence of many structural defects and give a very high static scattered intensity hindering the analysis of the spectra. The obtained values for swelling ratios 2 and 4 are in relatively good agreement with the ones found previously by spin labeling experiments.

The effect of the cosurfactant has been studied for a sample with a swelling ratio of 4 by the same technique. We again find that the role of the cosurfactant is indeed to lower the rigidity of the interfacial film. The results are shown on Fig. 11.4.

These experiments have allowed us to measure for the first time the rigidity constant of the interfacial film of microemulsions systems. These rigidities are

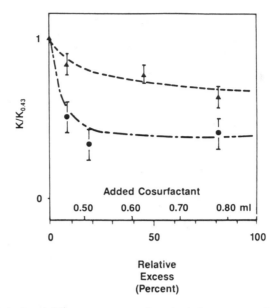

FIGURE 11.4. Relative rigidity constant as a function of excess cosurfactant for a phase with a swelling ratio of 4. K_{max} is the rigidity of the phase with the minimum cosurfactant amount (0.43 ml). Triangles (▲) are light scattering experiments, while circles (●) are the results of the previous spin labeling experiments.

very small (a little lower than kT) compared to the widely studied lecithin systems. We have studied lamellar systems, so we expect that the spontaneous curvature should be very small or even zero. We therefore concentrate on the flexibility and show experimentally that the cosurfactant lowers the rigidity of the interface as supposed by many authors.

III. Random Bicontinuous Microemulsions

A. Introduction

Since the early model of ordered bicontinuous structures generated by minimal surfaces proposed by Scriven [25], two models of random bicontinuous structures [26–28] in which the local curvature of the surfactant film strongly fluctuates have been published.

These random bicontinuous structures and the statistical configurations of the oil, water, and surfactant are described by simple mean-field models [26,28,29] in which the volume of the microemulsion is divided into elementary cells of mean size ξ, randomly filled by oil and water, the surfactant being distributed at the oil–water interface.

The models differ mainly by the way the random geometry is generated. In the original Talmon–Prager model [26], the cells are generated by a random Voronoi tesselation, and their size distribution is large. Contrary to Talmon and Prager, de Gennes et al. [28] and Widom [29] assume that the film stiffness forbids the curvature fluctuations at spatial scales smaller than ξ_K, the persistence length of the surfactant layer. This length is the basic size of the cells, which are taken to be identical and cubic.

In both models, in contrast to the well-known equation giving the radius R of, for example, water in oil spheres, one has

$$R = 3\phi_w/C_S\Sigma. \tag{8}$$

The mean size ξ of the oil and water elementary volumes, which is the length scale (~ 100 Å) of the random structure, is related to the microemulsion composition by the geometrical constraint

$$\xi = 6\phi_o\phi_w/C_S\Sigma. \tag{9}$$

Here, ϕ_o and ϕ_w are the oil and water volume fractions ($\phi_o + \phi_w = 1$), C_S is the surfactant concentration (number of molecules per unit volume), and Σ is the area per surfactant molecule in the film ($\Sigma \sim 60$ Å^2).

Equation (9) interpolates continuously between the case of water in oil droplets ($\phi_w \ll 1$) and the case of oil in water droplets ($\phi_o \ll 1$). In the models of random microemulsions, the average mean curvature of the film $\langle C \rangle$ (by convention positive for water/oil droplets) increases continuously with ϕ_o at constant interfacial area (C_S constant): $\langle C \rangle$ is negative for $\phi_o < 0.5$, positive for $\phi_o > 0.5$ and vanishes by symmetry when the microemulsions contain equal amounts of oil and water (inversion point, $\phi_o = \phi_w$).

In this respect, a particularly interesting case of a microemulsion system is the Winsor III microemulsion, which appears when an oil/surfactant/brine system separates into three phases. The middle-phase microemulsion is associated with extremely low interfacial tensions ($\sim 10^{-3}$ dynes/cm) and has many industrial applications in enhanced oil recovery and phase transfer. Its structural organization is not well understood. Several features are particularly interesting:

It is associated with extremely low interfacial tensions.

It appears in a range of salinity of the brine where the spontaneous curvature of the amphiphilic film (which is strongly dependent on electrostatic repulsions) could be very low.

If it also corresponds to highly flexible films, it could be good candidate for testing the random bicontinuous models that were recently proposed [26,28].

To test this prediction, different techniques have been used, such as conductivity measurements [30], self-diffusion coefficient measurements [31,32], electron microscopy [33], and scattering techniques [34–40]. However, as the measurements of transport coefficients only yield very indirect information on the microemulsion structure and as it has not yet been possible to obtain artifact-free electronmicroscopy pictures of the middle phases [41], the main information on these systems presently comes from the small-angle X-ray and neutron scattering experiments.

By using these techniques and the method of contrast variation with deuterated molecules [42] and studying a very representative and well-known system [43], we have obtained four main experimental results [36,37,39].

1. A well-defined surfactant film evidenced by the asymptotic behavior of the scattered intensities exists even in critical Winsor microemulsions.
2. The characteristic microemulsion size ξ, defined as a mean radius of curvature and drawn from the spectra in the intermediate range of scattering vector q, follows the prediction of Eq. (9) experimentally.
3. When $\phi_o = \phi_w$, the macroscopic concentration fluctuations of water and surfactant (measured from the scattering at zero angle for different contrasts) are not correlated.
4. The intensity scattered by the oil and water exhibits a pronounced peak at a given scattering vector q^* proportional to the surfactant concentration C_s.

Let us recall that in any scattering experiment the important parameter is the scattering vector q related to the scattering angle 2θ and wave length λ by

$$q = \frac{4\pi}{\lambda} \sin \theta. \tag{10}$$

By varying q, it is possible to test the structure at different length scales. In our experiments, q varies between 10^{-2} and $0.25 \, \text{Å}^{-1}$; that is, we explore the microemulsion structure between roughly 15 and 300 Å. Three different q ranges can be distinguished, in correlation with the characteristic sizes ξ around 100 Å of the microemulsions.

1. There is a large q range ($q > 0.1 \, \text{Å}$, small spatial scale $< 30 \, \text{Å}$) that corresponds to the effects due to the surfactant film. This range is called an asymptotic range since in this domain the scattered intensity follows general laws that are independent of the large-scale structure.
2. There is a medium q range that corresponds to the size of the water and oil domains.
3. There is a very small q domain that describes the large-scale structure of the microemulsions and depends on correlations between domains.

Another important element is the possibility of varying the contrast in neutron scattering techniques, that is, to selectively enhance the scattering power of the various domains [42] (oil, water, or film) of the system.

Assuming that a microemulsion reduces itself to three in-compressible geometrical parts: the oil, water, and film (volume fraction ϕ_o, ϕ_w, and ϕ_f, scattering length density n_o, n_w, and n_f), the intensity scattered by a unit volume of the sample is written [37]

$$i(q) = (n_w - n_o)^2 \chi_{ww}(q) + (n_f - n_o)^2 \chi_{ff}(q)$$
$$+ 2(n_w - n_o)(n_f - n_o)\chi_{wf}(q). \tag{11}$$

The partial structure factors are defined by the relation

$$\chi_{ij}(q) = \int d^3r \langle \delta\phi_i(0)\delta\phi_j(r)\rangle e^{i\mathbf{q}\cdot\mathbf{r}}, \tag{12}$$

where $\delta\phi_i(r)$ is the local fluctuation of the ith component volume fraction ($i = w, f$). The angle brackets mean average value.

The direct structure factors $\chi_{ww}(q)$ and $\chi_{ff}(q)$ can be measured in a single experiment by using, respectively, an oil–water contrast ($n_o = n_f \neq n_w$) and a film contrast ($n_o = n_w \neq n_f$). In contrast, a complete variation of the difference is necessary to determine the cross-structure factor $\chi_{wf}(q)$.

B. Results

Large-q Measurements

The signal is due to scattering length density variations at small scales. Asymptotic laws were established long ago, and experimentally, this domain has been widely investigated for checking the presence of interfaces in finely divided matter (measurement of specific area). It is particularly interesting in the case of microemulsions since the presence of a well-defined interface, which was clearly demonstrated in droplet microemulsions, is not evident in Winsor microemulsions, which are frequently in the vicinity of critical points [29,43].

These experiments revealed the existence of a well-defined surfactant film even in Winsor microemulsions. More precisely, the general laws of large-angle scattering intensity, $i(q)$, have been predicted by Porod [44] under the general form

$$i(q) = \frac{S}{V} f(q), \tag{13}$$

$\frac{S}{V}$ being the interfacial area per unit volume of the sample [$\frac{S}{V} = C_S \Sigma$, with C_S the surfactant concentration (number of molecules per unit volume) and Σ the area per surfactant molecule in the film ($\Sigma \sim 60$ Å^2)]. The function $f(q)$ depends on the type of contrast. One gets in the case of two densities (usual case of one finely divided component or D$_2$O, C$_7$H$_8$ contrast in neutron scattering),

$$i(q) = 2\pi(n_w - n_o)^2 C_S \Sigma q^{-4}; \tag{14}$$

in the pure film contrast,

$$i(q) = 2\pi(n_f - n_o)^2 C_S \Sigma d^2 q^{-2}; \tag{15}$$

and in the more complex contrast observed with X-rays,

$$i(q) \sim C_S \Sigma (A\, q^{-4} + B\, q^{-2}).\tag{16}$$

These various behaviors have been experimentally observed [36,37] in particular, the proportionality of the asymptotic intensity to C_S, which proves the existence of the interfacial film. The area per molecule Σ is found around 70 Å^2 ± 10 Å^2, and the thickness of the film is 10 Å + 5 Å.

Semilocal Scale ($q\xi \sim 1$)

Expressions [14–16] of Porod's law are valid as long as the interface looks flat at the scale defined by the scattering angle. Our idea is that the location of the q_1 crossover zone between the asymptotic behavior and the small-angle behavior gives the size scale (πq_1^{-1}) at which the film becomes curved, that is, an estimation of the persistence length of the film ξ_K introduced into the model.

Empirically, we have chosen as a definition of the crossover scattering vector q_1, the abcissa of the minimum of the $q^4 I$ versus q plot. The detailed discussion of the practical and theoretical limitations of this choice has been given in Ref. 36.

A very interesting point about this phenomenological procedure is that, for all the samples, the intensities in the medium q range can be written as

$$I(q) = I_o f(q/q_1),\tag{17}$$

thus proving that q_1 is a good choice to determine a characteristic length of the structure.

Recall that the main prediction of the models of bicontinuous microemulsions is [9]

$$\xi = \frac{6\phi_o \phi_w}{C_S \Sigma}.$$

In agreement with our assumption that q_1^{-1} reflects ξ, it was experimentally verified that q_1 is proportional to C_S. Moreover, $C_S q_1^{-1}$ is proportional to the product $\phi_o \phi_w \approx \phi_o, (1 - \phi_o)$; for example, Fig. 11.5 exhibits a parabolic shape around $\phi_o \sim .5$. At low water or oil volume fractions (ϕ_o or $\phi_w < 0.3$), the behavior is similar to that of spherical droplets. A semiquantitative calibration of q_1^{-1} is possible when observing an otherwise known system using this procedure.

The observed behavior indicates that as ϕ_o increases, one passes progressively from a structure made of distinct oil in water spheres to a bicontinuous structure described by the Talmon–Prager or de Gennes' model; the curvature of the film is inverted at $\phi_o = 0.5$, and as ϕ_o increases again, the reverse process takes place.

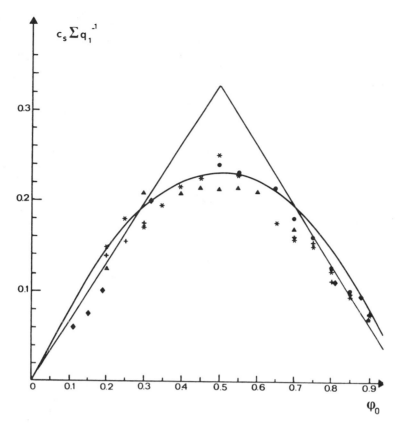

FIGURE 11.5. Product $C_s \Sigma q_1^{-1}$ plotted versus the toluene volume fraction ϕ_o. The straight line corresponds to $C_s \Sigma q_1^{-1} = 0.65$ (ϕ_w or ϕ_o). The parabola corresponds to $C_s \Sigma q_1^{-1} = 0.9\ \phi_o \phi_w$.

Very Small q Range

In this domain, large distance correlations or interactions are observed.

We previously noticed that the structure of the Winsor microemulsions is very similar (up to a scale factor that has been discussed in Ref. 45 at scales that are of the same order of magnitude or smaller than the characteristic size. The differences between the samples appear at larger scales ($q < 10^{-2}$ Å^{-1}).

In the inversion zone ($0.4 < \phi_o < 0.7$), the samples exhibit a diffuse band centered around q^*, the corresponding Bragg spacing being $2\pi q^{*-1}$ (Fig. 11.6). The quantity q^* does not move as one goes far away from the inversion point, but varies proportionally to the amount of surfactant C_S. The comparison of various systems reveals a strong correlation between the place of the peak and q_1.

In this inversion zone, the intensity scattered by a two-phase random oil and water partition in Voronoi polyhedra has been calculated by Kaler and

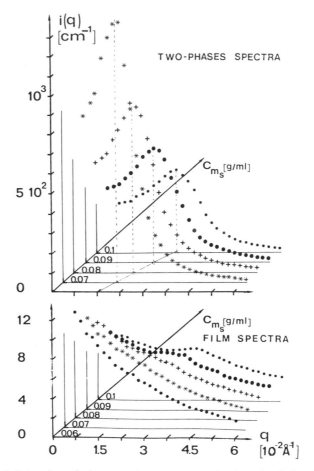

FIGURE 11.6. Intensity of the two-phases spectra (top) and of the film spectra (bottom) as a function of q and C_{m_s} [same q scale on the two plots, q^* (peak scattering vector) versus C_{m_s} in the q–C_{m_s} plane].

Prager [46]. As this partition is random, they predict that at small q the intensities approximately follow a Guinier law where the equivalent radius of gyration R_G is proportional to the characteristic length of the microemulsion; within this model, $R_G = 0.55\ \xi_K$. This prediction does not agree with the experimental results, which clearly exhibit a peak. Quantitatively, the peak is not very different from what it would be if the microemulsions were dispersions of droplets. In fact, the peak is clearly visible in the case of a water–oil contrast, but it disappears in the case of a pure film contrast (in contradiction with a droplet model; see Fig. 11.6). The reason for these effects is illustrated in Fig. 11.7.

The differences between the two structures appear either at the semilocal scale (studied in Ref. 36) or on the film geometry. The study of the film

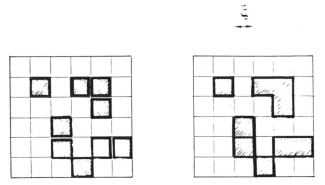

FIGURE 11.7. Parallel between a lattice-gas model of water in oil droplets and a model of random microemulsion with a well-defined length scale. If the surface and curvature energies of the interfacial film are neglected before the entropy of dispersion of the oil and the water, the statistical configurations and correlations of the oil and water are the same in the two models. The differences between the two models only appear at the local scale ($r < \xi$) or on the geometry of the interfacial film (thick line).

fluctuations and correlations through the two functions $\chi_{wf}(q)$ and $\chi_{ff}(q)$ is then particularly interesting. Here, we will focus mainly on the zero-angle limit of $\chi_{wf}(q)$, which is related to the film curvature.

Scattering at Zero Angle and Curvature of the Film

If $\delta\phi_w$ and $\delta\phi_f$ denote the fluctuations of the water and film volume fraction, respectively, in a macroscopic volume V, $\chi_{wf}(q)$ and $\chi_{ff}(q)$ are given in the thermodynamic limit of q going to zero by the two relations

$$\chi_{wf}(0) = V\langle\delta\phi_w\delta\phi_f\rangle, \tag{18}$$

and

$$\chi_{ff}(0) = V\langle\delta\phi_f^2\rangle. \tag{19}$$

Because of Schulman's condition, the area per surfactant in the interfacial film does not fluctuate in the first approximation. A surfactant (or film) fluctuation in a macroscopic volume V is proportional to the fluctuation of the interfacial area in the volume V. As first noticed by Widom [29] and developed in Refs. 37 and 39, $\chi_{wf}(0)$ and $\chi_{ff}(0)$ (calculable from the thermodynamical models of microemulsions [26,28,29] then have a simple geometrical interpretation.

For a microemulsion made of water in oil droplets, a water excess in a volume V means a droplets excess, hence a surfactant excess, that is, $\chi_{wf}(0)$ is positive. It is the reverse for oil in water microemulsions. In the past, this distinction (formulated in another way) has been used as evidence for the inversion of microemulsions [47]. Up to a numerical factor and to the first order in d (d is the thickness of the film, and $\langle C \rangle$ is the average mean curvature of the film), one gets

$$\chi_{wf}(0)/\chi_{ww}(0) \propto \langle C \rangle d, \qquad (20)$$

by geometrical considerations.

Widom's remark [29] enables the relation between $\chi_{wf}(0)$ and the film curvature to be generalized to the random microemulsions. In this case, one finds that the average curvature of the film and $\chi_{wf}(0)$ are both proportional to $(1 - 2\phi_w)$. As was predicted by Widom [29], when $\phi_o = \phi_w = 0.5$, the average curvature of the microemulsion vanishes, and the fluctuations of water and film are not correlated.

Figure 11.8 represents the experimental variations of the ratio $\chi_{wf}(0)/\chi_{ww}(0)$ as a function of ϕ_o (at constant C_S). This ratio increases progressively when ϕ_o increases, it changes its sign for $\phi_o = 0.5 \pm 0.03$. This directly shows that the film mean curvature is progressively inverted as the oil and water proportions in the samples are progressively inverted, in agreement with the predictions of the models of random bicontinuous microemulsions (the straight continuous line on Fig. 11.8 is calculated from the model of de Gennes et al.).

In order to appreciate the sensitivity of the method of contrast variations, it is interesting to compare the experimental results with a model of water in oil droplets. In this model,

$$\chi_{wf}(0)/\chi_{ww}(0) = 3d/R_w + 0(d/R_w)^2, \qquad (9)$$

[d is the film thickness, and R_w is given by Eq. (1).]. With $d = 10$ Å (see Ref. 26) and $\Sigma = 60$ Å^2, the value of $3d/R_w$ is 0.25 for the sample $\phi_o = 0.5$. This value is much higher than the almost zero experimental value. A description of the $\phi_o = 0.5$ sample in terms of a droplet model is clearly not possible. Let us now consider the sample that contains the largest proportion of oil, $\phi_o = 0.7$. If $\phi_o = 0.7$, $3d/R_w = 0.42$. This value is larger than the observed

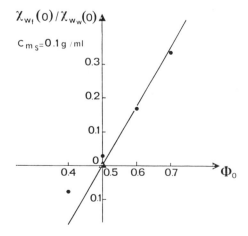

FIGURE 11.8. Variation of the ratio $\chi_{wf}(q)/\chi_{ww}(q)$ extrapolated at zero angle as a function of the oil volume fraction at constant surfactant concentration.

value of $\chi_{wf}(0)/\chi_{ww}(0) = 0.33$. The curvature of the interfacial film is smaller in this sample that it would be if the microemulsion was made of distinct droplets. This means that the structure of this microemulsion probably remains very connected.

IV. Conclusions

From these two sets of experimental results obtained with entirely different techniques, we checked the validity of recent ideas about microemulsions, in particular the role of low rigidity of the interfacial film when the spontaneous curvature is very small. First, the rigidity constant, measured for the first time, in lamellar birefringent microemulsions has been found to be several orders of magnitude smaller than that of lecithin systems. The role of the cosurfactant is clearly to decrease this rigidity. Second, it has been verified by structural experiments that the Winsor or middle-phase microemulsions exhibit the main features of bicontinuous models. Nevertheless, a characteristic length, the persistence length introduced by de Gennes, appears in the analysis of the scattering intensities, which seems to show a smaller degree of randomness than predicted by the theory.

Acknowledgments. This work has received partial financial support from P.I.R.S.E.M. (C.N.R.S.) under A.I.P. No. 2004.

References

1. P.G. de Gennes and C. Taupin, *J. Phys. Chem.* **86**, 2294 (1982).
2. A.M. Cazabat and M. Veyssié, ed., "*Colloïdes et Interfaces*," Summer School of the C.N.R.S., Aussois, 6–17 September 1983, Editions de Physique, Paris 1984.
3. J.S. Huang and M.W. Kim, *Phys. Rev. Lett.* **47**, 418 (1981).
4. A.M. Cazabat, D. Langevin, J. Meunier, and A. Pouchelon, *J. Phys. Lett.* **39**, 89 (1982).
5. M. Laguës, R. Ober, and C. Taupin, *J. Phys. Lett.* **39**, L-487 (1978).
6. For a recent review, see A.M. Bellocq, J. Biais, P. Bothorel, B. Clin, G. Fourche, P. Lalanne, B. Lemaire, B. Lenanceau, and D. Roux, *Adv. Colloid Interface Sci.* **20**, 167 (1984).
7. A. Pouchelon, D. Chatenay, J. Meunier, and D. Langevin, *J. Coll. Int. Sci.* **82**, 418 (1981).
8. D. Roux, Thesis of Doctorat d'Etat, University of Bordeaux I, 1984.
9. J.H. Schulman and J.B. Montagne, *Ann. N.Y. Acad. Sci.* **92**, 366 (1961).
10. S.A. Safran, L.A. Turkevich, and P.A. Pincus, *J. Phys. Lett.* **45**, L-69, (1984).
11. C.A. Miller, *J. Dispersion Science and Technology* **2**, 159 (1985).
12. M. Dvolaitzky, R. Ober, J. Billard, C. Taupin, J. Charvolin, and Y. Hendricks, *Comptes Rendus Acad. Sci.* **B295**, 45 (1981).
13. S.E. Friberg and C.S. Wohn, *Colloid and Polymer Sci.* **263**, 156 (1985).

14. W. Helfrich, *Z. Naturforschung* **28c**, 693 (1973).
15. L.J. Berliner, ed., *"Spin Labelling: Theory and Applications,"* Academic Press, New York, 1976.
16. M. Dvolaitzky and C. Taupin, *Nouv. J. Chim.* **1**, 355 (1977).
17. J.-M. di Meglio, M. Dvolaitzky, R. Ober, and C. Taupin, *J. Phys. Lett.* **44**, L-234 (1983).
18. W. Helfrich, *Z. Naturforschung* **33a**, 305 (1978).
19. J.-M. di Meglio, M. Dvolaitzky, and C. Taupin, *J. Phys. Chem.* **89**, 871 (1985).
20. R.M. Servuss, W. Harbich, and W. Helfrich, *Biochim. Biophys. Acta.* **436**, 900 (1976).
21. M.B. Schneider, J.T. Jenkins, and W.W. Webb, *Biophys. J.* **45**, 891 (1984).
22. J.-M. di Meglio, L. Paz, M. Dvolaitzky, and C. Taupin, *J. Phys. Chem.* **88**, 6036 (1984).
23. J.-M. di Meglio, M. Dvolaitzky, L. Léger, and C. Taupin, *Phys. Rev. Lett.* **54**, 1686 (1985).
24. F. Brochard and P.G. de Gennes, *Pramana* **1**, 1 (1975).
25. L.E. Scriven, in *"Micellization, Solubilization and Microemulsions,"* Vol. 2, p. 77, K.L. Mittal, ed., Plenum, New York, 1977; *Nature* **263**, 123 (1976).
26. Y. Talmon and S. Prager, *J. Chem. Phys.* **69**, 2984 (1978).
27. P.G. de Gennes and C. Taupin, *J. Phys. Chem.* **86**, 2294 (1982).
28. P.G. de Gennes, J. Jouffroy, and P. Levinson, *J. Physique* **43**, 1241 (1982).
29. B. Widom, *J. Chem. Phys.* **81**, 1030 (1984).
30. K.E. Bennett, J.C. Hatfield, H.T. Davis, C.M. Macosko, and L.E. Scriven, in *"Microemulsions,"* p. 65, I.D. Robb, ed., Plenum, New York, 1982.
31. M.T. Clarkson, D. Beaglehole, and P.T. Callaghan, *Phys. Rev. Lett.* **54**, 1722 (1985).
32. P. Guering and B. Lindman, *Langmuir* **1**, 464 (1985).
33. J. Biais, M. Mercier, P. Bothorel, B. Clin, B. Lalanne, and B. Lemanceau, *J. Microsc.* **121**, 169 (1981).
34. E.W. Kaler, K.E. Bennett, H.T. Davis, and L.E. Scriven, *J. Chem. Phys.* **79**, 5673 (1983).
35. E.W. Kaler, H.T. Davis, and L.E. Scriven, *J. Chem. Phys.* **79**, 5685 (1983).
36. L. Auvray, J.P. Cotton, R. Ober, and C. Taupin, *J. Physique* **45**, 913 (1984).
37. L. Auvray, J.P. Cotton, R. Ober, and C. Taupin, *J. Phys. Chem.* **88**, 4586 (1984).
38. A. de Geyer and J. Tabony, *Chem. Phys. Lett.* **124**, 357 (1986).
39. L. Auvray, Ph.D. Thesis, Université de Paris-Sud, France, 1985. Available on request.
40. J.N. Chang, R.T. Hamilton, J.F. Billman, and E.W. Kaler, paper presented at the American Chemical Society Meeting, Miami Beach, Miami, Florida, 1985.
41. Y. Talmon, in Proceedings of the 5th International Symposium on Surfactants in Solutions, Bordeaux, July 9th–13th, 1984, K.L. Mittal and P. Bothorel, eds.
42. H.B. Stuhrmann, *J. Appl. Crystallogr.* **7**, 173 (1974).
43. A.M. Cazabat, D. Langevin, J. Meuinier, and A. Pouchelon, *Adv. Coll. Int. Sci.* **16**, 175 (1982).
44. R. Kirste and G. Porod, *Koll. Z. u. Z. für Polym.* **184**, 1 (1962).
45. L. Auvray, Thesis of Doctorat d'Etat, University of Paris XI, 1985.
46. E.W. Kaler and S. Prager, *J. Coll. Int. Sci.* **86**, 359 (1982).
47. M. Laguës, R. Ober, and C. Taupin, *J. Phys. Lett.*, **39**, L-487 (1978).

12
Low Interfacial Tensions in Microemulsion Systems

D. LANGEVIN

We review the existing theories of interfacial tensions in microemulsions in the droplet and bicontinuous phases. Data are obtained on two model systems: brine-toluene-butanol-sodium dodecyl sulfate and brine-dodecane-butanol-sodium hexadecyl benzene sulfonate, and they are compared with these theories.

I. Introduction

Microemulsions are dispersions of oil and water made with surfactant molecules [1]. Contrary to emulsions, they are thermodynamically stable, because the characteristic size of the dispersion is very small, about 100 Å. It was shown theoretically by Ruckenstein [2] that thermodynamic stability arises from the fact that the surface energy, which is equal to the interfacial tension γ_{ow} times the total area between oil and water, can be compensated by the dispersion entropy, when γ_{ow} is sufficiently small; typically, $\gamma_{ow} \lesssim 10^{-2}$ dyn/cm. Such ultralow interfacial tensions are commonly reached by using a cosurfactant that is usually on alcohol. Let us recall that typical interfacial tensions between oil and water without surfactant are about $\gamma_{ow}^{o} \sim 50$ dyn/cm.

According to the spontaneous curvature of the surfactant film that covers the oil-water interfaces, the structure of the dispersion may be, as for emulsions, of the oil in water type (o/w: oil droplets in a water continuous phase) or of the water in oil type (w/o: water droplets in an oil phase). An evolution from the first type to the second can be obtained, for instance, by varying the temperature with nonionic surfactants or the salinity with ionic surfactants. In this chapter, we will restrict the discussion to ionic surfactants, and we will present a study of several model systems.

At low water salinity S, one obtains an o/w microemulsion that can coexist with excess oil. In such a case, the droplet radius has reached the spontaneous radius of curvature: the droplets cannot accommodate more oil without a high cost in curvature energy, and the excess oil is rejected in an excess phase

[3]. The interfacial tension between the microemulsion and the excess phase is very small $\gamma_{om} \lesssim 10^{-2}$ dyn/cm. It happens to be equal to the interfacial tension between the excess oil phase and the water continuous phase [4]. This is because the low tension is only due to the high surface pressure of the mixed surfactant–cosurfactant layer at the macroscopic flat interface γ_{ow}.

At larger salinities S, one obtains a w/o microemulsion that can coexist with excess water. In such a case, the droplet radius is again the spontaneous radius of curvature. The interfacial tension γ_{wm} is also equal to γ_{ow}, that is, less than or about 10^{-2} dyn/cm.

At intermediate salinities, a middle-phase microemulsion can coexist with both excess oil and water. Scriven proposed that it could be bicontinuous [5]. Several theoretical models have recently been elaborated to describe its properties, including Voronoi tesselation of polygons by Talmon and Prager [6] and cubic cells by de Gennes [7] and Widom [8]. When the oil or the water volume fractions ϕ_o or ϕ_w are small, these models are equivalent to the droplet models. The interconnections of both oil and water cubes or polygons over macroscopic distances are predicted for $0.2 \lesssim \phi_o$ (or $\phi_w) \lesssim 0.8$. The structure is then bicontinuous and evolves continuously from the o/w to the w/o structures between these limits. X-rays and neutron scattering data are in agreement with some of the predictions of the models.

The interfacial tensions between the microemulsion and the excess phases are still smaller than in the two-phase equilibria. One of these tensions, the larger one, is equal to the tension between the excess phases γ_{ow} [9]. Again the low tension is associated with the high surface pressure of the surfactant layer. The second tension, the lower one, decreases to almost zero close to the two-phase–three-phase boundaries. This was shown to be associated with the vicinity of critical end points in the phase diagram [9].

Let us mention that the low tensions associated with the surfactant film are the ones of interest in tertiary oil recovery [10]. Several theories predict that such tensions are of the order of kT/L^2, where L is the dispersion scale, whereas the low tensions associated to critical phenomena are of the order of kT/ξ_c^2, where ξ_c is the correlation length for concentration fluctuations. In order to gain information about the quantitative relationship between γ, L, and ξ_c, we have undertaken the study of two model systems with different surfactants.

In the following, we will briefly review the existing theories of interfacial tensions. We will then present the data obtained on the model systems and compare them with the theories.

II. Theoretical Models

A. Structural Models

The Talmon–Prager–de Gennes–Widom model can be used to describe droplet dispersions and bicontinuous structures as well. In the simplest version, given by de Gennes, the microemulsion volume is divided into consecutive cubes randomly filled with either oil or water. The interface between cubes of different type are covered by a surfactant film whose volume is, however, neglected. The linear size of the cube is determined by the persistence length of the interface ξ_K; at scales smaller than ξ_K, the interface is essentially flat, whereas as scales larger than ξ_K, it is strongly wrinkled. The quantity ξ_K is related to the curvature elastic modulus by

$$\xi_K = a \exp(2\pi K/kT), \tag{1}$$

a being a molecular length.

If ϕ_o and ϕ_w are the oil and water volume fractions, the average total area between oil and water per cube is $(6\phi_o\phi_w\xi_K^2)$, and the corresponding surface energy is $(6\phi_o\phi_w\xi_K^2\gamma_{ow})$. The entropy of dispersion per cube is $k(\phi_o \ln \phi_o + \phi_w \ln \phi_w)$. The statistics of the interface is thus reduced to a lattice-gas model. When

$$\gamma_{ow} \lesssim kT/\xi_k^2,$$

a single-phase microemulsion is thermodynamically stable. With $\xi_K \sim 100$ Å, it follows that $\gamma_{ow} \lesssim 10^{-2}$ dyn/cm.

The area per surfactant molecule, Σ, is generally reasonably constant whatever the spontaneous curvature is. It follows then that ξ_K is determined when the total number of surfactant molecules n_s is given. Writing that the total area is $n_s\Sigma$, and with $C_s = n_s/V$, V being the total volume,

$$\xi_K = \frac{6\phi_o\phi_w}{C_s\Sigma}. \tag{2}$$

A very similar expression holds for the average size of the polygons in the Talmon–Prager model: $5.82\phi_o\phi_w/C_s\Sigma$. In the limit of isolated oil or water droplets of radius R_o or R_w, one would have

$$R_o = 3\phi_o/C_s\Sigma \quad \text{or} \quad R_w = 3\phi_w/C_s\Sigma, \tag{3}$$

which are limiting cases of Eq. 2 for small ϕ_o or ϕ_w with $R = \xi_K/2$.

B. Interfacial Tension between a Microemulsion and an Excess Phase

Most existing calculations apply to microemulsions containing spherical droplets of radius R.

The simple description of Israelachvili [11] allows one to understand easily the nature of the contributing terms. Let us call μ_N the chemical potential of the surfactant molecule in an aggregate of N molecules:

$$\mu_N = \mu_N^0 + \frac{kT}{N} \ln \frac{X_N}{N},$$

where μ_N^0 is a standard chemical potential containing the interaction terms between the surfactant molecules in the droplet and X_N is the surfactant concentration fraction present in the droplet.

The interface between the microemulsion and the excess phase is covered by a surfactant layer that can be considered as an infinite aggregate. This aggregate is subjected to a tension. The surfactant chemical potential is, therefore,

$$\mu_s = \mu_\infty - \gamma \Sigma = \mu_\infty^0 - \gamma \Sigma.$$

The interfacial tension is finally obtained by writting that μ_N is equal to μ_s

$$\gamma = -\frac{1}{N\Sigma} kT \ln \frac{X_N}{N} + \frac{1}{\Sigma} (\mu_\infty^0 - \mu_N^0). \tag{4}$$

The first term is an entropic contribution, γ_e, arising from mixing. It is analogous to the expression derived by Ruckenstein on different thermodynamical basis [12].

The second term is a curvature contribution, γ_c, which has been calculated explicitly by other authors [13-15]. De Gennes has shown [7] that these calculations are equivalent to

$$\gamma_c = 2K/R \, R_0, \tag{5}$$

where R_0 is the spontaneous radius of curvature.

The order of magnitude of the two terms are the same; one has $N\Sigma = 4\pi R^2$, $R \sim R_0$, and $K \sim kT$; thus, $\gamma_c \sim \gamma_e \sim kT/R^2$. It should be noted that the tension ultimately depends on the droplets radius, although it does not change when the microemulsion is replaced by its continuous phase, that is, as the droplets are removed [4]. In fact, both γ and R are determined by the properties of the surfactant layer, and it is not surprising to find a relation between them.

Equation (4) does not contain any contribution from interaction between different droplets. This contribution is expected to be negligible as soon as one is sufficiently far from a critical point. This is certainly the case when γ is equal to γ_{ow}, which does not contain any interaction term.

When the microemulsion is bicontinuous and in equilibrium with both excess oil and water, there are no theoretical calculations of the interfacial tension. However, it can reasonably be expected that the interfacial tension will be of the order of kT/ξ_K^2.

C. Interfacial Tension between Two Microemulsions

The transition between two-phase and three-phase systems can sometimes be close to critical end points. Then, the microemulsion in the two-phase domain becomes very turbid and transforms into two turbid microemulsion phases. At the critical end point itself, turbidity diverges, and one interface in the three-phase region disappears without moving toward the top or bottom of the cell. Turbidity diverges because the correlation length for the concentration fluctuations ξ_c diverges. The interfacial tension goes to zero as

$$\gamma = \alpha k T_c / \xi_c^2, \tag{6}$$

T_c being the critical temperature.

In Ising models for simple fluids [16], if ε is the distance to the critical point, one has

$$\xi_c = \xi_0 \varepsilon^{-\nu} \qquad \nu = 0.63,$$

and

$$\gamma = \gamma_0 \varepsilon^\mu \qquad \mu = 2\nu = 1.26.$$

If the critical point is approached by varying the temperature, $\varepsilon = |T - T_c|/T_c$. ξ_0 and γ_0 are scale factors; they are related by [17]

$$\gamma_0 \xi_0^2 / k T_c = \alpha = 5.4 10^{-2}.$$

III. Experimental Procedure

A. Sample Composition

We have studied two model systems. The first one is a mixture of brine (47 wt.%), toluene (47 wt.%), butanol (4 wt.%), and sodium dodecyl sulfate (SDS, 2 wt.%). The brine is an aqueous solution of S wt.% sodium chloride. S was varied between 3.5 and 10. The salinities of phase separation have been found to be $S_1 = 5.4$ and $S_2 = 7.4$, at $T = 20°C$.

The second model system is a mixture of brine (56.83 wt.%), dodecane (38.19 wt.%), butanol (3.32 wt.%), and sodium hexadecyl benzene sulfonate (SHBS or Texas $\neq 1.666$ wt.%). This composition corresponds to equivalent amounts of oil and brine in volume. The brine is an aqueous solution of S wt.% sodium chloride. S was varied between 0.4 and 0.9. The salinities at phase separation have been found to be $S_1 = 0.52$ and $S_2 = 0.61$ at $T = 20°C$.

B. Experimental Techniques

Interfacial tensions have been measured with surface light scattering experiments. The frequency broadening of the scattered light is measured with a wave analyzer after a photon beating detection. The width of the spectrum is directly proportional to γ [18].

The characteristic sizes in microemulsion phases were measured with bulk light scattering techniques, both elastic and quasielastic [9,19]. In the two-phase regions, the microemulsion structure was that of droplets dispersed in a continuous phase. The composition of the continuous phase was found to be the extrapolation of the composition of the corresponding excess phases (either oil or water for water-in-oil or oil-in-water microemulsions) as salinity is varied [9]. By diluting the microemulsion phases, we were able to get the particle volume $v = \frac{4}{3}\pi R^3$ (from the scattered intensity), the hydrodynamic radius R_H from the correlation function of the scattered light, and sometimes the radius of gyration R_G (from the intensity angular anisotropy when $R_G \gtrsim 200$ Å). We also obtained two virial coefficients B and α for the osmotic compressibility and the diffusion coefficient. For hard spheres, $B = 8$ and $\alpha = 1.5$, respectively; for repulsive spheres, B and α are larger than these values; for attractive spheres, they are smaller.

In the three-phase region, dilution is not possible. A pseudocorrelation length has been deduced from quasielastic light scattering measurements with the equivalent Stokes law: $\xi = kT/6\pi\eta D$ $(q = 0)$. The persistence length ξ_K can be deduced from X-ray experiments [20,21].

Close to S_2 in the SDS system, critical behavior was evidenced from bulk light scattering experiments (angular anisotropy of both intensity and correlation functions). The correlation length for the concentration fluctuations ξ_c could then be deduced ($\xi_c = \xi$ in this case) [9].

IV. Experimental Data and Discussion

A. Sizes in Microemulsion Phases

The sizes and virial coefficients are reported in Table 12.1 for the two systems. Some of these data are plotted in Fig. 12.1, together with X-ray and neutron scattering data [22] obtained in the same systems.

TABLE 12.1.   Characteristic sizes, virial coefficients, and area per surfactant.[a]

System	S	R (Å)	R_H (Å)	R^* (Å)	ξ (Å)	ξ_k (Å)	B	α	Σ (Å^2)
SDS	3.5	85	95				5	− 1.5	74
	5	130	190				3	− 5	67
	5.2	155	240	233			0	− 6	69
	6.5				33	240 [20]			
	8	180	200	208			2	− 1.5	44
	10	125	150				6	0.5	43
SHBS	0.4	112	139						151
	0.5	203	276						134
	0.52	250	366	335					121
	0.565				49	400 [21]			102
	0.7	192	240						137
	0.9	129	141						107

[a]Defined according to text.

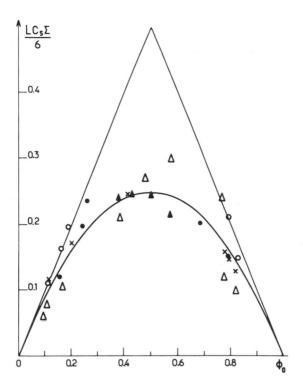

FIGURE 12.1. Characteristic length in the microemulsion phases, $LC_s \Sigma/6$, versus the oil volume fraction, ϕ_o. SDS system: open circle, light scattering data [9]; open triangle, X-ray data [20]; cross, neutron data [22]. SHBS system: closed circle, light scattering data [9]; closed triangle, X-ray data [21]. Straight lines correspond to Eq. (3) and $L = 2R$; the parabola corresponds to Eq. (2) and $L = \xi_K$.

It is seen that in the two-phase region, size L varies as ϕ_o or ϕ_w, as predicted by Eq. (3). In the three-phase region, there is a remarkable continuity between droplet radii and persistence length, ξ_K, measurements. ξ_K varies as $\phi_o \phi_w$, as predicted by Eq. (2). It was observed that the ξ_K values were unexpectedly constant over the three-phase domain. The area Σ being also constant, this happens because the product $\phi_o \phi_w$ is approximately constant. The measured values for ξ_K and Σ are also reported in Table 12.1.

A large difference between the L values for the two systems is observed. According to Eq. (1), the difference should be associated to larger elastic moduli in SHBS films than in SDS ones. This happens although the SHBS films contain more alcohol than the SDS ones, that is, three alcohol molecules and one alcohol molecule per surfactant, respectively. This was confirmed by recent measurements of K [23].

An X-ray study of a model system containing SHBS has been reported earlier [24]. The composition of the samples was the same in Ref. 24 as here, except for the alcohol, which was isobutanol instead of butanol, and temperature (25°C instead of 20°C). The sizes were found about 30% smaller. Although there is no information about the composition of the surfactant layer, it is expected that branched alcohols would lower the value of K; K/kT and ξ_K should therefore be smaller in this other model system.

A large difference between the different radii determinations is observed, especially close to the phase boundaries. This has been associated to droplet elongation in oil in water systems and to oil penetration in water in oil systems. An increasing oil penetration is known to be associated to increasing attractive forces between microemulsion droplets [25]. This is indeed the case here, where the virial coefficients are decreasing close to S_2 (see Table 12.1). In the SDS system, attraction is strong enough to give rise to critical behavior. The system is close to a critical end point. The second critical end point is much farther, since no critical behavior is observed close to S_1. The two end points are both far from S_1 and S_2 in the SHBS system, where the interactions are even more repulsive close to S_1. The nature of the attractive interactions giving rise to the end point close to S_1 is still very poorly understood.

B. Relation between Interfacial Tensions and Sizes

The measured interfacial tension versus salinity are plotted in Fig. 12.2; $\gamma^* = 4.5 \ 10^{-3}$ dyn/cm for the SDS system, and $\gamma^* = 8.7 \ 10^{-4}$ dyn/cm for the SHBS system. We have seen that the tensions $\gamma > \gamma^*$ are expected to be of the order of kT/L^2. The ratios $\gamma L^2/kT$ have been plotted in Table 12.2. As predicted, these ratios are of the order of unity. However, they systematically increase with salinity. Moreover, they are systematically larger in the SDS system than in the SHBS system.

Therefore, it has to be concluded that the ratio $\gamma L^2/kT$ is not universal. As explained, γ contains two kinds of terms, namely, an entropy term γ_E, which scales with kT/R^2 (the factor log ϕ being a slowly varying function of concentration), and a curvature term, which scales with K/RR_o. This second term is probably responsible for the variation of $\gamma R^2/kT$ with salinity and surfactant nature. A more quantitative test of the theories would require the knowledge of the curvature modulus K and of the spontaneous radius of curvature R_0. Experiments are in progress to determine these parameters in both systems.

C. Critical Behavior

There are very little data to compare with theory since critical behavior has been observed only close to S_2 in the SDS system. However, a partial comparison is in Table 12.3 with the tensions $\gamma < \gamma^*$. It is seen that $\gamma \xi_c^2$ is

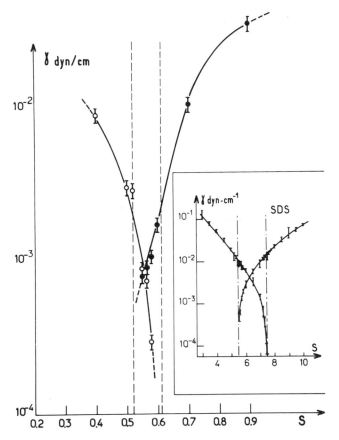

FIGURE 12.2. Interfacial tensions versus salinity for SHBS and SDS (insert) systems.

much smaller than predicted by theory. The comparison has been extended to samples that do not exhibit critical behavior (as evidenced from light scattering experiments) by assuming $\xi_c \sim \xi$. It has been shown that all the tensions $\gamma < \gamma^*$ vary as $\Delta\rho^4$, $\Delta\rho$ being the density difference between the coexisting phases [9]. This behavior is as predicted theoretically for critical behavior. Similarly, the correlation lengths ξ vary as $\Delta\rho^2$.[1] In the SHBS system, there were not enough experimental points to test these predictions. We assumed that they remain valid and have thus added the corresponding data in Table 12.3. It is seen that $\gamma\xi^2$ is still smaller than for the SDS system. We do not understand the origin of these differences.

We have recently undertaken a similar but more complete study of a micellar system containing SDS, butanol, and brine, where the critical point

---

[1] The distance ε to the critical point being unknown in these complicated mixtures, we have compared the variations of γ and ξ_c with another critical property: $\Delta\rho = \Delta\rho_0\varepsilon^\beta$, $\beta = 0.33$.

TABLE 12.2.   Scaling between the largest interfacial
tensions ($\gamma > \gamma^*$) and the characteristic lengths.

System	S	L	γ (dyn/cm)	$\gamma L^2/kT$
SDS	$S < S^*$		γ_{om}	
	3.5	R	7.5×10^{-2}	1.35
	5	R	2.0×10^{-2}	0.85
	5.2	R	1.4×10^{-2}	0.85
	$S = S^*$	ξ_K	4.5×10^{-3}	0.65
	$S > S^*$		γ_{wm}	
	8	R	2.3×10^{-2}	1.86
	10	R	7.5×10^{-2}	2.93
SHBS	$S < S^*$	R	γ_{om}	
	0.4		8.9×10^{-3}	0.28
	0.5		3.0×10^{-3}	0.31
	0.52		2.9×10^{-3}	0.46
	$S = S^*$	ξ_K	8.7×10^{-4}	0.35
	$S > S^*$	R	γ_{wm}	
	0.7		1.1×10^{-2}	0.95
	0.9		3.5×10^{-2}	1.46

TABLE 12.3.   Scaling between the lowest interfacial tensions
($\gamma < \gamma^*$) and the correlation lengths.

System	S	ξ_c (Å)	γ (dyn/cm)	$\gamma\xi_c^2/kT \times 10^2$
SDS	$S < S^*$		γ_{wm}	
	5.5	116	4.8×10^{-4}	1.6
	5.6	68	6.5×10^{-4}	0.8
	6	36	3×10^{-3}	1.0
	$S > S^*$		γ_{om}	
	6.5	30	3.5×10^{-3}	0.8
	7	46	1.5×10^{-3}	0.8
	7.3	106	4×10^{-4}	1.1
	7.4	228	8.7×10^{-5}	1.1
SHBS	$S < S^*$		γ_{wm}	
	0.55	46	8×10^{-4}	0.4
	$S > S^*$		γ_{om}	
	0.565	49	7.3×10^{-4}	0.4
	0.58	49	2.9×10^{-4}	0.2
Micelles [26]		$7\varepsilon^{-0.62}$	$0.48\varepsilon^{1.24}$	5.9
Pure fluids and Binary mixtures [27]				7–11
Theory [17]				5.4

was approached more carefully [26]. It was found that $\gamma\xi_c^2$ was still about two times smaller than the value found experimentally in most pure fluids and binary mixtures of small molecules [27]. These experimental values are larger by a factor of two than the theoretical ones [17], so the result for the micellar system is in excellent agreement with theory. It remains, however, that experimentally $\gamma\xi_c^2$ does not seem universal in surfactant systems and decreases with increasing characteristic size L.

This may indicate that the order parameter is not a scalar as in pure fluids or simple binary mixtures. It could be more complicated because of the presence of microscopic interfaces and could be analogous to the two-dimensional order parameters used to describe the smetic A phases close to critical points [28].

V. Conclusion

We have performed a detailed study of multiphase microemulsion model systems where, successively, an o/w microemulsion coexists with excess oil, a middle-phase microemulsion coexists with excess oil and water, and a w/o microemulsion coexists with excess water. The surfactants were SDS and pure alkyl benzene sulfonate (SHBS).

The interfacial tensions between the microemulsion and the excess phases were measured with surface light scattering methods. They were found to be ultralow and even lower in the SHBS system.

The sizes of the droplets have been measured in the o/w and w/o microemulsions with bulk light scattering methods. They are larger in the SHBS system. The difference remains in the three-phase domain, where sizes were taken from X-ray studies.

The relationship between the largest interfacial tension ($\gamma > \gamma^*$) and characteristic sizes has been quantitatively investigated. It has been shown that γ scales only approximately with kT/L^2. The ratio $\gamma L^2/kT$ varies with salinity and with the nature of the surfactant. It is proposed that these differences arise from curvature contribution to the interfacial tensions. Experiments to determine the surfactant film curvature properties on these systems are in progress to further test the theories.

The lowest interfacial tensions in the three-phase region ($\gamma < \gamma^*$) are likely to be associated with the vicinity of critical end points. The behavior of the systems differs markedly from those of simple fluids and binary mixtures of small molecules. The theoretically universal product $\gamma\xi_c^2/kT$ depends on the characteristic size L. Systematic experiments in which the critical end points can be better approached are needed to clarify this problem and to determine if a new kind of order parameter has to be introduced for these systems.

226   D. Langevin

Acknowledgments. The experiments presented in this paper have been performed in collaboration with the other members of the microemulsion group at the Ecole Normale Supérieure: O. Abillon, A.M. Cazabat, D. Chatenay, D. Guest, J. Meunier, and A. Pouchelon.

References

1. L.M. Prince, "*Microemulsions*," Academic Press, San Diego, California, 1977.
2. E. Ruckenstein and J. Chi, *J. Chem. Soc. Faraday Trans. 2* **71**, 1690 (1975).
3. S.A. Safran and L.A. Turkevich, *Phys. Rev. Lett.* **50**, 1930 (1983).
4. A. Pouchelon, D. Chatenay, J. Meunier, and D. Langevin, *J. Coll. Int. Sci.* **82**, 418 (1981).
5. L.E. Scriven, in "*Micellization, Solubilization and Microemulsions*," Vol. 2, p. 877, K.L. Mittal, ed., Plenum, New York, 1977.
6. Y. Talmon and S. Prager, *J. Chem. Phys.* **69**, 2984 (1978).
7. P.G. de Gennes and C. Taupin, *J. Phys. Chem.* **86**, 2294 (1982).
8. B. Widom, *J. Chem. Phys.* **81**, 1030 (1984).
9. A.M. Cazabat, D. Langevin, J. Meunier, and A. Pouchelon, *Adv. Coll. Int. Sci.* **16**, 175 (1982).
10. D.O. Shah, ed., "*Surface Phenomena in Enhanced Oil Recovery*," Plenum, New York, 1981.
11. J. Israelachvili, in "*Surfactants in Solution*," Vol. 4, p. 3, K.L. Mittal and P. Bothorel, eds., Plenum, New York, 1986.
12. E. Ruckenstein, in "*Surfactants in Solution*," Vol. 3, p. 1551, K.L. Mittal and B. Lindman, eds., Plenum, New York, 1984.
13. M. Robbins, in "*Micellization, Solubilization and Microemulsions*," Vol. 2, p. 713, K.L. Mittal, ed., Plenum, New York, 1977.
14. D.J. Mitchell and B.W. Ninham, *J. Phys. Chem.* **87**, 2996 (1983).
15. C. Huh, *J. Coll. Int. Sci.* **97**, 201 (1984).
16. B. Widom, in "*Fundamental Problems in Statistical Mechanics*," Vol. 3, p. 1, E.D.G. Cohen ed., North-Holland, Amsterdam, 1975.
17. E. Brézin and S. Feng, *Phys. Rev. B* **29**, 472 (1984).
18. D. Langevin, J. Meunier, and D. Chatenay, in "*Surfactants in Solution*," Vol. 3, p. 1991, K.L. Mittal and B. Lindman, eds., Plenum, New York, 1984.
19. D. Guest and D. Langevin, *J. Coll. Int. Sci.* **112**, 208 (1986).
20. L. Auvray, J.P. Cotton, R. Ober, and C. Taupin, *J. Physique* **45**, 913 (1984).
21. D. Guest, L. Auvray, and D. Langevin, *J. Phys. Lett.* **46**, L-1055, 1985.
22. A. de Geyer and J. Tabony, *Chem. Phys. Lett.* **113**, 83, 1985.
23. B.P. Binks, J. Meunier, O. Abellon and D. Langevin, *Langmuir* **5**, 415 (1989).
24. E.W. Kaler, H.T. Davis, and L.E. Scriven, *J. Chem. Phys.* **79**, 5685 (1983).
25. B. Lemaire, B. Bothorel, and D. Roux, *J. Phys. Chem.* **87**, 1023 (1983).
26. O. Abillon, D. Chatenay, D. Langevin, and J. Meunier, *J. Phys. Lett.* **45**, L-223 (1984).
27. M. Schneider, private communication, 1984; R. Moldover, *Phys. Rev. A* **31**, 1022 (1985).
28. P.G. de Gennes, "*The Physics of Liquid Crystals*," Clarendon Press, Oxford, 1974.

13
Structure and Properties of Three-Component Microemulsions Near the Critical Point

J.S. Huang, M. Kotlarchyk, and S.-H. Chen

We have used small-angle neutron scattering and light scattering to study the structure and nature of the phase transition near the lower critical point of a 3-component AOT microemulsion system. We have found that at both far and close to the critical point the system exists as an assembly of surfactant coated spherical water droplets dispersed in oil, characterized by a well-defined polydispersity. On approaching the critical point, the correlation length of the system diverges with a critical exponent of $v \simeq 0.75$ for a number of field variables, namely, temperature, pressure, alkane number, and salinity. This divergence can be attributed to increased correlations between the basic droplets. Furthermore, the critical exponents for the correlation length (v) and for the osmotic compressibility (γ) appear to have Ising-like values, with $\gamma = 2v$. The coexisting phases above the transition temperature are both water-in-oil microemulsions, differing only in the number density of droplets, and the nature of the phase transition is analogous to a simple liquid–gas coexistence. We have also determined that the critical phenomenon is driven by interdroplet attractions. Far below the cloud point temperature, the attraction is short-ranged. Near the transition, the interaction becomes progressively longer ranged, and its strength increases approximately linearly with temperature, consistent with the fact that the coexistence curve has a lower critical point.

I. Introduction

Critical phenomena [1,2,3] have been of great interest to the research community of condensed matter physicists over the past two decades or more. One of the most striking aspects of critical phenomena is the universality of the singular behavior in a host of largely unrelated physical systems. This is thought to be caused by the divergence of some of the susceptibilities of the system so that the thermodynamic properties of the system are dominated by the correspondingly large fluctuations. Since, near the critical point, these fluctuations have correlation length orders of magnitude larger than that of the intermolecular distance, the critical behavior becomes completely insensitive to the detailed nature of the intermolecular interactions. Rigorous scaling laws [4,5,6,7] are obeyed by a large class of critical systems, such as the simple liquid–gas coexistence, binary fluids mixtures, magnetic alloys, and quantum mechanical superfluids. There have been in the literature reports on observations of nonuniversal behavior in a number of surfactant systems that include several micellar solutions [8] and a microemulsion [9]. However, some of these results were later shown to be incorrect by more careful experiments [10]. We shall describe, in this article, a 3-component, oil-continuous microemulsion system whose critical behavior can be considered as normal, belonging to the Ising universality class. It is noted that there are other examples of microemulsions whose critical behavior can be considered as normal also, but these are all 4- or 5-component systems with possibly complicated structures [11,12].

From a fairly extensive small-angle neutron scattering (SANS) study, it is possible to understand the reason why this 3-component microemulsion system, containing sodium di-2-ethyl hexyl sulfosuccinate (AOT), water, and decane, should display a well-behaved critical phase transition. Nevertheless, even in this relatively simple 3-component microemulsion system, we shall see that there are a few more independent thermodynamic variables we can adjust, and there are also unexpected behaviors that do not seem to have any natural counterparts in simpler critical systems.

In many respects, the behavior of this AOT microemulsion is similar to that of a one-component liquid–gas system. This is perhaps a natural consequence of the stable droplet structure of this 3-component microemulsion. As we shall see in the later sections, the AOT microemulsion consists of microscopic water droplets, stabilized by a surfactant coating, dispersed in the oil phase. The mean radius of the droplets is uniquely determined by the water-to-surfactant ratio (volume to surface ratio) $X = [H_2O]/[AOT]$. The droplet system is found to be reasonably monodispersed with an average size spread of $\Delta R/R \simeq 0.3$. The properties of this fluid are dominated by the attractive interactions between the droplets, which behave like macromolecules in solution. However, there are additional degrees of freedom available to the microemulsions: the droplets could collide and exchange part of the interior phase or temporarily merge into a larger droplet and then split up again. So, if we compare this microemulsion with a macromolecular solution, we see that the equivalent molecular weight of the macromolecules is not a constant, nor is the instantaneous number concentration a constant. The interior liquid phase could provide an extra viscous damping to alter the dynamics of the solution, and the presence of a dilute but finite concentration of monomers in solution, in equilibrium with the interfacial surfactant molecules could contribute to the subtle complexities of the microemulsion system.

In the remainder of this chapter, we shall describe the experimental aspects in Section II and the observed scaling behavior of the AOT microemulsion system in Section III. In Section IV, we shall describe how we use the SANS technique to determine the structure of the AOT microemulsions. A brief discussion of the nature of the interaction potential that may be responsible for the observed phase transition near the critical point is presented in Section V. Section VI summarizes our findings.

II. Experimental Aspects

A. Sample Preparations

The composition of the system is denoted by 3-segmential numbers $X/Y/Z$, where Y/Z is the relative amount of water and oil (usually decane), respectively, in milliliters. A frequently used 3/5/95 composition designates a microemulsion containing 3 g of AOT dissolved in 100 cm³ of a 5/95 (by volume) mixture of water and decane.

AOT was obtained from the Fluka Chemical Company. It was purified in a hexane solution over a bed of activated carbon followed by methanol treatment [13]. The water used was twice distilled from deionized water from our own laboratory. The oils were a 99% gold label product purchased from Aldrich Chemical Company. The heavy water used contained 99.96% D_2O and 0.04% H_2O. With the exception of AOT, all other chemicals and distilled water were used as received.

B. Light Scattering

A conventional photon counting static/dynamic light scattering setup was employed in our experiments [14]. The major components used in the experiments included a Hewlett-Packard 9845B microcomputer, a Malvern K7025 128-channel multibit correlator, a PAR photon processor/counter, and a Spectra-Physics 124 15-mW He–Ne laser. The correlation length is measured by both the angular dissymmetry of the scattered intensity and by the Rayleigh linewidth Γ through a relation, $\xi = kTQ^2/6\pi\eta\Gamma$. Here, Q is the magnitude of the scattering wave vector, $Q = (4\pi/\lambda)(n \sin \theta/2)$, defined by the scattering angle θ, the wavelength λ, and the index of refraction of the medium, n. The Rayleigh linewidth $\Gamma = 1/\tau$, where τ is the measured correlation time of the intensity–intensity correlation function. The notation η stands for the viscosity of the solvent and k and T are the Boltzmann constant and the absolute temperature, respectively. The correlation length obtained through the static measurement is found to be in good agreement with that obtained through the dynamic measurement. Figure 13.1 shows a typical result of the intensity measurements. Here, the inverse of the intensity is plotted against Q^2. A straight line fit to the data yields the correlation length through the use of the Ornstein–Zernicke (OZ) relation $1/I(Q) \sim (1/\xi^2) + Q^2$.

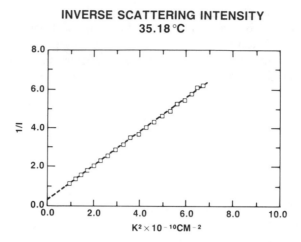

FIGURE 13.1. Inverse intensity is plotted versus K^2 for microemulsion 1°C below T_c. Here, $K = 4\pi/\lambda(\sin/2)$.

Because of the fact that AOT undergoes hydrolysis in higher temperatures, it is desirable to perform all of the meaurements near the critical temperature as rapidly as possible. For this reason, most of our data were obtained by swift, dynamic measurements. Figure 13.2 shows the comparison of the correlation length obtained by dynamic and static measurements. It is clear that both methods yield the same characteristic length.

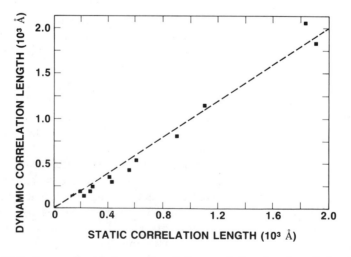

FIGURE 13.2. Comparison between the static correlation length and the dynamic correlation length.

C. Small-Angle Neutron Scattering

Small-angle neutron scattering experiments were performed using a 30 m SANS spectrometer belonging to the National Center for Small-Angle Scattering Research, which is located in the High-Flux Isotope Reactor of the Oak Ridge National Laboratory ($\lambda = 4.75$ Å). Alternatively, we also used the spectrometer at the High-Flux Beam Reactor of the Brookhaven National Laboratory ($\lambda = 5.28$ Å). Both spectrometers use a two-dimensional position-sensitive He^3 proportional counter to measure scattered neutron intensity as a function of the scattering angle. The data were corrected for detector sensitivity, transmission, solvent scattering, and background [13]. The final intensity spectra were converted to an absolute differential scattering cross section, $d\Sigma/d\Omega$ (cm^{-1}), by use of well-calibrated standard samples [15]. The detailed procedures for fitting the data to models are rather involved and can be found in cross references mentioned in subsequent references. In this chapter, we give only a qualitative summary of the procedures necessary for interpreting the results.

III. Critical Scaling Behavior

We have found that the attractive interaction potential between the droplets is approximately proportional to the droplet size [16], so a higher water to surfactant ratio (larger size) always yields a lower critical temperature. This is due to the fact that the AOT system exhibits a lower critical solution temperature, and the stronger attractive potential between the larger droplets results in the lowering of the phase separation temperatures. In the interest of surfactant stability, we have chosen to study mainly the microemulsion system with a large mean radius at a high water to surfactant molar ratio of $X \simeq 41$. The critical temperature for this 3/5/95 system in decane is found to be 36.01°C [17] if H$_2$O is used, 43°C if D$_2$O is used. The critical scaling behavior is described in the following sections.

A. The Coexistence Curve

It is known that the AOT microemulsion exhibits phase separation upon heating. The temperature at which the phase separation occurs is called the cloud point. The cloud point is a function of $X = [\text{D}_2\text{O}]/[\text{AOT}]$ and the volume fraction of the dispersed phase (water plus surfactant). For $X = 41.15$, the inverted coexistence curve is shown in Fig. 13.3. A lower critical point is observed at $\varphi = 0.08$. The coexisting phases above the cloud point temperature are oil-continuous microemulsions containing a different number density of the droplets, as we shall discuss later. The droplets in the coexisting phases have the same mean radii and the same degree of polydispersity. We view this cloud point transition as similar to that of the

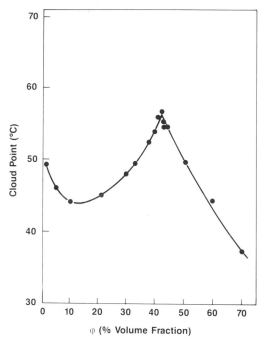

AOT/D$_2$O: 3/5 IN DECANE

FIGURE 13.3. Coexistence curve of AOT microemulsion with $X = 24.5$.

liquid–gas phase transition in simple fluid systems. Here, the microemulsion droplets play the role of the molecules in a single-component system. The critical index that characterizes the coexistence curve is $\beta \simeq 0.31$, similar to that found in simple fluids. The reason for an inverted coexistence curve is that the attractive interactions between the droplets increase as the temperature increases. Attractive interactions are thought to be caused by the branched surfactant tails [16,18,19].

At the dispersed phase volume fraction $\varphi \simeq 0.41$ and above, a new phase appears at higher temperatures. This phase boundary is probably due to the influence of an impending liquid–solid transition. The structure of the new phase has not been determined, but it is likely to be a lamella phase. However, the structure of the homogeneous low-temperature phase has been determined to be consisting of a dispersed droplet phase of water (or heavy water) in decane.

B. Divergence of the Correlation Length

A critical phenomenon is always accompanied by a divergence of the correlation range of the order parameter fluctuations. It is not, however, a priori clear what is a good choice of the order parameter for the microemulsion system and what are the best field variables with which to approach the critical point. The Ising-like coexistence curve (the left branch in Fig. 13.3) suggests that either the volume fraction of the dispersed phase or the droplet concentration may be serviceable as an adequate order parameter. Then, the droplet–droplet correlation length as measured by scattering experiments will reflect the divergence of the susceptibility of the order parameter at the critical point.

The correlation length is measured in most cases by dynamic light scattering. Simple power law divergence is generally found near the critical point such that $\xi \propto \delta Z^{-\nu}$, where $\delta Z \equiv Z - Z_c$. Here, Z represents various field parameters, such as the temperature, pressure, or carbon chain length of the oil phase, and Z_c is the critical value of Z where the correlation length ξ diverges. Figure 13.4 shows the divergence of ξ with respect to the relative temperature $t = T_c - T$. The straight line fit to the data on a log–log plot depicts a power law divergence characterized by the critical exponent $\nu = 0.75$ for $T < T_c$ in the one-phase region and $\nu = 0.65$ in the lower phase of the two-phase region [20].

For AOT microemulsions, it has been shown [21] that the alkyl carbon number of the oil phase modulates the interactions between the droplets. A set of measurements of the divergence of ξ in different oils characterized by the alkyl carbon number N, as a function of temperature, is obtained (Fig.

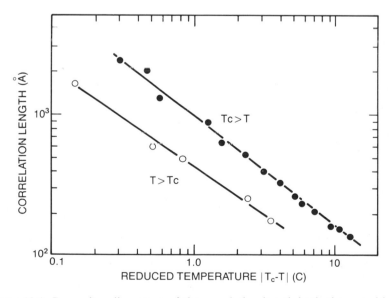

FIGURE 13.4. Power law divergence of the correlation length both above and below T_c; $\nu = 0.75$ for $T < T_c$ in the single phase, and $\nu = 0.65$ for $T > T_c$ in the lower phase or the 2-phase coexistence.

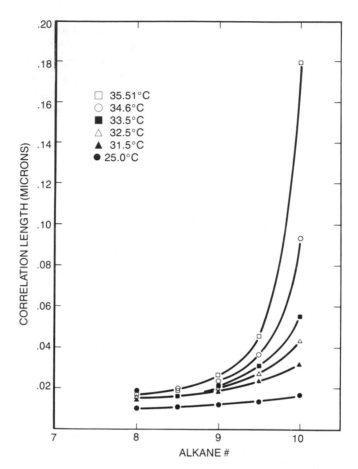

FIGURE 13.5. Divergence of the correlation length as function of the alkyl carbon number of the oil phase at various temperatures.

13.5). It is found that this family of curves apparently diverges at different critical temperature $T_c(N)$. We can plot the correlation length ξ obtained in the different oils as a function of their respective relative temperature $T - T_c(N)$, then all of the data fall on the same line on the log–log scale, as shown in Fig. 13.6. The best-fit slope of the data shown in Fig. 13.6 is again $v \simeq 0.75$. It is readily seen from Fig. 13.5 that at constant temperature ξ is a function of N. We can define a critical carbon number $N_c(T)$ to be the value of N such that ξ diverges at temperature T (Fig. 13.7). A plot of ξ as a function of $N_c - N$ is shown in Fig. 13.8. Here, a different scaling behavior is observed: $\xi \propto (N_c - N)^{-v_N}$, with $v_N \simeq 0.75$, also. This suggests that the carbon number of the oil phase plays a function similar to that of the more familiar field variable T.

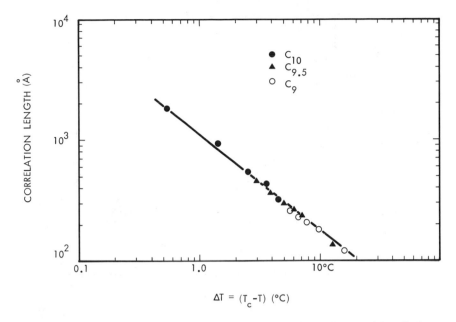

FIGURE 13.6. Scaling with respect to temperature for various alkanes of the oil phase. The slope $v_0 \simeq 0.75$.

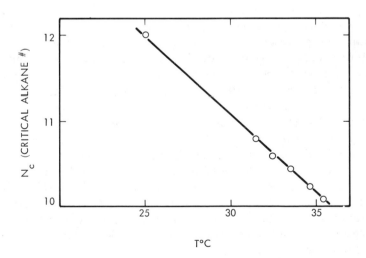

FIGURE 13.7. The temperature dependence of the critical alkyl number of the oil phase.

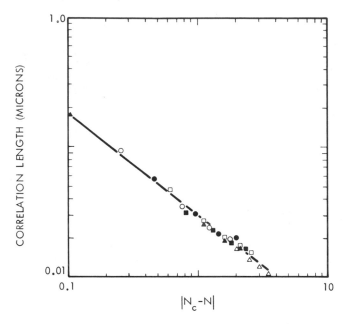

FIGURE 13.8. Scaling of ξ with respect to $|N_c - N|$ for different temperatures: closed triangles represent $T = 35.51°C$; open circles represent $T = 34.61°C$; closed circles represent $T = 33.5°C$; open squares represent $T = 32.5°C$; closed squares represent $T = 31.5°C$; and open triangles represent $T = 25.0°C$.

There are other field variables one can choose: pressure, the salinity of the aqueous phase, and the molar ratio of the water (internal phase) to AOT. These parameters all couple to the interaction strength between the droplets and are therefore capable of driving the system toward criticality. Changes in the hydrostatic pressure of the system serve to change the critical temperature [22]. If we assume, to the first order, that the shift of the critical temperature is linear in applied pressure, then at constant temperature, the correlation length will diverge as $(P_c - P)^{-v_p}$, with $v_p = v$. Figure 13.9 shows the scaling behavior of ξ with respect to δ_p. The best-fit slope $v_p \simeq 0.7$ is somewhat lower than, but consistent with, that found for the temperature index of 0.75. The salinity of the aqueous phase in this system turns out to be a very sensitive parameter, for the salt tolerance of the AOT microemulsion in the single-phase regime is low. In other systems, such as the alkyl xylene sulfonates, we have found that the correlation length scales with the relative salinity similar to that of the temperature scaling [21]. For the AOT system, the critical temperature of the microemulsion is lowered by the addition of sodium chloride, but apparently the correlation length would just rescale with respect to the new value.

It is interesting to point out that values of v_z for these different choices of the field variables Z are apparently very similar, that is, $v_z \simeq 0.75$. For the

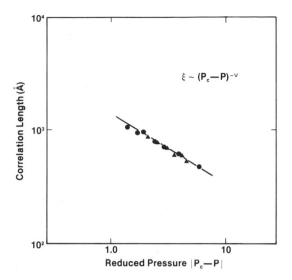

FIGURE 13.9. Scaling of ξ with respect to $|P_c - P|$ at two different temperatures. Circles are for $T = 26.3°C$, and triangles are for $T = 28.2°C$.

scattering experiments, where the temperature is the variable, the closest approach to the critical point is $\delta T/T_c \simeq 5 \times 10^{-5}$, typically an order of magnitude worse than what is obtained in critical binary fluid mixtures. For other scaling parameters, $\delta N/N_c \simeq 5 \times 10^{-2}$, and $\delta P/P_c \simeq 6 \times 10^{-2}$; both seem to be far away from the critical point, yet an almost identical power law divergence is observed for these variables. Furthermore, the dependence of ξ on the relative temperature is roughly consistent with the value obtained for an Ising system. It should also be noted that the prefactor of the correlation length is an order of magnitude larger than that obtained in a simple binary fluid, so it is harder for the microemulsion system to be in the asymptotic regime of the critical point according to the Ginsburg criteria.

C. Divergence of the Shear Viscosity

Recently, Berg et al. [23] have measured the behavior of the shear viscosity of the AOT microemulsion very near the critical point. The measurements were made in the AOT/D_2O/decane system with a 3/5/95 composition. The shear viscosity is measured by the damping of the oscillations of a cylinder suspended by a thin quartz fiber. The cylinder containing the microemulsion is temperature regulated to better than 0.001°C. It is found that the critical part of the shear viscosity behaves like $\eta \propto (T_c - T)^{-y}$, with $y = 0.03$. This value is lower than, but still close enough to, the value of 0.042 for the critical index observed in a critical binary fluid. However, the temperature dependence of the noncritical background viscosity is quite different from the simpler liquid mixtures. Attempts to describe the background viscosity enhancement by considering a hard-sphere model with a temperature-dependent attractive interaction were only partly successful [23]. A viscosity anomaly in another microemulsion system was also observed [24], but the measurements were taken relatively far from the critical point. Given the nature of the weak divergence of the shear viscosity at T_c, the critical behavior could not be obtained in that system.

D. Dynamic Scaling Behavior and Droplet Structure

We have studied the dynamics of the fluctuations of the 3/5/95 system and found that the dynamic scaling holds for this microemulsion [25]. However, the scaling function that describes the universal dependence of the linewidth Γ/AQ^2 as a function of the dimensionless wave vector $Q\xi$ differs from the expression given by the mode–mode coupling theory [26] by a constant numerical factor of 1.2. Figure 13.10 shows the dynamic scaling behavior of

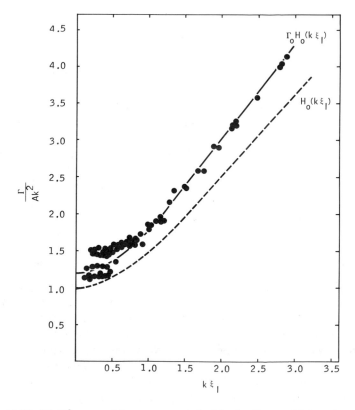

FIGURE 13.10. Γ/Ak^2 versus $k\xi$ are compared with the Kawasaki function $H_0(k\xi_I)$. The solid line represents the best-fit function $\Gamma_0 H_0(K\xi_I)$ with $\Gamma_0 = 1.20$.

the microemulsion. The quantity $A = kT/(6\pi\eta\xi)$ is a measured quantity, and the dashed line represents the mode–mode coupling theory of the Kawasaki function. A more recent light scattering experiment and its analysis by Rouch et al. [51] identified the source of the numerical factor 1.2 between the experimental data and the calculation of the mode–mode coupling theory. If one takes into account the so-called background effect in the mode–mode coupling theoretical calculation, then this additional factor of 1.2 comes naturally. This is one more evidence that near the critical point the characteristic length as measured by static and dynamic light scattering should be identified with the correlation length between the droplets and not the mean droplet radius.

From the observed scaling between the compressibility and the characteristic length ($\gamma \simeq 2\nu$), we may deduce that the scattering intensity due to the droplet density fluctuation, $I \propto t^{-\gamma} \propto t^{-2\nu} \propto \xi^2$, is consistent with the fact that ξ is the correlation length. For, if the characteristic length is identified with the droplet radius R, then we expect that $I \simeq R^3$, which definitely lies outside of the experimental errors.

IV. Droplet Structure

Before discussing the microstructure of microemulsions near the critical point, it is appropriate to discuss the structure of water in oil (W/O) AOT microemulsions in general. AOT forms reverse micelles in hydrocarbon solvents [27–33]. In W/O microemulsions, the basic micellar aggregate simply solubilizes the water by incorporating it into the hydrophilic micellar core. Previously performed light scattering experiments [32,34] have established a picture of AOT W/O microemulsions consisting of spherical water cores, each coated by a monolayer of AOT, immersed in a continuous oil phase. SANS studies have been performed to confirm this picture of reverse micelles with water cores [35,36].

The composition variable that determines the size of the surfactant-coated water droplets is the molar ratio $X = $ [water]/[AOT]. Dilution with oil and variations in temperature have only weak effects on the resultant droplet dimensions [13]. In addition, an important result from the analysis of the SANS data is that polydispersity of the droplets is an integral part of the W/O microemulsion structure. This polydispersity is also primarily determined by the variable X [13,37].

From the SANS data, one can measure both the mean droplet radius $\bar{R}(X)$ and a polydispersity index $p(X) = \sigma_R/\bar{R}$, where σ_R is the root-mean-square deviation from the mean radius. By analyzing various SANS spectra using a histogram polydispersity analysis [38], it was found that the Schultz distribution is an appropriate functional form for modeling the microemulsion polydispersity [37]. However, in no way should the Schultz distribution be construed as the unique functional form.

For SANS experiments, the microemulsions were composed of D_2O and protonated oil. In this way, the neutrons see essentially only the water core because the tails of the AOT molecules are indistinguishable from the oil. Under this condition, one can analyze each SANS spectrum to get the two parameters R and p. Again, we emphasize that R and p turn out to be functions of the parameter X. In particular,

$$(1 + 2p^2)\bar{R} = \frac{3V_w}{a_H} X + \frac{3V_H}{a_H}.$$

Here, V_w is the specific volume of a D_2O molecule $\simeq 30 \text{ Å}^3$, V_H is the volume of an AOT head group, and a_H is the area subtended by an AOT head group. The latter equation can be derived from a volume conservation condition for surfactant and water molecules plus a packing constraint that each surfactant head group maintain an optimal area. Figure 13.11 shows a plot of $(1 + 2p^2)R$ versus X for a series of microemulsions containing a fixed total volume fraction of water plus surfactant ($\phi = 0.05$). The resulting straight line gives the values $a_H = 60 \text{ Å}^2$ and $V_H = 103 \text{ Å}^3$, corresponding to a head group radius of approximately 2.3 Å, which is a very reasonable result. One should note that the linearity of the relation is satisfied only if one includes the polydispersity factor $(1 + 2p^2)$.

Naturally, one now asks how the microemulsion droplet structure is affected when the composition and temperature approach the critical point. To answer this question, SANS measurements were obtained in both the one-phase and two-phase regions near the critical point [13,39].

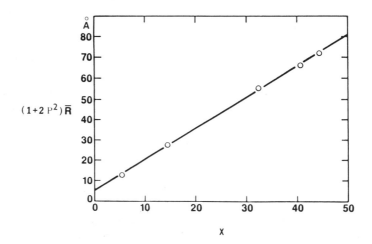

FIGURE 13.11. $(1 + 2p^2)R$ plotted against the molar ratio $X = $ [water]/[AOT]. The straight-line fit corresponds to a constant packing area per surfactant head group of $a_H = 60 \text{ Å}^2$.

In the one phase region near the critical point, the divergence of the correlation length ξ has already been discussed in view of the light scattering data. A crucial point to ascertain is whether this length corresponds to the droplet size or to a characteristic length of the concentration fluctuations of the droplets. SANS is an ideal probe for determining the answer to this question because the scattering wave vector spans values in the range $\xi^{-1} < Q < R^{-1}$.

The single phase 3/5/95 microemulsion ($X = 40.8$, $\phi = 0.0746$) was studied at temperatures below the cloud point. Figure 13.12 shows the small Q behavior of the SANS spectra from a temperature just below T_c to far below T_c. One clearly sees a rapid increase in the forward scattering intensity signaling the onset of long-range correlations when one approaches T_c. At large Q, however, the scattering curves become virtually indistinguishable, indicating the invariant underlying length scale dictated by the droplet structure previously discussed. The data can be successfully modeled by assuming that well-defined polydisperse droplets continue to exist near the critical point, and that these droplets develop long-range correlations as the critical point is approached. The pronounced forward scattering is modeled with a modified OZ form [40] (parameterized by the isothermal osmotic compressibility of the droplets, χ, and the correlation length of the droplet concentration fluctuations, ξ), while the scattering at large Q is that of polydisperse spheres. As with the light scattering data, Fig. 13.13 shows that χ exhibits a power law behavior having an exponent $v = 0.72 \pm 0.04$. Furthermore, the compressibility χ also exhibits a power law behavior characterized by an exponent $\gamma = 1.61 \pm 0.09$ [39]. In actuality, the precise values of the extracted exponents depend sensitively on the details of the model used for the scattering data. For example, if, in addition to OZ scattering, it is assumed that hard-sphere droplet interactions are significant at small Q, then the exponents become $v = 0.62 \pm 0.01$ and $\gamma = 1.20 \pm 0.06$ [41], values very close to that obtained by light scattering. In either case, however, it appears that neutron scattering gives critical exponents that are Ising-like, satisfying the scaling relation $\gamma = 2v$.

Light scattering is the more reliable way to extract the critical exponents. However, only by measuring the neutron scattering is it certain that the critical scattering is caused by increased correlations between the underlying submicroscopic droplets. This is made clear by looking at the results extracted from the scattering curves for a large number of samples at temperatures and volume fractions near the critical point; in Fig. 13.14, the values of T and ϕ are indicated in relation to the cloud point curve for X at the fixed value of 40.8. The different values of ϕ were obtained by diluting the 3/5 ratio of AOT/D$_2$O with varying amounts of oil. Figure 13.15 shows the resulting mean droplet size and the polydispersity as a function of ϕ and T. As previously stated, both R and p are approximately independent of concentration and temperature for a fixed value of X. Also, it is seen that the polydispersity index near the critical point is between 0.25 and 0.30. (The

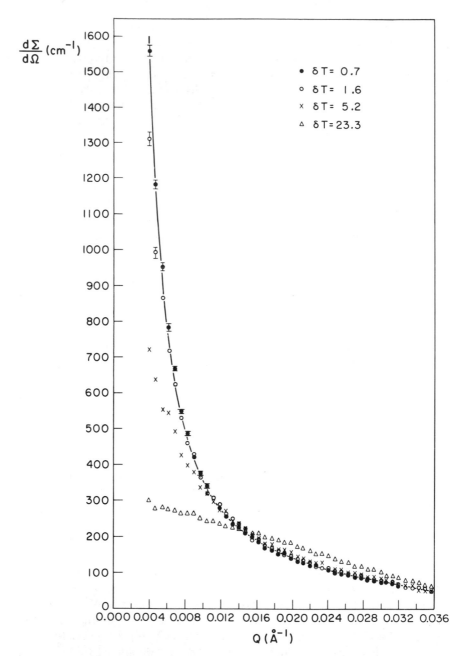

FIGURE 13.12. Measured neutron differential cross sections versus Q for a sample with composition $AOT/D_2O/decane = 3/5/95$, at various temperatures, δT, below the critical temperature.

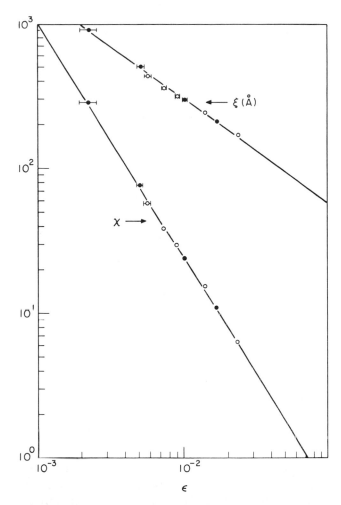

FIGURE 13.13. Critical parameters ξ and χ versus reduced temperature $\varepsilon = \delta T/T_c$ from the SANS analysis. The slopes give the exponents ν and γ.

parameter $\Delta\rho$ is the difference in neutron scattering length density between the water core and the continuous hydrocarbon medium. It is extracted from the absolute intensity scale of the neutron data and should not depend on composition or temperature).

We also performed SANS measurements on the 3/5/95 microemulsion heated in the two-phase region at a temperature about 1° above T_c. In this case, the single-phase microemulsion ($\phi = 0.0746$) separates into two phases; a separate SANS spectrum was obtained for each of the two resultant phases. The two spectra were fitted using the procedure described previously to obtain the droplet parameters and the critical parameters. The results can be

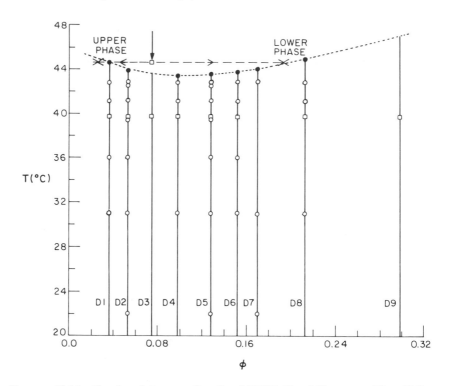

FIGURE 13.14. Cloud point curve for the $AOT/D_2O = 3/5$ system ($X = 40.8$) at different droplet volume fractions ϕ. The phase separation experiment at $\phi = 0.0746$ is also illustrated.

summarized as follows. The nature of the cloud point transition is a one-phase microemulsion to two-phase microemulsion transition. The difference in the two phases is simply the number density of the microemulsion droplets. The size and polydispersity of the droplets do not change in the phase transition, that is, the values are $R \simeq 44$ Å and $\rho \simeq 0.3$ for both phases. As proposed by Safran and Turkevich [42], this is the evidence that shows the so-called bending elastic energy of the surfactant interface is much larger than either kT or the interdroplet interaction energy. When this bending energy is large, it is predicted that the coexisting phases should consist of a high and low density of identical droplets, analogous to a binary mixture phase separation. Furthermore, with the aid of the measured phase volumes, we were able to calculate the volume fractions of the upper phase ($\phi = 0.024$) and the lower phase ($\phi = 0.193$). These values are in agreement with the phase boundary determined by the cloud point measurements at different droplet volume fractions in the one-phase region (see Fig. 13.14). Therefore, the droplet volume fraction is the proper order parameter of the phase transition.

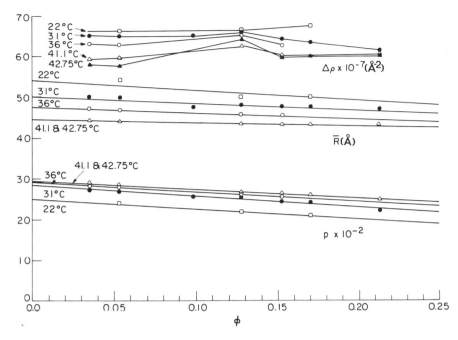

FIGURE 13.15. Concentration dependence of the mean water-core radius (R), poly-dispersity (p), and contrast factor ($\Delta\rho$) at various temperatures.

V. Interdroplet Interactions

It seems that the cloud point in these three-component microemulsions is analogous to a liquid–gas phase transition. The role of the atoms in the liquid–gas system is played by the surfactant-stabilized water droplets in microemulsions. A major difference, however, between the two systems is that the microemulsion coexistence curve exhibits a lower consolute temperature, while the liquid–gas system has an upper consolute temperature. The reason for the latter behavior is that the atoms in a simple fluid interact through the Lennard–Jones potential with a fixed attractive well depth (ε) and range independent of temperature. To drive the one-phase system toward two phases, the interaction strength, given by ε/kT, must exceed a critical value; therefore, one must decrease the temperature to produce a phase separation. It thus seems reasonable to postulate that the microemulsion one-phase to two-phase transition can be effected if there exists an attractive droplet–droplet interaction characterized by a temperature-dependent attractive interaction strength that increases with temperature.

Away from the critical point, it has been demonstrated that attractive interactions between AOT reversed micelles and AOT w/o microemulsion droplets can be represented by an interdroplet potential consisting of a hard core plus an attractive square well [16]. Analysis of SANS data with interparticle structure factors derived by Sharma and Sharma [43] shows the range of the attraction to be about 2.4 Å and the well depth to be a linear function of the droplet size. These results imply that the attraction may be due to the mutual interpenetration of the tips of the surfactant tails, the penetration length being limited by the branched double-chain structure of the AOT molecule. This result is consistent with a theory developed by Lemaire et al. [18] and light scattering experiments performed by Brunetti et al. [19] for a four-component w/o system, for which it is proposed that the interpenetration of the surfactant tails is limited by the length of the cosurfactant. Recently, however, Pincus and Safran [44] have suggested that the details of the surfactant molecules may not be responsible for the attractive mechanism. Instead, they propose that the short-range attraction is caused by the fact that the concentrated regions of surfactant at the droplet interfaces cause a perturbation in the overall dilute surfactant concentration in the bulk, which is near the value of the critical micelle concentration (c.m.c). When droplets approach each other, an attractive force arises because the perturbations in the surfactant concentration become correlated, resulting in a lower free energy of the system.

Upon approaching the critical point, the latter perturbation theory predicts that the range of the attractive interaction should diverge. Instead of modeling SANS spectra with the previously described OZ model, which is phenomenological in nature, it is possible to fit the data with a microscopic model that uses structure factors calculated for a given interaction range and well depth [13,41,45,46]. Structure factors calculated using either the so-called mean spherical approximation [47] or the optimum cluster theory [48] result in an interaction that becomes long-ranged near the critical point, but the range does not sensitively depend on the precise value of $\delta T/T_c$. This result was obtained by assuming that the form of the potential $U(x)$ consists of a hard core of diameter σ, determined by the diameter of the droplet water core plus surfactant coating and an attractive Yukawa tail:

$$U(x)/kT = \begin{cases} -\gamma \exp(-kx)/x & x \geq 1 \\ \infty & x < 1 \end{cases},$$

where $x = r/\sigma$. The range parameter k is found to be approximately 1.4 near the critical point. The actual range of the attraction is $\sigma/k \simeq 130\ \text{Å}/1.4 = 93\ \text{Å}$. The effective well depth γ, on the other hand, turns out to be a linearly increasing function of temperature [45]; thus, the interaction strength increases with temperature, as required for a lower consolute point. Furthermore, at a given value of ϕ, by extrapolating the linear increase in γ to the point where the calculated osmotic compressibility diverges, one can locate a point on the spinodal curve, which lies above the cloud point curve [45]. The resulting spinodal curve for the 3/5 system is displayed in Fig. 13.16. Actual spinodal decomposition has been directly observed in AOT microemulsions [49], as well as in a four-component system by light scattering [50].

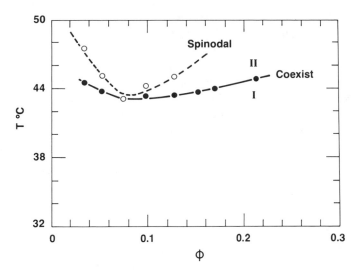

F<small>IGURE</small> 13.16. Experimentally measured cloud points and theoretically predicted spinodal curve for AOT/D$_2$O = 3/5 system.

VI. Summary

The AOT three-component microemulsion system exhibits a phase transition characterized by a lower critical point. Light scattering and neutron scattering have provided an understanding of the structure and properties of the system near the critical point. Both far from and close to the phase transition, the system exists as an assembly of surfactant-coated spherical water droplets in oil, characterized by a well-defined polydispersity. Upon approaching the critical point, the correlation length of the system diverges with a critical exponent of $v \simeq 0.75$ for a number of field variables, namely, temperature, pressure, and the alkyl carbon number of the oil. This divergence can be attributed to increased correlations between the basic droplets. Furthermore, the critical exponents for the correlation length and the osmotic compressibility appear to be Ising-like, with $\gamma = 2v$.

The phase transition is a one-phase microemulsion to two-phase microemulsion transition. The coexisting phases above the cloud point temperature are both w/o microemulsions, differing only in the number density of droplets. The phase transition is analogous to a simple liquid–gas phase transition.

The critical phenomenon is driven by interdroplet attractions. Far below the cloud point temperature the attraction is short-ranged. Near the cloud point, however, the attraction becomes long-ranged, and its strength increases linearly with temperature. This accounts for the fact that the coexistence curve has a lower consolute point.

Acknowledgments. We acknowledge important contributions to this review by M. W. Kim. The viscosity measurements near the critical point were carried out at the National Bureau of Standards by R. Berg and M. Moldover. We have benefited by many discussions with S.A. Safran, L. Turkevich, M. Robbins, and J. Bock. We have also received experimental assistance from J. Sung, W. Gallagher, and D. Schneider. Neutron beam times generously granted by the Oak Ridge National Laboratory and the Brookhaven National Laboratory are greatly appreciated. Research support for S.H.C. was from an NSF grant administered through the Center for Materials Science and Engineering at MIT.

References

1. M.E. Fisher, *Reports on Progress in Physics*, **30**, 615 (1967).
2. H.E. Stanley, "*Introduction to Phase Transition and Critical Phenomena*," Oxford Univ. Press, London and New York, 1971.
3. C. Domb and M.S. Green, "*Phase Transition and Critical Phenomena*," Vols. 1-6, Academic Press, San Diego, California, 1974.
4. B. Widom, *J. Chem. Phys.* **43**, 3895 (1965).
5. R.B. Griffiths, *Phys. Rev.* **158**, 176 (1967).
6. L.P. Kadanoff, *Rev. Mod. Phys.* **39**, (1967).
7. B.I. Halprerin and P.C. Hohenberg, *Phys. Rev.* **117**, 952 (1969).
8. M. Corti and V. Degiorgio, *Phys. Rev. Lett.* **45**, 1045 (1980); M. Corti and V. Degiorgio, *Phys. Rev. Lett.* **55**, 2005 (1985); K. Hamano, T. Sato, T. Koyama, and N. Kuwahara, *Phys. Rev. Lett.* **55**, 1472 (1985).
9. R. Dorshow, F. de Buzzaccarini, C.A. Bunton, and D.F. Nicoli, *Phys. Rev. Lett.* **47**, 1336 (1981).
10. R. Strey and A. Pakusch, in "*Surfactants in Solution*," Vol. 4, K.L. Mittal and P. Bothorel, eds., Plenum, New York, 1987; G. Dietler and D.S. Cannell, *Phys. Rev. Lett.* **60**, 1852 (1988).
11. O. Abillon, D. Chatenay, D. Langevin, and J. Meunier, *J. Physique Lett.* **45**, L-233 (1984).
12. A.M. Cazabat, D. Langevin, J. Meunier, and A. Pouchelon, *J. Physique Lett.* **43**, L-89 (1982).
13. M. Kotlarchyk, S.-H. Chen, J.S. Huang, and M.W. Kim, *Phys. Rev.* **A29**, 2054 (1984).
14. B. Chu, "*Laser Light Scattering*," Academic Press, San Diego, California, 1974.
15. G.D. Wignall and F.S. Bates, *J. Appl. Cryst.* **20**, 28 (1987).
16. J.S. Huang, S.A. Safran, M.W. Kim, G.S. Grest, M. Kotlarchyk, and N. Quirke, *Phys. Rev. Lett.* **53**, 592 (1984).
17. The critical temperature tends to be different from different batches of surfactants. A variation of 2°C is typical.
18. B. Lemaire, P. Bothorel, and D. Roux, *J. Chem. Phys.* **87**, 1023 (1983).
19. S. Brunetti, D. Roux, A.M. Bellocq, G. Fourche, and P. Bothorel, *J. Chem. Phys.* **83**, 1028 (1983).
20. J.S. Huang and M.W. Kim, *Phys. Rev. Lett.* **47**, 1462 (1981).

21. J.S. Huang and M.W. Kim, "*Physics of Amphiphiles: Micelles, Vesicles, and Microemulsions,*" p. 303, V. Degiorgio and M. Corti, eds., North-Holland, Amsterdam, 1985.
22. M.W. Kim, J. Bock, and J.S. Huang, *Phys. Rev. Lett.* **54**, 46 (1985).
23. R.F. Berg, M.R. Moldover, and J.S. Huang, *J. Chem. Phys.* **87**, 3687 (1987).
24. A.M. Cazabat, D. Langevin, and O. Sorba, *J. Physique Lett.* **43**, L-505 (1982).
25. M.W. Kim and J.S. Huang, *Phys. Rev.* **B26**, 2703 (1982).
26. K. Kawasaki, *Ann. Phys. (NY)* **61**, (1970).
27. P. Ekwall, L. Mandell, and K. Fontell, *J. Coll. Inter. Sci.* **33**, 215 (1970).
28. J.B. Peri, *J. Coll. Inter. Sci.* **29**, 6 (1969).
29. T. Assih, F. Larche, and P. Delord, *J. Coll. Inter. Sci.* **89**, 35 (1982).
30. Y.C. Jean and H.J. Ache, *J. Am. Chem. Soc.* **100**, 6320 (1978).
31. J.H. Fendler and E.J. Fendler, "*Catalysis in Micellar and Macromolecular Systems,*" Academic Press, San Diego, California, 1975.
32. M. Zulauf and H.F. Eicke, *J. Phys. Chem.* **83**, 840 (1979).
33. M. Kotlarchyk, J.S. Huang, and S.-H. Chen, *J. Phys. Chem.* **89**, 4382 (1985).
34. R.A. Day, B.H. Robinson, J.H.R. Clarke, and J.V. Doherty, *J. Chem. Soc., Faraday Trans. I* **75**, 132 (1979).
35. M. Kotlarchyk, S.-H. Chen, and J.S. Huang, *J. Phys. Chem.* **86**, 3273 (1982).
36. C. Cabos and P. DeLord, *J. Appl. Cryst.* **12**, 502 (1979).
37. M. Kotlarchyk, R.B. Stephens, and J.S. Huang, *J. Phys. Chem.* **92**, 1533 (1988).
38. R.B. Stephens, *Appl. Phys.* **61**, 1348 (1987).
39. M. Kotlarchyk, S.-H. Chen, and J.S. Huang, *Phys. Rev. A* **28**, 508 (1983).
40. L.S. Ornstein and F. Zernicke, *Proc. Ned. Akad. Sci.* **17**, 793 (1914).
41. S.-H. Chen, T.L. Lin, and M. Kotlarchyk, in "*Surfactants in Solution,*" Vol. 4, K.L. Mittal and P. Bothorel, eds., Plenum, New York, 1987.
42. S.A. Safran and L.A. Turkevich, *Phys. Rev. Lett.* **50**, 1930 (1983).
43. P.V. Sharma and K.C. Sharma, *Physica (Utrecht)* **89A**, 213 (1977).
44. P.A. Pincus and S.A. Safran, *J. Chem. Phys.* **86**, 1644 (1987).
45. S.-H. Chen, T.L. Lin, and J.S. Huang, in "*Physics of Complex and Supermolecular Fluids,*" p. 285, S.A. Safran and N.A. Clark, eds., Wiley (Interscience), New York, 1987.
46. S.-H. Chen, *Physica* **137B**, 183 (1986).
47. J.L. Lebowitz and J.K. Percus, *Phys. Rev.* **144**, 251 (1966).
48. H.C. Andersen and D. Chandler, *J. Chem. Phys.* **57**, 1918 (1972).
49. H.M. Lindsay and M.W. Kim, unpublished result.
50. D. Roux, *J. Physique* **47**, 733 (1986).
51. J. Rouch, A. Safouane, P. Tantaglia, and S.-H. Chen, *J. Phys. C. Condensed Matter* **1**, 1773 (1989); *J. Chem. Phys.* **90**, 3756 (1989); *Prog. in Colloid and Polymer Sci.* **79**, 279 (1989).

14
Field Variables and Critical Phenomena in Microemulsions

D. ROUX and A.M. BELLOCQ

Some general features of field variables and critical phenomena in multicomponent systems are discussed. Critical behavior of surfactants in solution is specially examined, and results for a ternary system composed with pentanol, water, and sodium dodecyl sulfate (SDS) are presented. The study of the phase diagram leads us to evidence of a critical point and an associated field-like variable. The critical behavior is studied using the temperature (path I) and the field variable (path II) as variables. In both cases, an Ising-like behavior is found.

I. Introduction

Observations of critical phenomena in micellar solutions [1,2] and microemulsion [3–6] have been reported in the last few years. In most cases, it seems possible to interpret the critical behavior of these mixtures as a liquid–gas-like critical phenomenon [7–9]. The structure of the medium can be described as a solution of interacting aggregates dispersed in a continuous phase made mostly of water (for the case of normal micelles in water) or oil (for the case of inverted micelles in oil). Attractive interactions exist between these aggregates and, if they are large enough, they could lead to a first-order phase separation between a micellar-rich phase and a micellar-poor one. For some appropriate values of the interaction energy (comparing to kT), this phase transition leads to a critical point where the two phases merge.

It is well established that, on the one hand, the critical point of the liquid–gas phase transition belongs to the same universality class as the three-dimensional Ising model [10]. However, on the other hand, the phase separation in a binary fluid mixture at the critical solution point is, in many ways, similar to the liquid-vapor phase transition in a single-component fluid near the critical point [11]. For mixtures with three or more components, liquid–liquid phase separation also seems similar to the Ising model [12].

The goal of investigations of critical behavior of surfactant solutions is to identify the differences, if any exists, between these critical points and the liquid–gas critical point of a pure fluid. Some experimental results [3,13–15] are in agreement with the Ising-like behavior, while others indicate a more complex behavior [4,9,16]. In this chapter, we address the question of whether the micellar solutions and microemulsion systems belong to the Ising universality class.

In comparing the critical behavior in surfactant solutions with those of pure fluids or binary mixtures it is essential to know the path taken to approach the critical point. Indeed, usually microemulsions are quaternary or quinary mixtures, and the number of independent variables required to describe the system is large. An interesting way to control the path of approach to the critical point for microemulsions was recently discovered [15–17]. For these mixtures, a simple ratio between compositions, namely, the water to surfactant ratio, behaves as a field variable. This property allowed us to approach the critical point, at a constant temperature, by using the water to surfactant ratio as a variable.

In what follows, we review some generalities about field variables and their relation with the phase diagram (Section II) and about the critical behavior (Section III) in multicomponent systems. Then, in Section IV, we present some new results on a ternary mixture (water, SDS, and pentanol). This system has been chosen as a model microemulsion for its simplicity.

II. Generalities About Field Variables and the Phase Diagram

A system with n components may be characterized by $(n + 1)$ independent variables, the pressure P, the temperature T, and the molar densities per unit volume ρ_i $(1 \leqslant i \leqslant n - 1)$. All these variables are intensive, but, following Griffith and Wheeler's definition [18], some are fields (P and T) and others are densities (ρ_i). The fields must take the same values in the two phases that coexist. However, densities usually take different values in the different phases in equilibrium.

Phase diagrams are generally represented with a set of mixed variables. For a pure fluid, we can choose from the set, P, T, and ρ (the density). Only two of them are necessary to represent the phase diagram since the equation of states links these three variables. Figure 14.1 gives the three possible representations of the phase diagram corresponding to the three projections of the surface representing the coexisting region. For systems with more than one component, the phase diagram is drawn at constant vapor pressure, ignoring the vapor phase. This leads to the elimination of one degree of freedom. For example, a two-component mixture requires basically three independent variables to represent the phase diagram. In order to have a two-dimensional representation, the phase diagram is drawn at constant vapor pressure using T and the concentration of one component as variables. Only

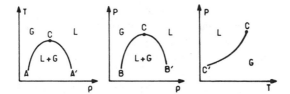

FIGURE 14.1. Three possible representations of the phase diagram of a pure fluid. Here, G corresponds to the gas phase, and 1 corresponds to the liquid phase. C is the critical point, and the lines ACA', BCB', and $C'C$ are the projections of the coexisting line.

the liquid phases are represented. For a three-component mixture, four variables are required to describe the complete phase diagram (P, T, and two concentrations). Usually, the phase diagram is represented by a cut at constant vapor pressure and constant temperature. This two-dimensional cut of the phase diagram is generally drawn in a triangle for reasons of symmetry (Fig. 14.2). The case of a four-component mixture is more complex. Indeed, five variables are now required to represent the phase diagram, and even by keeping the vapor pressure constant and fixing the temperature, the phase diagram must be drawn in a three-dimensional space (three concentrations may be used as variables). As before, for reasons of symmetry, the phase diagram is generally represented in a tetrahedron. If the phase diagram is complex, which is generally the case for a four-component mixture, this kind of representation is difficult to visualize. A way to simplify the phase diagram representation is to turn back to a two-dimensional picture keeping one concentration, or more usually the ratio between two concentrations, constant. This representation gives a misleading idea of the multiphase equilibria. Indeed, in a region of the phase diagram where two phases coexist, the

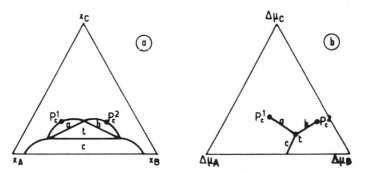

FIGURE 14.2. Example of a phase diagram for a ternary mixture, ABC, in (a) the density space, x_A, x_B, x_C, and (b) the field space, $\Delta\mu_A$, $\Delta\mu_B$, $\Delta\mu_C$. The field variables are related to the chemical potential of A, B, and C; a, b, and c are two-phase regions, t is a three-phase region, and P_c^i are critical points.

phases that are in equilibrium are generally not in the plane corresponding to the cut of the three-dimensional phase diagram; this would not be the case if a field was kept fixed. It is important to notice that fixing a field variable (temperature or pressure) is not equivalent to fixing a density. For more than four components, it is impossible to obtain an exact representation of the phase diagram since, even in a three-dimensional cut of the phase diagram, the phases that coexist are usually not in this three-dimensional representation. However, the use of field variables allows a simplification of the phase diagram representation, since fixing a field leads to a reduction of one for the number of variables of the system. Moreover, the replacement of densities by fields reduces the dimensionality of the region of phase coexistence. For example, let us consider a ternary mixture. In a density space, all the regions of existence and coexistence of phases appear as surfaces whatever their variance; but, in a field space, the one-phase regions appear as surfaces, two-phase regions as lines, and three-phase regions as points (Fig. 14.2). In this representation, the dimensionality of each region is equal to its number of degrees of freedom. Another advantage of using fields instead of densities is the possibility of approaching a critical point with a field variable other than temperature [18].

III. Critical Behavior in Multicomponent Systems

When a critical point corresponding to a second-order phase transition is approached using a field h as the variable, the divergence of the fluctuations leads to a singularity of the thermodynamic properties that follows a power law function of the reduced distance ε from the critical point ($\varepsilon = (h - h_c)/h_c$, where h_c is the value of the field h at the critical point). For example, the divergence of the compressibility K_T is characterized by the exponent γ. The exponent v is characteristic of the divergence of the correlation length of the order parameter fluctuations, and β is characteristic of the shape of the coexisting curve in the h, ρ representation [10]. The critical exponents associated with these singularities depend only upon the universality class of the critical point (i.e., on the space dimensionality and the number of variables of the order parameter). These exponents are obtained along special paths approaching the critical point. For example, the unique path approaching the pure fluid liquid–gas critical point and giving the characteristic exponents of the critical point is the critical isochore. (This path maintains a constant density and allows changes in temperature). The liquid–gas critical point of the pure fluid belongs to the Ising universality class; consequently, the values of the exponents γ, v, and β are equal to 1.24, 0.64, and 0.32, respectively.

In the case of liquid–liquid critical point of a two-component mixture, compressibility is replaced by osmotic compressibility, and density is replaced by concentration. In this case, the path approaching the critical point equivalent to the isochore path for the pure fluid corresponds to varying the temperature at fixed concentration of the components and at constant vapor pressure. The liquid–liquid critical point of a binary mixture belongs to the same universality class as the liquid–gas critical point of a pure fluid; therefore, the same values for the critical exponents are found [11]. In these two cases, we approach the critical point by fixing only one density and by varying the temperature. That condition entirely fixes the path in the pure fluid case; however, in the case of binary mixtures, another field (the pressure) must be controlled.

For a three-component mixture, the situation is more complex. First, the order parameter is not clearly defined, because of the multiple choices of densities. Moreover, there exist several ways to approach the critical point. Indeed, the path equivalent to the critical isochore is one in which two fields (namely, the pressure and one chemical potential) and one density are fixed and the temperature is varied. Following this path, we expect to obtain Ising values for the critical indices. In practice, instead of holding two fields and one density constant, the experimentalist fixes two densities and controls one field (the pressure). Keeping the second density fixed leads to a renormalization of the critical exponents along the experimental path [19]. The renormalized values of the exponents γ', ν', and β' are 10% higher than the Ising values [they are multiplied by the coefficient $1/(1 - \alpha)$, with α equal to 0.11]. Another possible way to approach this critical point is to keep two fields (pressure and temperature) and one density constant and to vary a chemical potential. In this case, because of the nonlinear relation between the chemical potential and the reduced temperature, we also expect renormalized exponents [20]. Any path different from these three leads to a more complex behavior involving crossover functions.

For more than three components, nothing is very clear. However, Griffiths and Wheeler have proposed that we may expect the good exponents (i.e., Ising exponents for the liquid–gas-like phase transition) to follow a path that is asymptotically parallel to the coexisting curve (in the field space) and not parallel to the critical line [18]. Consequently, there are two possibilities to obtain the characteristic exponents of the critical point studied. The first one is to fix enough constant fields in order to reduce the number of variables until one reaches a situation equivalent to a known case (namely, binary or ternary mixture). The second one is to approach a critical point along a path satisfying the Griffiths and Wheeler criterion. In these two cases, the knowledge of one or more chemical potentials is required. The problem is to be able to experimentally control fields other than temperature or pressure.

A. Experimental Results on Critical Behavior of Surfactants in Solution

The experimental study of critical points of surfactants in solution began in the early 1980's [2–6]. Prior to this, the considerable increase of turbidity as the coexistence curve was approached was interpreted as an increase in the size of the aggregates. Several authors have published experimental measurements of the critical exponents. In most cases, the exponents measured are v and γ; sometimes β is also given. The choice of these exponents (rather than others) is due to the use of the light scattering technique for measuring the critical behavior. Table 14.1 summarizes the main results obtained for several systems. The mixtures are classified according to the number of components. Values of theoretical exponents are also given. The first point we may emphasize is that most of the values are close to the Ising exponents (renormalized or not). However, three series of measurements give very different exponents. The first one is the work of Dorshow et al. [4], who have

TABLE 14.1. Experimental critical exponents for several binary, ternary, quaternary, and quinary mixtures containing surfactant.[a]

Mixture	Exponent		
	v	γ	β
Ising model [10]	0.64	1.24	0.32
Renormalized Ising model [19]	0.72	1.40	0.36
Binary mixture			
C1OE4, water [13]			0.36
C6E3, water [9]	0.63	1.25	
C8E4, water [9]	0.57	1.15	
C12E6, water [9]	0.53	0.97	
C12E8, water [9]	0.44	0.92	
Ternary mixture			
AOT, water, decane [3] $T_c = 36°C$	0.75	1.22	
AOT, water, decane [14] $T_c = 43°C$	0.72	1.61	
AOT, water, decane [17] $T_c = 26°C$	0.76	1.30	
AOT, water, decane [17] $T_c = 30°C$	0.71	1.25	
AOT, water, decane[b] [17] $T_c = 25°C$	0.61	1.26	0.4
AOT, water, decane[c] [24]	0.70	1.50	
AOT, water, decane [25]	0.67	1.65	
CPBr, water, NaClO3 [26]	0.6		
Quaternary mixture			
SDS, water, butanol, NaCl [27]	0.62		
SDS, water, pentanol, dodecane for	(16,23)		
different points of a critical	.64 − .2	1.2 − .4	0.40
line and with different paths			
Quinary mixture			
CTAB, water, NaBr, butanol,			
octane [4]	1.13	2.24	2.4

[a] Values are taken from the literature.
[b] The critical point is approached using a chemical potential as the variable.
[c] The critical point is approached using pressure as the variable.

obtained exponents that are twice the values of the Ising model. Such values are expected when the critical point line is approached tangentially [21]. Further investigations are required in order to verify if the path followed corresponds to this special approach. In the two other series of experiments [9,16], even if a path problem exists in one case [16], the surprisingly low values of the exponents are probably due to more complex behavior. Besides these cases, another problem exists. Indeed, in molecular ternary mixtures where no aggregation exists renormalized exponents are found [12]. In surfactant solutions, because of the existence of aggregates, a pseudobinary description of the structure may be given. Does this microscopic structure lead to binary-like behavior (i.e., without Fisher's renormalization) or is the behavior strictly similar to that of a molecular multicomponent mixture? The analysis of the results of Table 14.1 does not allow us to answer this question. Indeed, the theoretical difference is only of 10%, and the experimental data are not accurate enough.

IV. Study of a Ternary Mixture: Water, SDS, and Pentanol

In order to understand better the relationship between chemical potential and composition in a simple case, we have studied a ternary mixture made of water, SDS, and pentanol. This ternary mixture corresponds to a limiting case of a microemulsion (i.e., without oil). Moreover, a critical point exists in the phase diagram, and the study of the critical behavior would allow one to obtain information about the value of the critical indices measured along different paths.

Usually, a chemical potential is a function of the other variables; $\mu = f(P, T, \rho_i)$, and at constant pressure and temperature it is a function of the composition: $\mu = f(\rho_i)$. By definition, the function f must take the same value in every phase in equilibrium with each other. For example, in a two-phase equilibrium (phases 1 and 2) $f(\rho_i^1) = f(\rho_i^2)$, where ρ_i^j is the density of the component i in phase j (1 or 2 in our example). Near a critical point, this function may take a simple form [28], but a general expression valid in a large region of the phase diagram would be very interesting. In earlier articles, we have presented results that indicate the water to surfactant ratio allows one to control a chemical potential [15,16,17] in a quaternary microemulsion mixture. Indeed, it was evidenced that this ratio takes the same value in several phases (two or three) in equilibrium in the oil-rich part of the phase diagram.

The phase diagram at constant temperature ($T = 25°C$) of the ternary mixture consisting of water, SDS, and pentanol is given in the Fig. 14.3. This phase diagram was experimentally established with a technique described elsewhere [22]. Four one-phase regions are seen. One is a liquid isotropic

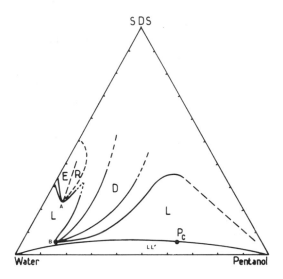

FIGURE 14.3. Phase diagram at $T = 25°C$ of the ternary water–pentanol–SDS system. L is an isotropic phase; E, R, and D are mesomorphic phases. The dashed boundaries have not been accurately determined. The points A and B are azeotropic-like points; P_c is a critical point.

phase (L); at 25°C, this phase extends continuously from the water corner to the pentanol corner. This property is characteristic of microemulsion behavior. The liquid phase is known to be a micellar phase in the water-rich part of the phase diagram. The three other one-phase regions are mesophases. One (phase D) is lamellar and the two others are, respectively, hexagonal (phase E) and rectangular (phase R). These different one-phase regions are separated from each other with polyphasic equilibria. In particular, there exists a large region where two isotropic liquid phases are in equilibrium (region LL'). This region ends with a critical point P_c, where the two liquid phases in equilibrium merge. Other special points are the points denoted by A and B. At these points, it seems clear that an azeotropic equilibrium takes place. Careful measurements of the tie lines on each side of B point indicate that they become shorter and shorter until they disappear at point B. At this point, the lamellar and liquid phases in equilibrium have exactly the same compositions. The evolution of this phase diagram as a function of temperature was reported elsewhere [23]. The main feature of this evolution is the splitting of the L region into two domains, L_1 and L_2, as the temperature is decreased. This separation at $T = 20°C$ takes place through an indifferent state. Figure 14.4 summarizes this behavior using a qualitative field representation of the phase diagrams.

The shape of the coexisting curve, which limits the LL' region, was precisely measured by preparing a large number of samples and visually determining if they were situated in the one-phase or two-phase regions. This

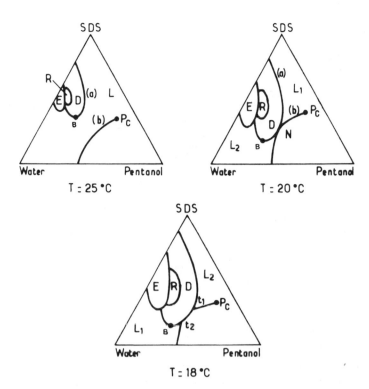

FIGURE 14.4. Schematic phase diagrams of the water–pentanol–SDS system at different temperatures in the field space. At $T = 20°C$, the two coexisting lines (a) and (b), limiting L and D regions, touch one each other producing an indifferent state at the point N. Below $T = 20°C$, this indifferent state gives two three-phase equilibria t_1 and t_2.

determination was made after storing the samples for several days in order to achieve a complete separation of the phases. Then, the positions of several tie lines in the LL' region around the critical point P_c were determined. Precise measurements were obtained by using density measurements. First, densities of several one-phase mixtures along the coexisting curve of the LL' region were determined using a high-precision picnometer (relative densities are measured with a precision better than 10^{-5}). The experimental variation of the density d along the coexisting curve was found to be a linear function of the alcohol concentration X_a:

$$d = d_w + aX_a, \qquad (1)$$

where $d_w = 1.0012$ (at 25°C), and $a = 0.19689$.

Then, two-phase equilibria were prepared in the LL' region, and the density of both coexisting phases was measured. From Eq. (1) and the knowledge of the coexisting curve, we were able to locate precisely the tie

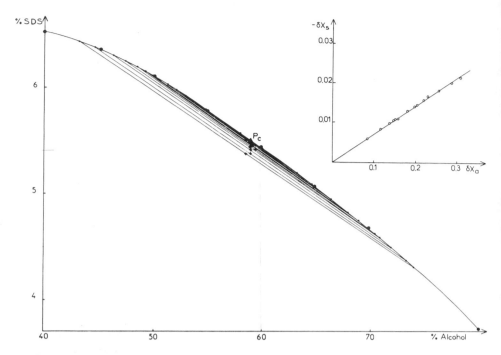

FIGURE 14.5. Magnification of the two liquid-phase LL′ regions around the critical point P_c. The segments represent the tie lines, the full points correspond to the visual determination of the coexisting curve, the crosses are the positions of the middle of the tie lines, and the triangle is the position of the critical point. In addition, we have plotted the difference of the surfactant concentration between the two phases in equilibrium: δX_s as a function of the difference of the alcohol concentration and δX_a in the same two-phase equilibrium. The linear behavior of these two differences indicates that the tie lines are parallel.

lines. Figure 14.5 gives the positions of these tie lines around the critical point. It appears that around the critical point, the tie lines seem to be parallel; this also holds for mixtures far from the critical point. This remark was made several years ago for another system by Zollweg [28]. In addition, the middle points of the tie lines are situated on a straight line that is perpendicular to the tie lines. This direction is defined by a relation between two concentrations. Let us choose, for example, X_s (surfactant concentration) and X_a (the alcohol concentration). From experimental behavior, we have noticed that a linear relation exists between δX_s and δX_a (Fig. 14.5):

$$\delta X_s = -A\delta X_a$$

with $A = +0.07$, $\delta X_s = X_s^1 - X_s^2$, and $\delta X_a = X_a^1 - X_a^2$, where X_a^j and X_s^j are the alcohol and surfactant concentrations in the upper phase ($j = 1$) and lower phase ($j = 2$).

Consequently, for each two-phase equilibrium, the function $Y(X_a, X_s)$ is conserved in each phase in equilibrium:

$$Y(X_a^1, X_s^1) = Y(X_a^2, X_s^2) \quad \text{with} \quad Y(X_a, X_s) = AX_a + X_s.$$

The direction parallel to the tie lines corresponds to the density conjugated to the field Y. The order parameter O of the phase transition is given by the function $O(X_a, X_s) = -X_a/A + X_s$.

It is possible to redraw the phase diagram as a function of the field Y and the order parameter O at constant temperature. Figure 14.6 shows this phase diagram representation. The critical point P_c is obviously an extremum of the coexisting curve. However, it is more surprising that the points A and B are also extrema, which indicates that the variable Y seems to behave as a field even far from the critical point P_c. Note that the indifferent state that appears at lower temperature is also nearly perpendicular to this direction [23]. In fact, a precise analysis shows that Y is not exactly constant on each tie line;

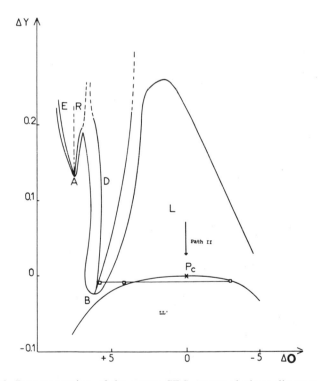

FIGURE 14.6. Representation of the water–SDS–pentanol phase diagram at $T = 25°$ using the field Y and the order parameter O variables: $\Delta Y = Y(X_s, X_a) - Y(X_s^c, X_a^c)$, and $\Delta O = O(X_s, X_a) - O(X_s^c, X_a^c)$ (see text). The notation of the regions is the same as in Fig. 14.3. The squares correspond to the positions of the three phases of the indifferent state evidenced at $T = 20°C$.

especially far from the critical point, it would be better to consider that $Y(X_a, X_s)$ is an approximate expression for a field variable and that a more complex expression exists. The determination of a more accurate and, consequently, more complex formula for Y is not probably of great interest; moreover, it requires more precise analysis.

This knowledge of an approximate expression for a chemical potential has allowed us to approach the critical point P_c along two different paths. The first one is the classical path: all the concentrations are fixed at their critical values, and we have used a change of temperature to approach the critical point. The critical point, being an upper one, was approached by decreasing the temperature (path I). The second path keeps the temperature constant and varies the concentrations along a straight line (of direction $X_s = X_a/A$), while fixing O at its critical value $O(X_a^c, X_s^c)$, where X_a^c and X_s^c are the alcohol and surfactant critical concentrations (path II). This path is indicated in Fig. 14.6. For each path, the distance of the critical point is measured with a reduced variable ε:

$$\varepsilon_t = (T - T_c)/T_c, \qquad \text{following the path I,}$$

and

$$\varepsilon_y = (Y - Y_c)/Y_c, \qquad \text{following the path II,}$$

where Y_c is the value of the function $Y(X_a, X_s)$ at the critical point.

We have determined the value of the exponents γ and ν by following path I, using the light scattering technique and turbidity measurements. Figure 14.7 shows a log–log plot of the light scattered at zero wave vector [$I(0)$ is a direct measurement of the osmotic compressibility and diverges with the exponent γ]. The experimental value we have obtained is $\gamma = 1.27 \pm .10$. Because of the large turbidity of the sample, it was not possible to get very close to the critical point. In order to approach closer to the critical point, we performed turbidity measurements. Results are given in Fig. 14.8, and the whole curve is fitted with the expression [29]

$$\tau = K\chi_T \frac{2\alpha^2 + 2\alpha + 1}{\alpha^3} \ln(1 + 2\alpha) - \frac{2(1 + \alpha)}{\alpha^2}, \qquad (2)$$

where K is a constant, k_0 is the wave vector of the light, $\alpha = 2(k_0\xi)^2$, and χ_T is the osmotic compressibility. We used the power law divergence of χ_T and ξ according to the relations

$$\chi_T = \chi_T^0 \varepsilon_t^{-\gamma}, \qquad \xi = \xi_0 \varepsilon_t^{-\nu}.$$

The fit with the expression of the turbidity given in Eq. (2) involves four parameters: K, ξ_0, ν, and γ. Even with precise measurements of the turbidity, the least-square fit with the four free parameters does not converge. In order to obtain the best values of the exponents ν and γ, we decided to fix γ to the Ising value (1.24) or to the Fisher renormalized Ising value (1.38) and to let the other parameters free. The best fit was obtained with the nonrenormalized

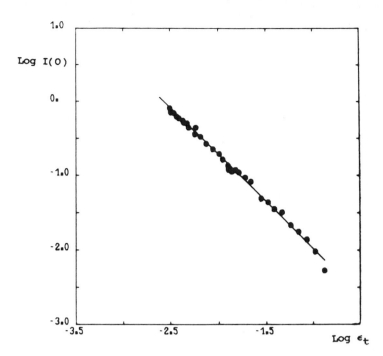

FIGURE 14.7. Log–log plot of the scattered light at zero angle $I(0)$ (which measured the osmotic compressibility) as a function of the reduced temperature ε_t along path I. The critical index γ is equal to 1.27.

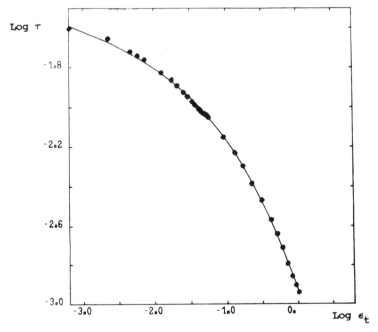

FIGURE 14.8. Log–log plot of the turbidity as a function of the reduced temperature ε_t. The full line corresponds to the fit with Eq. (2) and with the values of γ, ν, and ξ_0 equal to 1.24, 0.63, and 21 Å, respectively.

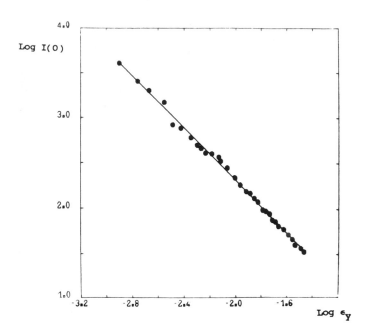

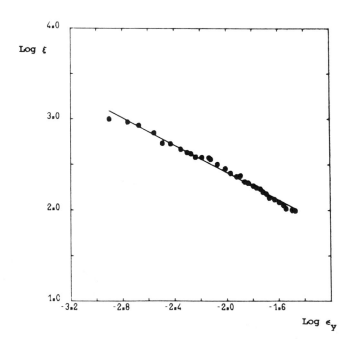

FIGURE 14.9. Log–log plot of the scattered light at zero angle $I(0)$ and of the correlation length ξ as a function of the reduced field variable ε_y along path II (shown in Fig. 14.6). The critical indices γ and ν are equal to 1.45 and 0.75, respectively.

Ising value $\gamma = 1.24$ and gives $v = 0.63$ and $\xi_0 = 21$ Å. The value of γ is close to the index obtained by light scattering, and this confirms that the exponents measured far from the critical point $(T - T_c \geqslant 1°C)$ are in accordance with the values obtained near the critical point $(0.01°C \geqslant T - T_c \leqslant 1°C)$. The critical exponents obtained indicate that this ternary mixture seems to behave like a binary mixture rather than a molecular ternary mixture. However, the precision of the experiment is not sufficient to definitively conclude that the exponents are not renormalized. More accurate measurements are still in progress.

Along the second path (path II), we have measured the exponent β in addition to the exponents γ and v. Figure 14.9 shows in a log–log plot the results of the measurements of the osmotic compressibility and the correlation length as functions of the reduced distance from the critical point ε_y. Figure 14.10 gives in a log–log plot the density difference between the two phases in equilibrium as the critical point is approached. The experimental behavior clearly follows a power law relation similar to the one observed with the temperature variation approach. The respective values of γ, v, and β are 1.45 ± 0.10, $.75 \pm 0.08$, and 0.38 ± 0.03. The accuracy of γ and v has the same order of magnitude as the preceding approach (path I); however, the value of

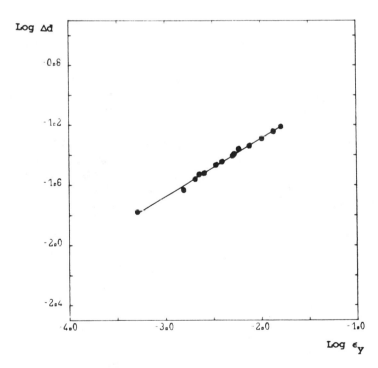

FIGURE 14.10. Log–log plot of the density difference between several two-phase equilibria around the critical point P_c. The value of the exponent β is 0.38.

the exponent β is more precise. These results indicate that following this path Fisher's renormalization is obtained, which is in accordance with the theoretical predictions for a ternary mixture [20]. This behavior confirms the field-like behavior of the function Y.

V. Conclusion

The existence of field variables related to composition has been previously observed in a quaternary mixture [15,16]. In this mixture, the field variable was the water to surfactant ratio (denoted by X). The use of this ratio as a field variable was also extended to the ternary mixture AOT, water, and decane [17]. The relation between the ratio X and the field variable Y for the ternary mixture SDS, water, and pentanol is not obvious. However, we may remark that another way to express the variable Y is to use the water concentration (X_w). We may then define another field variable:

$$Z(X_a, X_s, X_w) = \frac{X_w + (1 - A)X_a}{X_s + AX_a}.$$

Variable Z is strictly equivalent to the variable Y. Using the relation between the concentrations ($X_s + X_a + X_w = 1$), one has $Z = 1/Y - 1$. This expression of the field emphasizes the role of the water to surfactant ratio. In microemulsions, it is possible to relate the field character of this ratio to a characteristic length. Indeed, because of the microscopic structure of the medium, which can be described as a dispersion of water droplets in a continuous phase of oil, the water to surfactant ratio fixes the size of the particles. In the ternary mixture studied here, the structure is probably more complex; moreover, the partioning of the alcohol among the continuous phase, the interfacial film, and the water region could be the origin of the introduction of parameter A.

The critical indices measured following the two paths indicate an Ising-like behavior. Consequently, the critical point of this ternary mixture belongs to the same universality class as the critical point of a pure fluid or a binary mixture. The question of whether Fisher's renormalization is observed or not is still open as it requires the determination of the exponents with a precision better than 10%.

Acknowledgments. It is a pleasure to thank M. Maugey and O. Babagbeto for their technical assistance. The authors are also grateful to B. Pouligny for his contribution to the turbidity measurements.

References

1. M. Corti and V. Degiorgio, *Opt. Commun.* **14**, 158 (1975).
2. M. Corti and V. Degiorgio, *Phys. Rev. Lett.* **45**, 1045 (1980).
3. J.S. Huang and M.W. Kim, *Phys. Rev. Lett.* **47**, 1462 (1981).
4. R. Dorshow, F. de Buzzaccarini, C.A. Bunton, and D.F. Nicoli, *Phys. Rev. Lett.* **47**, 1336 (1981).
5. G. Fourche, A.M. Bellocq, and S. Brunetti, *J. Coll. Int. Sci.* **89**, 427 (1982).
6. A.M. Cazabat, D. Langevin, J. Meunier, and A. Pouchelon, *J. Phys. Lett.* **43**, L-89 (1982).
7. D. Roux, A.M. Bellocq, and M.S. Leblanc, *Chem. Phys. Lett.* **94**, 156 (1983).
8. S.A. Safran and L.A. Turkevich, *Phys. Rev. Lett.* **50**, 1930 (1983).
9. M. Corti, C. Minero, and V. Degiorgio, *J. Phys. Chem.* **88**, 309 (1984).
10. H.E. Stanley, "*Introduction to Phase Transitions and Critical Phenomena*," Oxford University Press, London and New York, 1971.
11. D. Beysens, Nato Adv. Study Inst., Ser. 82, Vol. 72, pp. 25–62 (1982).
12. K. Ohbayashi and B. Chu, *J. Chem. Phys.* **68**, 5066 (1978).
13. J.C. Lang, R.D. Morgan, *J. Chem. Phys.* **73**, 5849 (1980).
14. M. Kotlarchyk, S.-H. Chen, and J.S. Huang, *Phys. Rev.* **A28**, 508 (1983).
15. D. Roux, A.M. Bellocq, *Phys. Rev. Lett.* **52**, 1895 (1984).
16. A.M. Bellocq, P. Honorat, and D Roux, *J. Physique* **46**, 743 (1985).
17. P. Honorat, D. Roux, and A.M. Bellocq, *J. Phys. Lett.* **45**, L-961 (1984).
18. R.B. Griffiths and J.C. Wheeler, *Phys. Rev. A* **2**, 1047 (1970).
19. M.F. Fisher, *Phys. Rev.* **176**, 257 (1968).
20. B. Widom, *J. Chem. Phys.* **46**, 3324 (1967).
21. R.G. Johnston, N.A. Clark, P. Wiltzius, and D.S. Cannel, *Phys. Rev. Lett.* **54**, 49 (1985).
22. A.M. Bellocq and D. Roux, in "*Surfactants in Solution*," Vol. 4, K.L. Mittal and P. Bothorel, eds., Plenum, New York, 1987.
23. A.M. Bellocq and D. Roux "*Microemulsions*," S. Friberg and P. Bothorel, eds., CRC Press, Boca Raton, p. 33, 1987.
24. M.W. Kim, J. Bock, and J.S. Huang, *Phys. Rev. Lett.* **54**, 46 (1985).
25. C. Toprakcioglu, J.C. Dore, B.H. Robinson, A. Howe, and P. Chieux, *J. Chem. Soc., Faraday Trans. 1* **80**, 413 (1984).
26. J. Appel and G. Porte, *J. Phys. Lett.* **44**, L-689 (1983).
27. O. Abillon, D. Chatenay, D. Langevin, and J. Meunier, *J. Phys. Lett.* **45**, L-223 (1984).
28. J.A. Zollweg, *J. Chem. Phys.* **55**, 1430 (1971).
29. V.G. Puglielli and N.C. Ford, *Phys. Rev. Lett.* **25**, 143 (1970).

15
On the Structure of Five-Component Microemulsions

N.J. CHANG, J.F. BILLMAN, R.A. LICKLIDER, and
E.W. KALER

A family of 5-component microemulsions made with sodium 4-(1′-heptyl-onyl)-benzenesulfonate, iso-butyl alcohol, dodecane, and NaCl brines of various concentrations has been studied with small-angle neutron scattering (SANS) and quasielastic light scattering (QLS) methods. Both techniques show that when the water volume fraction, Φ_w, is either less than 0.20 or greater than 0.75, the size of the well-defined droplets (water in oil or oil in water) increases nearly linearly as the volume fraction of the dispersed phases increases. The water in oil droplets are spherical and polydispersed in size, while the oil in water droplets are prolate with an aspect ratio of approximately 3:1. For the intermediate aqueous volume fractions between 0.20 and 0.75, both the QLS and SANS results indicate that the length scales in the microemulsion phase vary approximately in proportion to $\Phi_w(1 - \Phi_w)$. These observations favor a model of bicontinuous structure in the intermediate regime.

I. Introduction

Microemulsions are of interest because they solubilize materials as dissimilar as oil and water into a single, thermodynamically stable, liquid phase. Their potential for industrial applications, as well as their intrinsic interest as a unique arrangement of oil and water on a colloidal scale, has fueled a tremendous expansion of the scientific study of microemulsions in the past decade. Despite this study, there remain fundamental unanswered questions about the structure of microemulsions, especially when the microemulsion contains comparable amounts of oil and water. An obstacle to understanding is that the phase behavior patterns and structure changes observed in microemulsion systems depend on many variables, with, for example, temperature, electrolyte concentration, and the type of hydrocarbon being just a few. Elucidation of the mechanisms by which these variables affect the structure of microemulsions has proved to be a persistent problem. Our purpose is to examine the structure changes in a model system with the goal of producing an unambiguous picture of the arrangement of oil and water in samples containing a range of oil/water ratios.

The minimum number of components necessary to form a microemulsion is three: oil, water, and surfactant. In practice, however, the easiest way to produce microemulsions containing equal or comparable amounts of oil and water is to add both a cosurfactant (usually a low molecular weight alcohol) to modify the surfactant-rich film and an electrolyte to moderate the interactions between surfactant head groups. The resulting five-component or Winsor microemulsions [1] are often of low viscosity and can produce ultralow interfacial tensions with the excess oil or water phases with which they are in equilibrium.

In general, the experimental evidence from a variety of systems points to the presence of at least three kinds of microemulsion microstructure, each existing over a range of composition [1-7]. As the volume fraction of oil solubilized in a microemulsion increases from near zero to near one, the structure evolves from an arrangement of oil-swollen micelles or droplets in water [an oil/water (o/w) microemulsion] to one of water-swollen inverted micelles in oil [a water/oil (w/o) microemulsion]. The intermediate structure is likely to be a bicontinuous one, in which both oil and water span the sample [7]. The surfactant in all of these structures is arranged as a film at the internal interfaces between the oil and water domains.

The measurements reported here are made on samples containing the surfactant sodium 4-(1'-heptylnonyl)benzenesulfonate (SHBS), iso-butyl alcohol (iBA), dodecane, water, and NaCl. The concentration of NaCl in each sample is varied to control the amounts of oil and water solubilized into the microemulsion phase. In a simple view, changing the electrolyte concentration controls the curvature of the surfactant film, and thus the structure, in the following way. At low salt concentrations, the microemulsion contains oil droplets dispersed in water, and the microemulsion phase exists at equilibrium with excess oil. Electrical interactions between the ionic surfactant head groups are moderated by their counterions and the electrolyte. As the salinity increases, the head-group repulsions are further screened, the curvature of the surfactant film decreases, and the droplets swell. Continued addition of salt leads eventually to microemulsion phases that contain equal amounts of oil and water and often coexist with both excess oil-rich and water-rich phases. Further increases in salinity lead to the formation of microemulsions containing inverted structures (droplets of water dispersed in oil), and these phases are usually found at equilibrium with excess water. Of particular interest are the intermediate samples that contain equal or nearly equal amounts of oil and water and are stabilized by surfactant films that are necessarily of very low, or zero, mean curvature [7].

In order to unravel the complex changes in the morphology of microemulsions with changes in composition, we have used the complementary techniques of small-angle neutron scattering (SANS), light scattering (LS), and quasielastic light scattering (QLS). SANS and LS probe the time-averaged size and shape and arrangement of oil and water domains in the microemulsions, while the QLS signal depends on fluctuations of the local dielectric constant of the sample with time. Thus, the QLS measurement contains important information about relaxation processes in the sample, as well as information about the size and shape of the oil and water domains [8]. The SANS measurements were taken with an instrument capable of resolving the scattering present at low values of scattering angle and is thus more accurate than the small-angle X-ray scattering data reported previously [3].

When the volume fraction of droplets is less than approximately 25%, the SANS and LS results reported in this Chapter are well represented by models of the microemulsion as a dispersion of droplets of either oil in water or water in oil interacting with coulombic or hard-sphere repulsions, respectively. For samples with compositions outside that range, the results suggest the presence of a bicontinuous structure. These observations are consistent with the results of QLS measurements when proper account is made of the interplay of hydrodynamic forces and interparticle thermodynamic interactions in modulating the fluctuations that produce the recorded relaxation times. Numerous theories for the interpretation of the QLS signal from concentrated colloidal solutions have been presented in the literature [9–14]. There is not, however, complete agreement among them, and they are applicable only under restricted conditions. We have used the theory of Felderhof [13], as extended by Fijnaut [14], to interpret the results of quasielastic light scattering from water in oil and oil in water microemulsions. This model, which is correct only to first order in the volume fraction of the droplets, allows arbitrary thermodynamic interaction potentials to be used. The hydrodynamic interactions are modeled in a low-density approximation, wherein only 2-body interactions are considered. The results calculated from this model fit our experimental data well, and more importantly, yield estimates of the dimensions of the microemulsion droplets that are in excellent agreement with those measured with SANS.

Section II describes in a heuristic way the physical quantities accessible for measurement with scattering techniques, while Section III contains experimental details. The principal results and their discussion are in Section IV, and they are organized into sections on oil-rich, water-rich, and midrange microemulsions.

II. Scattering Techniques

Light scattering, small-angle scattering, and recently, quasielastic light scattering or photon correlation spectroscopy have become standard methods for the study of colloidal suspensions in general and microemulsions in particular. SANS and LS measure the static or time-averaged structure of the material in a microemulsion, and the very different wavelengths of light (5000 Å) and thermal neutrons (4.5 Å) means that the length scales in the sample probed with the two radiations may be different. In contrast, in a QLS experiment, the time correlations of the scattered light are measured, so the signal depends directly on the motions of the scattering centers in the sample. Qualitatively, the dimensions probed by radiation of wavelength λ are inversely proportional to the magnitude of the scattering vector $\mathbf{q}$, which is

$$q = (4\pi n/\lambda) \sin(\Theta/2),$$

where n is the refractive index of the scattering medium and Θ is the scattering angle. Experimental limitations dictate that q usually varies from approximately $3 \times 10^{-3}\,\text{Å}^{-1}$ to $0.5\,\text{Å}^{-1}$ for SANS, and $5 \times 10^{-4}\,\text{Å}^{-1}$ to $3 \times 10^{-3}\,\text{Å}^{-1}$ for LS and QLS.

The quantitative methods used to treat the data when the microemulsion contains droplets are described in the following, with examination first of SANS and LS and then of QLS.

A. Small-Angle Neutron Scattering

Analysis of neutrons scattered by a sample yields information about the distribution of scattering centers (atoms) within the sample. The following treatment follows that of Ref. (15) and is phrased in terms of neutron scattering, but with appropriate substitutions, the expressions hold as well for light scattering. The methods used to understand the problems of interparticle interactions and polydispersity are stressed. When the scattering sample does not contain particles, as is the case for a bicontinuous microemulsion, a more general treatment of the problem is necessary [2].

In the absence of multiple scattering, the measured intensity $I(q)$ is proportional to the coherent differential scattering cross section, which for a scattering volume divided into N_p cells (each of which contains a single particle and solvent) is given by [15]

$$\frac{d\Sigma}{d\Omega}(\mathbf{q}) = \left\langle \sum_{N=1}^{N_p} \sum_{M=1}^{N_p} \exp[i\mathbf{q}\cdot(\mathbf{R}_N - \mathbf{R}_M)] \sum_{i=1}^{n(N)} \sum_{j=1}^{m(M)} b_{iN}\, b_{jM} \exp[i\mathbf{q}\cdot(\mathbf{r}_{iN} - \mathbf{r}_{jM})] \right\rangle,$$

$$(1)$$

where b_{iN} and b_{jM} are the averaged bound scattering lengths of atoms i in cell N and j in cell M; $\mathbf{r}_{iN}$ and $\mathbf{r}_{jM}$ are the locations of atoms i and j relative to the centers of mass of the cells N and M; $\mathbf{R}_N$ and $\mathbf{R}_M$ are the locations of the centers of mass for cells N and M relative to an arbitrary origin; and $n(N)$ and $m(M)$ are the number of atoms in cells N and M, respectively. The angular brackets indicate an ensemble average. It is convenient to recast this expression in terms of the scattering length densities, where the scattering length density in cell N at position $\mathbf{r}$ is

$$\rho_N(\mathbf{r}) = \sum_{i=1}^{n(N)} b_{iN}\, \delta(\mathbf{r} - \mathbf{r}_{iN}),$$

and r_{iN} is the position of atom i in cell N. Combining this definition with Eq. (1) yields

$$\frac{d\Sigma}{d\Omega}(\mathbf{q}) = \left\langle \sum_{N=1}^{N_p} \sum_{M=1}^{N_p} \exp[i\mathbf{q}\cdot(\mathbf{R}_N - \mathbf{R}_M)] \right.$$

$$\left. \times \int_{\text{cell }N} dr_1 \int_{\text{cell }M} dr_2 \, \rho_N(\mathbf{r}_1)\rho_M(\mathbf{r}_2) \exp[i\mathbf{q}\cdot(\mathbf{r}_1 - \mathbf{r}_2)] \right\rangle. \quad (2)$$

The observable small-angle scattering is due to the difference in scattering length densities between the particle and the solvent in cells N and M, while the scattering from the solvent will be observable only at $q = 0$. The contribution of the particle can be written in terms of the form factor $F_N(q)$ as

$$F_N(\mathbf{q}) = \int_{\text{particle }N} dr[\rho(\mathbf{r}) - \rho_s] \exp(i\mathbf{q}\cdot\mathbf{r}),$$

where ρ_s is the average scattering length density of the solvent. Introduction of this expression into Eq. (2) and rearranging yields, for $q \neq 0$,

$$\frac{d\Sigma}{d\Omega}(\mathbf{q}) = \sum_{N-1}^{N_p} \langle|F_N(\mathbf{q})|^2\rangle + \left\langle \sum_{N=1}^{N_p} \sum_{\substack{M\neq N \\ M=1}}^{N_p} \exp[i\mathbf{q}\cdot(\mathbf{R}_N - \mathbf{R}_M)]F_N(\mathbf{q})F_M(\mathbf{q}) \right\rangle. \quad (3)$$

Thus, the scattering from a collection of particles is made up of the sum of two terms: the first depending on intraparticle scattering and the second on interparticle scattering. The intraparticle scattering term is simply the average of the square of the particle form factor and can be calculated easily for particles of any geometry. This term depends upon both the shape of the particle and the distribution of atomic species within it. The interparticle term, however, can be evaluated in a closed form only if assumptions about the correlations between the spacing of the particles and their sizes and orientations are made. An important example is the case in which the sample contains a monodisperse population of spherical scatterers, the situation for which Eq. (3) can be simplified to the familiar form

$$\frac{d\Sigma}{d\Omega}(\mathbf{q}) = \langle|F(\mathbf{q})|^2\rangle\left\{1 + \left\langle \sum_{N=1}^{N_p} \sum_{\substack{M\neq N \\ M=1}}^{N_p} \exp[i\mathbf{q}\cdot(\mathbf{R}_N - \mathbf{R}_M)] \right\rangle\right\}, \quad (4)$$

where the quantity in braces is called the structure factor $S(q)$.

Clearly, the relative positions of particles N and M depend directly on the interparticle potentials, so we digress briefly to discuss the models of interparticle potentials used here. Two potentials for which the structure factors have been calculated are considered: a hard-sphere (HS) potential [16],

$$u(r) = \begin{cases} \infty & r < \sigma, \\ 0 & r > \sigma, \end{cases}$$

where σ is the particle diameter, and a repulsive coulomb (RC) potential [17,18],

$$
u(r) = \begin{cases} \infty & r < \sigma \\[2ex] \dfrac{\pi \Psi_0^2 \, \varepsilon_0 \varepsilon \sigma^2}{r} \exp[-\kappa(r - \sigma)] & r > \sigma, \end{cases}
$$

where Ψ_0 is the surface potential, κ is the inverse of the Debye screening length, ε_0 is the permitivity of free space, and ε is the solvent dielectric constant. The HS model represents the spheres as impenetrable and noninteracting beyond the hard-sphere radius. In the RC or macroion model, the impenetrable spheres carry a surface charge, and the counter ions and solvent are treated as a uniform, electrically neutralizing background [18].

The important influence of a polydispersed population of scatterers on the measured intensity can also be treated. A rigorous calculation of $I(q)$ for the HS potential acting between particles of various sizes is available [19]. For other potentials and particle shapes, a variety of decoupling approximations may be made in which the polydispersity and interaction problems are treated independently. For example, the scattering from a system of polydispersed spherical particles can be calculated with the decoupling assumption that there is no correlation between interparticle separation and particle size. This assumption allows simplification of the expression for the scattering cross section [Eq. (3)] to

$$
\frac{d\Sigma}{d\Omega}(\mathbf{q}) = \langle |F(\mathbf{q})|\rangle^2 + |\langle F(\mathbf{q})\rangle|^2 \left\langle \sum_{N=1}^{N_p} \sum_{\substack{M \neq N \\ M = 1}}^{N_p} \exp[i\mathbf{q} \cdot (\mathbf{R}_N - \mathbf{R}_M)]\right\rangle. \tag{5}
$$

It is important to note that, for a system of polydispersed spheres, application of the decoupling assumption results in an expression for $I(q)$ that departs significantly at low q from the $I(q)$ calculated with the rigorous solution. However, in the limit of small polydispersity or in a dilute system, this decoupling assumption should be valid. It has been argued [20] that this decoupling approximation is a better approximation for the macroion model than for the HS model because the electrostatic repulsion incorporated in the former model prevents the close approach of charged particles, and therefore reduces the correlation of sphere size and separation that is present in the concentrated polydisperse HS system.

Finally, the model of the system of monodispersed, nonspherical interacting particles can be decoupled with assumption that there is no correlation in the separation between particles and their orientation. This assumption allows the expression for the scattering cross section, Eq. (3), to be simplified to the same form as Eq. (5). Again, this decoupling assumption will most likely be valid for dilute solutions, and for the reasons given above, the approximation should prove better for the RC potential than for the HS potential.

B. *Quasielastic Light Scattering*

In homodyne QLS experiments, the autocorrelation function of the scattered intensity $g^{(2)}(q, t)$ is observed directly. In the absence of multiple scattering, it is related to the normalized scattered electric field correlation function $g^{(1)}(q, t)$ via the Siegert relation:

$$|g^{(2)}(q, t)| = C_1[1 + \beta|g^{(1)}(q, t)|^2], \tag{6}$$

where C_1 and β are the baseline and an instrument constant, respectively [21]. When the sample contains particles or droplets, $g^{(1)}(q, t)$ is related to the dynamic structure factor $S(q, t)$ through [22]

$$S(q, t) = S(q, 0)|g^{(1)}(q, t)|. \tag{7}$$

In a system of N identical spherical particles, $S(q, t)$, which is essentially the correlation function of the qth spatial Fourier component of the particle number density, is defined by

$$S(q, t) = N^{-1}\left\langle \sum_{i=1}^{N} \sum_{j=1}^{N} e^{i\mathbf{q}\cdot\mathbf{r}_i(0)} e^{-i\mathbf{q}\cdot\mathbf{r}_j(t)} \right\rangle, \tag{8}$$

where $\mathbf{r}_i(t)$ is the position of the center of particle i at time t. $S(q, t)$ is the dynamic analogy of $S(q)$ introduced in Eq. (4).

From Eqs. (6)–(8), it is clear that by beginning with an equation describing fluctuations in particle number densities $S(q, t)$, and in turn $g^{(2)}(q, t)$, may be determined in terms of the mutual interactions of the particles. An equation describing such fluctuations has been derived [13,14] under the assumption that in an interacting colloidal system the Brownian motion of the particles can be described by an N-particle Smoluchowski equation for the probability distribution $P_N(r_1, \ldots, r_N, t)$ of particles in configuration space. Considering only 2-body interactions yields a relation between the measured q-dependent D_{app} and D_0 [14], namely,

$$D_{app}(q) = D_0\{1 + \phi_H[S_1(q) + H_1 + H_2(q)]\}, \tag{9}$$

where $S_1(q)$ and $H_1 + H_2(q)$ are complicated functions of the potential of interaction and respectively account for the thermodynamic and hydrodynamic interactions. D_0 is equal to $k_B T/6\pi\eta R_H$, where R_H is the hydrodynamic radius of the particle, and η is the solvent viscosity. The hydrodynamic volume fraction ϕ_H is $(4\pi/3)\bar{\rho}R_H^3$, where $\bar{\rho}$ is the bulk number density of droplets in the solution.

III. Methods and Materials

The microemulsions studied here are formed with SHBS, iso-butyl alcohol (reagent grade, Matheson), dodecane (99% pure, Sigma), and NaCl brines containing different ratios of H_2O (double-distilled, deionized) and D_2O

(99.9% D, Norell). The SHBS was purchased from the University of Texas and was recrystallized three times from mixtures of heptane and acetone before use. The preparation of microemulsion samples is described elsewhere [23].

Small-angle neutron scattering experiments were performed with the 30-meter SANS camera at the National Center for Small-Angle Scattering Research at Oak Ridge National Laboratory. The neutron wavelength was 4.75 Å with $d\lambda/\lambda$ of 6%. Scattered intensities were measured for values of q from 0.005 to 0.025 Å^{-1}. The samples were held in 1 mm quartz cuvettes at 25°C. The data were corrected for detector sensitivity and background counts, and an absolute intensity determination was made using a calibrated aluminum standard.   The electrical conductivity and light scattering measurements are described elsewhere [23,24]. The intensity autocorrelation function $g^{(2)}(q, t)$ was analyzed using the method of cumulants [25], in which the logarithm of the normalized $g^{(2)}(q, t)$ was fitted to a polynomial equation using a nonlinear least-squares fitting routine. The fitted equation is

$$\ln|g^{(2)}(q, t)|_{\text{normalized}} = \Gamma_0 - \Gamma_1 t + \Gamma_2(t^2/2!), \qquad (10)$$

where Γ_0 is ideally zero. The first cumulant, Γ_1, is equal to $2q^2 D_{\text{app}}(q)$. In the absence of interparticle interactions, Γ_2, the second cumulant, is related to the variance of the particle size distribution.

IV. Results and Discussion

A. Oil-Rich Microemulsions

When the volume fraction of water in a microemulsion is low, often [2–5,23], but apparently not always [26], the microemulsion contains swollen inverted micelles or droplets of water dispersed in oil. In Winsor microemulsions, such as the ones studied here, these structures occur when the electrostatic repulsions between head groups are diminished by the addition of sufficient electrolyte to the aqueous component. In this case, the important forces controlling the interactions between the droplets are basically only hard-sphere repulsions [27] due to the impermeability of aqueous cores and van der Waals attractive forces between aqueous cores. In some cases, however, there seem to be additional short-range attractive forces arising from the mutual solubility of the surfactant tails.

Several theories aimed at predicting the shape and size or size distribution of microemulsion droplets have been put forth [28,29]. Important results that the theories must predict, as illustrated in this chapter and elsewhere, is that w/o droplets are spherical and the population has a significant polydispersity. A further complication is that as the volume fraction of water increases, there is evidence, particularly from electrical conductivity data, that droplets begin to merge and fuse into a bicontinuous network [2–4]. The structure of these bicontinuous phases and their relation to the droplet structures from which they evolve are discussed in the following.

Small-Angle Neutron Scattering

We have modeled the water in oil microemulsions in this system as dispersions of hard spherical particles for ϕ_w ranging from 0 to 0.35. In fitting the models to data, we found that the results are not sensitive to the details of the particle form factor. Thus, no account was made of any excess scattering from the surfactant-rich film, and the scattering length density in the droplet was assumed to be uniform. The fits of the models to the data with the assumptions of both a monodisperse population of droplets and a polydisperse one are shown in Fig. 15.1.

The monodispersed hard-sphere model provides a good fit of the experimental data for aqueous volume fractions less than 0.15. The hard-sphere structure factor was calculated in the Percus–Yevick approximation [30]. This model has two parameters: the volume fraction of the dispersed phase and the radius of the spherical particles. The volume fraction of the hard spheres was set equal to the measured water volume fraction, and the radius was determined by a least-squares fit to the data.

Fitting the data to the rigorous polydispersed hard-sphere model of van Beurten and Vrij [19], which is also calculated in the Percus–Yevick approximation, extends the range of agreement between model and SANS data to water volume fractions as high as 0.35 (Fig. 15.1). There are three parameters in this model; the volume fraction of particles, the average particle radius R, and the standard deviation μ of the (assumed) Gaussian distribution of particle radii. The results of this model were fit to the experimental results by again setting the volume fraction of the particles equal to the experimentally determined water volume in the microemulsion phase and then adjusting the average particle radius and the standard deviation of the distribution of particle radii until a visual best fit was obtained. Table 15.1 summarizes the results of the hard-sphere models for ϕ_w ranging from 0.1 to nearly 0.35. Both the droplet radii and their polydispersity grow with increasing ϕ_w. Note that the fit of the model to the intensity of scattered light at low q (filled symbols in Fig. 15.1) is not good for samples with higher ϕ_w. This failure is discussed further in Section IV.C, and the light scattering data were not used in the fitting procedure.

The other models discussed previously, namely the decoupled polydispersed hard-sphere model and a model of the particles as monodispersed hard ellipsoids, fit the experimental data no better than the monodispersed hard-sphere model, that is, they fail above water volume fractions of 0.15.

Quasielastic Light Scattering

Plots of the logarithm of the normalized intensity autocorrelation functions measured at a 90° scattering angle from microemulsions at various water

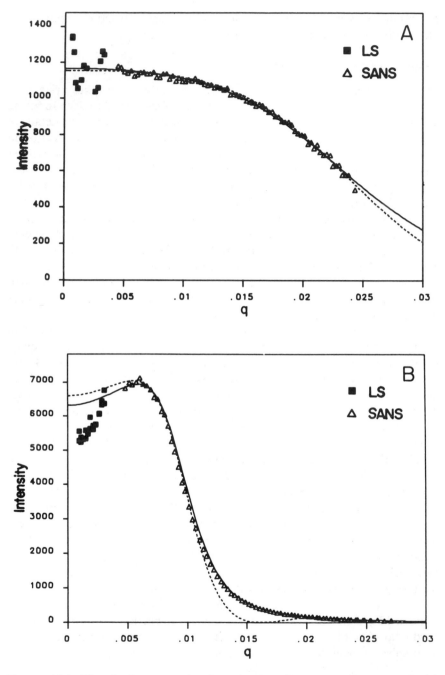

FIGURE 15.1. The absolute scattering intensity (cm^{-1}) versus q as measured with SANS (open triangles) from water in oil samples. The dashed curves in (A) and (B) are fits of the hard-sphere model assuming a monodisperse population, and the solid curves are fits of the hard-sphere model including a polydisperse population. The

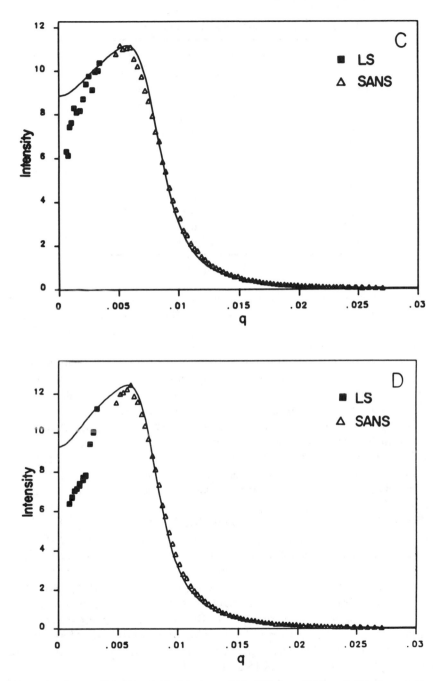

compositions are (A) $\phi_w = 0.09$, (B) $\phi_w = 0.19$, (C) $\phi_w = 0.33$, and (D) $\phi_w = 0.41$; q has units of Å^{-1}, and the absolute intensities in (C) and (D) are ×1000. The closed squares at lower q values are measured with light scattering.

TABLE 15.1. Dimensions of water in oil droplets.

Salinity (g NaCl/100 ml water)	ϕ_w	R (Å)	μ (Å)
Monodisperse population			
1.4	0.09	85.4	
1.2	0.10	95.0	
1.0	0.11	107	
0.9	0.13	120	
0.8	0.15	144	
Polydisperse population			
0.8	0.15	120	31
0.68	0.19	120	68
0.64	0.22	150	80
0.63	0.24	185	95
0.57	0.33	223	131

compositions are linear [23] when the water volume fraction is less than 15%. As the water composition increases, the second cumulant deviates increasingly from zero and the plot of $\ln g^{(2)}(q, t)$ becomes curved. This is probably a result of both an increase in polydispersity of particle sizes and an increase in the strength of interactions between particles [10]. To model these interactions, we make use of the theory of Felderhof and Fijnaut outlined in Section III. As in the analysis of the SANS data, the inverted swollen micelles are approximated as uniformly scattering hard spheres.

This theory has been used to calculate a q-dependent apparent diffusion coefficient for comparison with experiment [23]. The fitted parameters are the hard-sphere radius R_{hs} and the hydrodynamic radius R_H. The results are that R_H grows nearly linearly as the water volume fraction increases (Fig. 15.2) and that the radius of the inverted swollen micelles (or droplets) grows from 60 Å to 300 Å.

The hydrodynamic volume fraction of droplets ϕ_H used in Eq. (9) includes the volume fractions of water ϕ_w, of surfactant ϕ_s, and of alcohol ϕ_a that are not in the continuous phase. In a previous study of the same microemulsion system [2], it was found, using gas chromatography and hyamine titration, that the microemulsion phase contains approximately 4 wt.% iBA at all salinities and that a negligible amount of surfactant is solubilized in any nonmicroemulsion phase. Here, we make the further assumption that the iBA is equally distributed between the droplet and continuous phase in the microemulsion in estimating ϕ_H. Therefore, $\phi_H = \phi_w + \phi_s + 0.5\,\phi_a$. For these samples, we take $\phi_s = 0.015$ and $\phi_a = 0.03$.

In matching the predictions of the theory to data, the parameters R_{hs} and R_H are not fitted independently, but are related by a volume fraction constraint. Thus, in this model, R_H includes the thickness of the surfactant-rich and alcohol-rich boundary of the inverted swollen micelle, while R_{hs} is

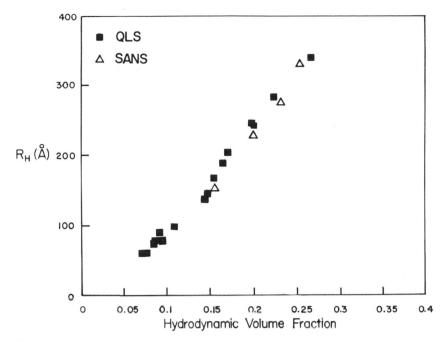

FIGURE 15.2. The hydrodynamic radius measured with QLS as a function of the volume fraction of droplets in water in oil microemulsions. The triangles are calculated from the SANS data, including the effect of polydispersity.

approximately a measure of the dimension of the water-rich core alone. Based on the relative volume fractions of water and of surfactant and alcohol in the droplet, R_H is set as R_{hs} plus 10 Å. For droplets of the size measured here, this is a minor effect.

The hard-sphere model used here is very simple, and the predictions for droplets interacting with a triangular potential [14] and with an attractive square-well potential [31] have also been calculated. The results do not differ significantly from those of the hard-sphere model, however, because the range of interaction is far less than the radius of the microemulsion droplets.

The droplet sizes measured with the two methods can now be compared. Note first that the QLS measurements yield the z-averaged diffusion coefficient, which can be used to calculate an average hydrodynamic radius R_H through the Stokes–Einstein equation, while fitting the SANS results yields the number-averaged radius $\langle R \rangle_N$ and the standard deviation μ in the assumed Gaussian size distribution. Comparison of these averages is straightforward. For a Gaussian distribution,

$$\frac{kT}{6\pi\eta\langle D\rangle_z} = \langle R\rangle_N \frac{[1 + 10\delta_N + 15\delta_N^2]}{[1 + 6\delta_N + 3\delta_N^2]}(1 + \delta_N),$$

where $\delta_N = \mu^2/\langle R \rangle_N^2$, and the left-hand side of the equation is the average radius measured with QLS. The agreement between the QLS measurement and results calculated from the SANS data (Table 15.1) with this equation is good (Fig. 15.2) and suggests that the corrections for hydrodynamic interactions and the analysis of polydispersity are correct.

The roughly linear growth of the droplets with increasing water volume fraction is in accord with a particularly simple model. This model assumes that there is a constant total interfacial area between water and oil in the microemulsion, since the total amount of surfactant in the microemulsion phase is fixed. The interfacial area is $3\phi_H V_{me}/R_H$, where V_{me} is the volume of the microemulsion phase. V_{me} remains roughly constant as the salinity changes [23], and thus R_H is predicted to rise in proportion to ϕ_H, as observed in Fig. 15.2. Application of this simple model indicates that the area per surfactant head group is approximately 80 Å^2, with no account made for the area occupied by alcohol.

Using the droplet dimensions measured with SANS, a more detailed calculation of surfactant head-group area can be made by taking into account the polydispersity in droplet size distribution and the small changes measured in the volume of the microemulsion phase as ϕ_w increases. The results are presented in Table 15.2. The average value of the head-group area calculated this way is roughly 80 Å^2, and it is also reassuring that the area per surfactant head group increases as the concentration of screening electrolyte decreases.

B. Water-Rich Microemulsions

When small amounts of oil are solubilized into a microemulsion, the swollen micelles are thought to be roughly spherical and, for microemulsions containing ionic surfactant, to interact strongly through screened coulombic interactions. These considerations have been used to formulate the models used to represent the scattering from oil in water microemulsions.

TABLE 15.2. Surfactant head-group areas in water in oil microemulsions.

Salinity (g NaCl/100 ml water)	ϕ_w	Head-group area (Å^2)
0.8	0.15	73
0.68	0.19	77
0.66	0.22	78
0.63	0.24	76
0.57	0.33	86
0.555	0.41	90

Small-Angle Neutron Scattering

We have modeled oil in water microemulsions as dispersions of charged prolate ellipsoids interacting through screened coulomb potentials. As in the hard-sphere models, this model neglects the distribution of atomic centers within the particles and so treats the scattering length density of ellipsoids as homogeneous. The structure factor was calculated using the solution of the Ornstein–Zernike equation obtained in the mean spherical approximation by Hansen and Hayter for the screened coulomb potential [32]. Comparisons of the scattering curves predicted using the charged ellipsoid model and those measured are presented in Fig. 15.3. Charged sphere models, with or without polydispersity, fail to fit the experimental data for q greater than approximately 0.015 Å^{-1}. The aspect ratio of the ellipsoid was found to be constant at 3:1, and the dimensions of the droplets are listed in Table 15.3.

The failure of the spherical macroion model including polydispersity probably indicates that because the surface charges of the particles are highly screened the coulombic repulsions are not great enough to preclude the correlation of neighboring particles. This is consistent with the small Debye screening length (from 10 to 15 Å) present in these microemulsions. However,

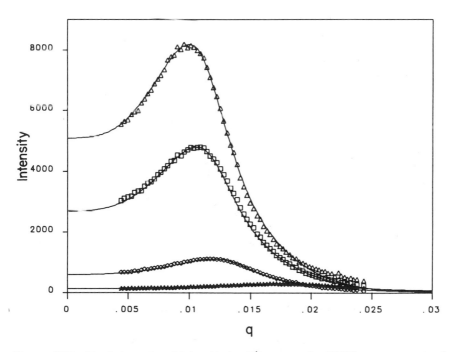

FIGURE 15.3. Absolute scattered intensity (cm^{-1}) versus q for SANS measurements of oil in water microemulsions. The solid curves are fits of the macroion model discussed in the text. From bottom to top, the data are from samples containing $\phi_{oil} = 0.15$, 0.16, 0.20, and 0.22, respectively; q has units of Å^{-1}.

TABLE 15.3. Dimensions and head-group areas for oil in water droplets.

Salinity (g NaCl/100 ml water)	ϕ_{oil}	a (Å)	b (Å)	Head-group area (Å²)
0.3	0.15	260	87	120
0.35	0.16	363	121	91
0.4	0.20	422	141	98

the coulombic repulsions are certainly not negligible, since all the hard-particle models fail to fit the data in this range.

Quasielastic Light Scattering

Use of Eq. (9) allows numerical calculation of $D_{app}(q)$. The hydrodynamic volume fraction is taken as $\phi_H = 1 - \phi_w = 0.5\,\phi_a$. The theoretical and experimental results for $D(q)$ are again in agreement [23], and the radius rises in proportion to ϕ_H (Fig. 15.4). Note that in this microemulsion system it is not possible to reach lower volume fractions of dispersed phase by simply

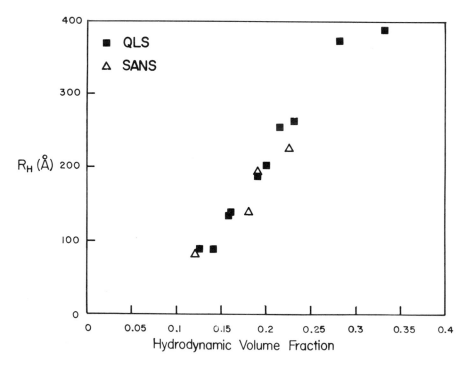

FIGURE 15.4. The hydrodynamic radius measured with QLS plotted as a function of volume fraction of droplets in oil in water microemulsions. The triangles are the calculated values from the SANS data.

varying the brine salinity, for even in the sample without added salt, the microemulsion phase contains only 90 % water.

Again, we can compare the results of the SANS and QLS measurements. The SANS data indicate that the swollen micelles are prolate ellipsoids. The equivalent hydrodynamic radius of a prolate ellipsoid is calculated using the equation [33]

$$R_H = \frac{b[(a/b)^2 - 1]^{1/2}}{\ln\{(a/b) + [(a/b)^2 - 1]^{1/2}\}}, \tag{11}$$

for the ratio of the ellipsoid axes $a/b > 1$. As in the water in oil case, the agreement between the two measurements is good (Fig. 15.4).

The areas occupied by the surfactant molecules in the oil in water system are listed in Table 15.3. Note that the salinity is lower and the head group areas are larger than those found in the water in oil microemulsions (Table 15.2). In water in oil microemulsions, the surfactant head groups are buried inside the droplets, while in oil in water microemulsions, the surfactant head groups are facing out. Elementary consideration of the shape and packing arrangements of SHBS molecules in a curved monolayer would suggest that the head groups on the surface of an oil in water droplet would be further apart than those on the surface of a water in oil droplet, and such is the case. The partitioning of alcohol into the surfactant-rich film, which is poorly understood and is not accounted for in the above calculation, also must play a role in determining the surfactant head-group area.

C. Midrange Microemulsions

For intermediate values of the salinity, these microemulsions contain comparable amounts of oil and water. The microemulsions evolve continuously from water-rich ones as salinity rises to oil-rich ones as salinity further increases. Current theories hold that the oil and water in these systems are organized into bicontinuous structures.

Theoretical models of bicontinuous microemulsion structures [34,35] show that a characteristic thickness of a phase domain, ξ, is given by

$$\xi \sim \frac{\phi_{oil} \, \phi_{water}}{C_S \Sigma}, \tag{12}$$

where C_s is the surfactant concentration, and Σ is the area of each surfactant head group. The constant of proportionality is 5.82 for the Talmon–Prager Voronoi model [34], and it is 6 for the de Gennes–Taupin model [35]. In the dilute microemulsions discussed previously, the droplet radius varies linearly as ϕ_H changes, which implies that the total interfacial area in the microemulsion phase remains roughly constant. Since the surfactant concentration is also nearly constant in each microemulsion, when the total composition in the microemulsion phase changes, the product $C_s\Sigma$ remains approximately a constant. With these assumptions, $\xi \sim \phi_{oil} \, \phi_w$.

Small-Angle Neutron Scattering

The measured SANS patterns begin to deviate from the predictions of the models used above for microemulsion droplets as the volume fraction of water increases toward 0.3 (Fig. 15.1). The trends are most conveniently discussed for the evolution from the water in oil microemulsions.

The disagreement between the hard-sphere model for water in oil micro-emulsions and the experimental results is most obvious at low values of q and is even more apparent when the intensity of scattered light (in arbitrary units) is presented on the same scale (Fig. 15.1). Since the intensity of the scattered light is artificially set at the value predicted by the hard-sphere theory for water in oil microemulsions, it falls much more rapidly than predicted as q decreases. In particular, $S(q = 0)$, which is proportional to the isothermal (osmotic) compressibility of the microemulsion, is much lower than predicted by the model. This is significant because a reason for the failure of the HS model in concentrated dispersions could be the growing importance of attractive interactions between droplets as their concentration increases. Such attractions would, however, make the dispersion more compressible and so raise $S(q = 0)$ above the expected hard-sphere value, rather than depress it as is observed.

We can extract a crude numerical estimate of the correlation length in the microemulsion from the SANS data by assuming that the value of q of the maximum intensity (q_{max}) is the signature of a preferred length in the sample. Thus, we expect $\xi \sim 2\pi/q_{max}$. Plotting $1/q_{max}$ as a function of composition (Fig. 15.5) shows that it follows precisely the same trend with composition as D_{app}^{-1}. In the region in Fig. 15.5 where ϕ_w lies between 0.25 and 0.7, the value of $2\pi/q_{max}$ is approximately 1200 Å.

Quasielastic Light Scattering

The apparent diffusion coefficients, measured at 90° from samples contained in 2 mm cells, have been recorded at various water compositions (Fig. 15.5). It is provocative to note the symmetry between o/w and w/o microemulsions. From the overall results, it is apparent that the radius of swollen micelles, or inverted swollen micelles, increases from approximately 60 Å to 300 Å as the upper or lower phase microemulsions approach middle phase microemulsions. In the intermediate range, where water composition in the microemulsion phase varies from 25% to 70%, the apparent diffusion coefficient remains nearly constant. Under the assumption that D_{app}^{-1} is proportional to ξ, the data imply that the characteristic size of phase domains varies only slightly in this range, which is qualitatively consistent with the theory of bicontinuous structures. The existence of water-continuous paths in the microemulsion is also consistent with the electrical conducitivity data [23,36].

It is interesting to compare the magnitude of the length scales that correspond to $2\pi/q_{max}$ and to the diffusion coefficient. The first dimension is

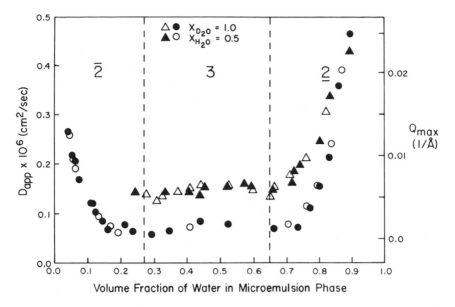

FIGURE 15.5. The apparent diffusion coefficient (measured at $\theta = 90°$, circles) and $1/q_{max}$ (triangles) plotted as a function of the water volume fraction.

about 1200 Å, but inserting the measured value of the diffusion coefficient into Stokes' law yields a value of R_H of ca. 400 Å. The smooth evolution of D_{app} and $1/q_{max}$ from and into the regions where droplets exist as the amount of oil increases is clear. The absolute value of the curvature of the surfactant film decreases as the compositions approach those in the plateau, and therefore the intermediate region must contain structures with curvature near zero.

V. Conclusions

The results of both SANS and QLS measurements consistently indicate that there are three types of microemulsion structures present in this five-component system. When the volume fraction of water is less than about 0.20, the neutron scattering from these water in oil systems is well represented by models of the dispersion as a population of polydisperse hard spheres. The surfactant head group area in these samples increases as the salinity decreases. The droplets grow as their volume fraction increases. In the other extreme, examination of oil in water microemulsions shows that the droplets therein are not spherical, but instead are prolate with an aspect ratio of 3:1. The droplets in those samples interact through screened coulombic forces and also grow as the volume fraction of oil increases.

When the microemulsions contain comparable amounts of oil and water, neither of the detailed models for oil in water or water in oil microemulsions fits the data. In particular, the osmotic compressibility of the samples is lower than predicted by the hard-sphere model. Qualitatively, the length scales corresponding to the maximum in SANS intensity and the inverse of the diffusion coefficient vary roughly as the product $\phi_w\phi_{oil}$, which is the prediction of the models of bicontinuous structures. This view is also consistent with the measured electrical conductivities reported elsewhere [23,36].

In conclusion, we observe that the structures present when these microemulsions contain small amounts of either oil or water are fairly well represented by conventional models. The challenge for both experimentalists and theorists is to understand the nature of the bicontinuous structures into which the droplets evolve.

Acknowledgments. This research was supported by the National Science Foundation under grants CPE-8307188 and CBT-8351179 and by a Faculty Career Initiation grant from the Shell Companies Foundation. We are grateful to G.D. Wignall and J.S. Johnson of the NCSASR and Oak Ridge National Laboratory for their help with the SANS experiments and to J.B. Hayter for a copy of the computer program used to calculate the macroion structure factor. A portion of this research was performed at the National Center for Small-Angle Scattering Research. The NCSASR is funded by National Science Foundation Grant No. DMR-77-244-59 through Interagency Agreement No. 40-636-77 with the U.S. Department of Energy under contract number DE-AL05-840R21400 with Martin Marietta Energy Systems, Inc.

References

1. P.A. Winsor, "*Solvent Properties of Amphiphilic Compounds,*" Butterworths, London, 1954.
2. E.W. Kaler, K.E. Bennett, H.T. Davis, and L.E. Scriven, *J. Chem. Phys.* **79**, 5673 (1983).
3. E.W. Kaler, H.T. Davis, and L.E. Scriven, *J. Chem. Phys.* **79**, 5685 (1983).
4. L. Auvray, J.-P. Cotton, R. Ober, and C. Taupin, *J. Physique* **45**, 913 (1984).
5. A.M. Cazabat and D. Langevin, *J. Chem. Phys.* **74**, 3148 (1981).
6. M. Zulauf and H.F. Eicke, *J. Phys. Chem.* **83**, 480 (1979).
7. L.E. Scriven, *Nature* **263**, 123 (1976), and in "*Micellization, Solubilization, and Microemulsions,*" p. 877, K. Mittal, ed., Plenum, New York, 1977.
8. B. Chu in "*Laser Light Scattering,*" Academic Press, San Diego, California, 1974.
9. G.K. Batchelor, *J. Fluid Mech.* **74**, (1976).
10. B.J. Ackerson, *J. Chem. Phys.* **64**, 242 (1976).
11. A.R. Altenberger and J.M. Deutch, *J. Chem. Phys.* **59**, 894 (1973).

12. S. Hanna, W. Hess, and R. Klein, *Physica* **111A**, 181 (1982).

13. B.U. Felderhof, *J. Phys. A: Math. Gen.* **11**, 929 (1978).

14. H.M. Fijnaut, *J. Chem. Phys.* **74**, 6857 (1981).

15. M. Kotlarchyk and S.-H. Chen, *J. Chem. Phys.* **79**, 2461 (1983).

16. L. Blum and G. Stell, *J. Chem. Phys.* **71**, 1 (1979).

17. E.J. Verwey and J.Th.F. Overbeek, in "*Theory of the Stability of Lyophobic Colloids*," Elsevier, Amsterdam, 1948.

18. J.B. Hayter and J. Penfold, *Molec. Phys.* **42**, 109 (1981).

19. P. van Beurten and A. Vrij, *J. Chem. Phys.* **74**, 2744 (1981).

20. J.B. Hayter and J. Penfold, *Colloid & Polymer Sci.* **261**, 1022 (1983).

21. E. Jackman, in "*Photon Correlation and Light Beating Spectroscopy*," H.Z. Cummins and E.R. Pike, eds., Plenum, New York, 1974.

22. P.N. Pusey, *J. Phys. A: Math. Gen.* **8**, 1433 (1975).

23. N.J. Chang and E.W. Kaler, *Langmuir* **2**, 184 (1986).

24. N.J. Chang and E.W. Kaler, *J. Phys. Chem.* **89**, 2996 (1985).

25. D.E. Koppel, *J. Chem. Phys.* **57**, 4814 (1972).

26. S.J. Chen, D.F. Evans, and B.W. Ninham, *J. Phys. Chem.* **88**, 1631 (1984).

27. D.J. Cebula, R.H. Ottewill, J. Ralston, and P.N. Pusey, *J. Chem. Soc. Faraday Trans. I* **77**, 2585 (1981).

28. E. Ruckenstein and R. Krishan, *J. Colloid and Interface Sci.* **75**, 476 (1980).

29. S.A. Safran, this volume.

30. N.W. Ashcroft and J. Lekner, *Phys. Rev.* **145**, 83 (1966).

31. R.V. Sharma and K.C. Sharma, *Physica* **89A**, 213 (1977).

32. J.P. Hansen and J.B. Hayter, *Molec. Phys.* **46**, 651 (1982).

33. F. Perrin, *J. Phys. Rad.* **7**, 1 (1936).

34. Y. Talmon and S. Prager, *J. Chem. Phys.* **69**, 2984 (1978).

35. P.G. de Gennes and C. Taupin, *J. Phys. Chem.* **86**, 2294 (1982).

36. K.E. Bennett, J.C. Hatfield, H.T. Davis, C.W. Macosko, and L.E. Scriven, in "*Microemulsions*," I.D. Robb, ed., Plenum, New York, 1982.

16
Transport Properties of Microemulsions

M.W. KIM and W.D. DOZIER

A study of transport properties of microemulsions is presented. The measurements of self-diffusion, mutual diffusivity, and electric conductivity were performed on oil continuous microemulsions as a function of the internal phase volume fraction (water) with the fixed molar ratio of water to surfactant. The properties of these microemulsions in dilute and concentrated regimes are discussed with a theoretical model.

I. Introduction

A three-component oil-continuous microemulsion containing water, oil, and AOT (sodium 2-di-ethylhexyl sulfosuccinate) has been studied extensively for its static structure [1-3] at low and high internal phase concentration regimes. It has been found that this thermodynamically stable, clear microemulsion consists of water droplets covered with surfactants in an oil-continuous medium. An average droplet radius [4] is in the range of 40 to 60 Å, depending on the water to AOT molar ratio. The average droplet size and the size distribution function of droplets remain constant for dispersed phase (surfactant and water) volume fractions between 0.5 and 50%, regardless of temperature variations near room temperature (20 ~ 45° C), as long as the relative amounts of water and surfactant are kept constant. Furthermore, the static properties of this microemulsion system have been successfully described by a hard-sphere liquid with a short-range square-well attractive interaction [5,6]. Thus, this system is a well-characterized system to study the transport behavior of microemulsion systems in low and high concentrations of dispersed droplets. Since the interaction between the droplets is known from the static measurements, it will be interesting to see the effect of this attractive interaction on the dynamic properties of microemulsion droplets. For this purpose, the self-diffusion and mutual diffusion constant have been measured in a dilute concentration regime. The experimental results will be discussed with regard to theoretical [7-10] predictions. Furthermore, the electrical conductivity has also been measured in dilute and dense concentration regimes. Since the internal aqueous phase is the only electrical conductor, the external oil phase being a insulator, a possible conduction mechanism is proposed in the dilute concentration regimes. The conductivity shows a sharp increase at $\phi = 16\%$ as the dispersed phase volume is increased. This may be an indication of the formation of a bicontinuous phase [11] or the formation of a percolation cluster [12] by the water droplets. A water soluble dye [13] was used to distinguish these two processes by measuring the self-diffusion constant [14] near the threshold volume concentration. Since the dye molecules are trapped inside of the water droplets, the dyed self-diffusion constant will directly provide the droplet self-diffusion constant in a dilute concentration. However, the dye molecule will be free from the confined water droplet if a bicontinuous phase develops, and the dye molecule will diffuse much more rapidly through the continuous water channel. It is the purpose of this study to discuss these transport properties of microemulsions in dilute and dense concentration regimes of the dispersed phase.

This chapter is organized in the following manner: Section II is an experimental section. Section III is devoted to experimental results and a discussion. Conclusions are given in Section IV.

II. Experimental

A. Sample Preparation

A purified aerosol-OT surfactant, spectral grade decane, and double-distilled water were used in this study. AOT microemulsions used in this study contained a 3/5 weight ratio of AOT surfactant to water in various volumes of decane. The stock solution for the dilute regime study was a 4 vol % dispersed component (water plus surfactant), and other samples were made by diluting this stock with decane containing 200 ppm AOT [15]. The stock solution for the dense regime study was either a 32 vol % or 60 vol % dispersed component.

B. Mutual Diffusion Constant

Dynamic light scattering [16] was done using an He–Ne laser ($\lambda = 6328$ Å) with a Malvern goniometer and correlator (M3000) at a scattering angle of 45°. The sample was temperature controlled by a water bath at a temperature of 25°C within $\pm$ 10 mK.

The mutual diffusion constant $D(\phi)$ at a volume fraction ϕ was obtained by the cumulant analysis of the intensity autocorrelation function:

$$C(\tau) = \langle I(t)I(t + \tau)\rangle \sim Ae^{-2Dq^2\tau} + B. \tag{1}$$

C. Self-Diffusion Constant

The samples for this measurement contained 0.3 mg/cm^3 congo red in an aqueous solution. Congo red is water soluble and is insoluble in decane. This was checked by adding decane to a solution of congo red in water. The two liquids immediately separated, and there was no visible dye in the decane. Also, the interface exhibited the normal miniscus one would expect between water and decane. Congo red is a dye of the azobenzene family [13], which exhibits photochromism due to the cis–trans isomerization about the nitrogen double bond.

The self-diffusion constant was measured via forced Rayleigh scattering [14,17]. A laser beam was split and recombined in the sample, giving a pattern of interference fringes. An argon-ion laser operating at 514.5 nm and a power of 100–400 mW was used. This sets up a spatially periodic density of excited dye molecules. A second laser (He–Ne) can then be Bragg scattered from what is effectively a diffraction grating. Since the first laser beam is only flashed on the sample a short time (50–250 ms), the pattern begins to fade as the dye molecules diffuse and the density of excited molecules smears out. The diffracted spot from the second laser decays in intensity as this occurs, with the following time dependence:

$$I(t) \propto (e^{-Dq^2t} + \alpha)^2 + \beta^2, \tag{2}$$

where α is the noise mixing coherently with the signal, and β is the background that mixes incoherently. This technique gives a direct measurement of the self-diffusion constant of the dye. The idea in this study is to use a dye that will stay in the water and then observe the diffusion of the dye. As long as the water is contained in droplets, the dye will be trapped inside, and we will see the diffusion of the droplets by watching the dye. This method is reminiscent of the typical way of measuring self-diffusion in solids by a radioactive tracer. It has been used previously to study polymers [14,17], oil in water microemulsions [18], and charged polystyrene spheres [19] in a colloidal liquid.

D. Electric Conductivity

Conductivity measurements were carried out with a Hewlitt–Packard LRC meter (model 4274A) in an electrode cell with a cell constant of 1.0 cm^{-1}. Temperature during the measurements were regulated to ± 0.1 K in a range (15–40°C) for the various dispersed phase volumes from 2 vol% up to 60 vol%. In a dilute concentration, the conductivity can be expressed as

$$\sigma = \frac{N(Ze)^2}{6\pi R\eta}, \tag{3}$$

where N is the charge carrier density, e is the electron charge, Z is a number of changes per carrier, η is a medium viscosity (solvent), and R is the radius of the charge carrier. The temperature resistance coefficient

$$\alpha = \left(\frac{1}{R}\right)\left(\frac{dR}{dT}\right)$$

was estimated. The temperature resistance coefficient exhibits a maximum at a certain temperature, which marks the position of the percolation threshold. The dispersed volume dependence shows very similar behavior, which can be expressed as $\sigma \sim |\phi - \phi_c|^{-t}$ or $\sigma \sim |\phi - \phi_c|^{-s}$, depending on the nature of the percolation [20–22].

III. Results and Discussion

A. Mutual Diffusion Constant $D(\phi)$

The mutual diffusion coefficient [9] $D(\phi)$ at a volume fraction ϕ is given by

$$D(\phi) = \frac{k_B T}{S(\phi)f(\phi)}, \tag{4}$$

where $f(\phi)$ denotes the mutual friction coefficient, and $S(\phi)$ is the long wavelength limit of the static structure factor. Since we are interested here only in the linear regime in ϕ, we expand $f(\phi) = f_0(1 + k_f\phi)$ and $S(\phi) = 1 + K_s\phi$ and obtain

$$D = D_0[1 - (k_f + K_s)\phi] = D_0[1 + k_d\phi], \tag{5}$$

where D_0 is the limit of $D(\phi)$ as $\phi \to 0$, and k_d is the second virial coefficient of $D(\phi)$ and $S(\phi)$, respectively.

In water in oil microemulsions of AOT, water and decane, it has been shown that the interparticle interaction is short-range and attractive and can be represented by a model potential of a hard core plus an attractive square well. This attraction is said to be due to the partial overlap of the tails of the AOT surfactant. For such a potential, it has been shown that k_d [10] is

$$k_d = 1.45 + [e^\varepsilon - 1]\left[-8\lambda^3 - 18\lambda^2 - 12\lambda \right.$$
$$\left. + \frac{15}{8(1 + \lambda)} - \frac{9}{64(1 + \lambda)^3} - \frac{75}{256(1 + \lambda)^4} - \frac{369}{256}\right]. \tag{6}$$

The dynamic light scattering was done using an He–Ne laser and Malvern goniometer at a scattering angle of 45°, the sample being immersed in a water bath at a temperature of 25°C. At this angle, the scattering wavevector $\mathbf{q}$ has a magnitude of 7.6×10^4 cm^{-1}. The number density of droplets in the samples

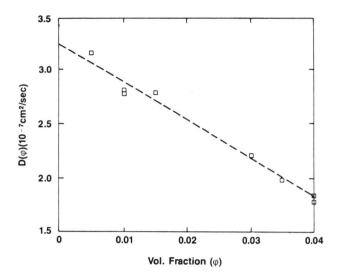

FIGURE 16.1. Plot of the mutual diffusion constant $D(\phi)$ as a function of the volume fraction (ϕ) of minor components (water and surfactant).

studied varied from 10^{16} to 10^{17} cm^{-3}, giving a typical q_{max} for the first peak in the structure factor of about 10^8 cm^{-1}, so the $q = 0$ approximation is justified. The autocorrelation function of the scattered light intensity was then collected by a Malvern correlator, and the result was analyzed using a 2-cumulant fit.

A plot of $D(\phi)$ is given in Fig. 16.1. The slope of this line yields $k_d = 11 \pm 0.5$. The intercept gives $D_0 = 3.2 \times 10^{-7}$ cm^2/s, which corresponds to a hydrodynamic radius of 75 Å. We can use Eq. (5) to calculate k_d from the neutron scattering [5] results $\varepsilon = 3.82$ and $\lambda = 0.02$. The resulting k_d from this calculation is -10, which shows in Fig. 16.1 as a dashed line. This line agrees quite well with the experimental data.

B. Self-Diffusion Constant

The definition of the self-diffusion constant [9] is

$$D_s = \lim_{t \to \infty} \frac{w(t)}{t} = \int_0^\infty V(t)\, dt, \tag{7}$$

where $w(t)$ is the mean-square displacement of the tagged particles:

$$w(t) - \tfrac{1}{6}\langle |\mathbf{r}(t) - \mathbf{r}(0)|^2 \rangle, \tag{8}$$

and $V(t)$ is the velocity $[V_1(t)]$ autocorrelation function of tagged particles:

$$V(t) = \tfrac{1}{3}\langle V_1(t)V_1(0) \rangle. \tag{9}$$

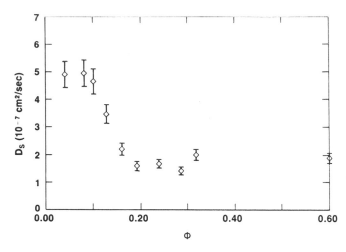

FIGURE 16.2. Plot of the self-diffusion constant $D_s(\phi)$ versus the volume fraction (ϕ) of minor components.

It is important to note that the interaction forces always reduce the value of D_s independently of the sign of the interaction forces, that is, at low ϕ.

$$D_s(\phi) = D_s(0)(1 + k_s\phi), \qquad \text{with} \qquad k_s < 0. \qquad (10)$$

The self-diffusion coefficient for several volume fractions of minor components is shown in Fig. 16.2. Measurements were taken for several fringe spacings , and the relaxation rates have been plotted against q^2 ($q = 2\pi/d$; d is the fringe spacing). The slope of this line is the diffusion constant ($1/\tau = D_s q^2$). One of these plots is shown in Fig. 16.3. For small ϕ, D_s is about 6×10^{-7} cm^2/s, which corresponds to a droplet radius of about 50 Å (using the Stokes–Einstein formula for the free diffusion of spheres and the viscosity of decane). As ϕ is increased the diffusion constant decreases more rapidly from ϕ from 10–16% and is then apparently independent of ϕ above 16%. For all ϕ, we observe a q^2 dependence of the relaxation rate, indicating diffusive behavior.

At low ϕ, D_s might be expected to decrease because of the small attractive interaction between the droplets as ϕ increases. However, D_s does not continue to decrease with ϕ, as one would expect if the interactions were dominating the process.

The result that is most interesting and difficult to explain is the lack of concentration dependence of D_s at the high volume fraction. It should be pointed out again that since we are directly measuring only the diffusion of the dye, the curve in Fig. 16.2 represents an upper bound for the self-diffusion constant of the water droplets. Clearly, the dye molecules (radius $\sim$ 20 Å) must move at least as quickly as droplets having twice their radius. Some kind of exchange [23] of dye between droplets cannot be ruled out complete-

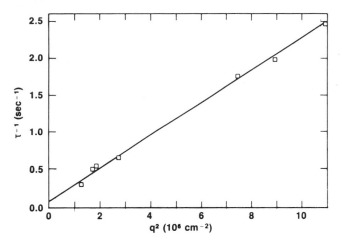

FIGURE 16.3. Graph of τ^{-1} as a function of the scattering vector (q^2).

ly. When droplets collide, their contents may interchange. Also, the exchange of dye molecules solubilized in the decane by surfactant is also possible. Under such assumptions, though, one would not expect D_s to be independent of ϕ.

At high ϕ, droplets may form a percolation cluster. The droplets may detach from the cluster and perform random motion until the cluster reforms. In this process, τ^{-1} is the characteristic frequency of the cluster reorganization, and l is the characteristic length (travel distance before recapture on the cluster, expected to be about one droplet diameter). The self-diffusion constant is then $D_s \sim l^2/\tau$. D_s is thus independent of ϕ, because τ is a function mainly of the strength of interaction between droplets, and l is just the droplet size. However, for small ϕ, the ϕ independence on D_s is very hard to explain with the known attractive potential. It seems to be a canceling out of effects from the interaction and hydrodynamics in the concentration regime that is not yet theoretically understood. It is a very challenging problem for the theorist to predict such behavior.

C. Electric Conductivity

As stated, the electrical resistance was measured as a function of ϕ at a frequency of 1 kHz. The normalized conductivity as shown in Fig. 16.4 is constant at a dilute concentration regime. At low ϕ, the conductivity is linear with ϕ, in agreement with a conduction mechanism, as given in Eq. (3). There are several possibilities for a charge carrier. One is for the ions to move through the decane, which takes too much energy. A second possibility is for the droplet to be charged. To make a charged droplet, it is only necessary for the droplets to exchange surfactants without their counterions. A charged

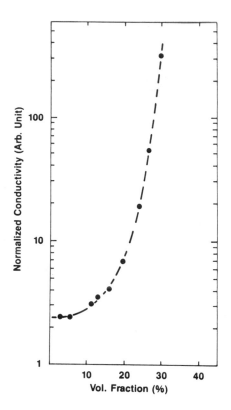

FIGURE 16.4. Normalized conductivity is plotted versus ϕ.

surfactant has a faster mobility than the droplets because of its much smaller size. The measured conductivity values gives Z to be one, so that there is, on average, one charged AOT per droplet exchanged. Therefore, the charged surfactant is a possible charge carrier. This charged surfactant could be responsible for an extra shell [24] around the droplet to explain the low percolation threshold volume fraction (ϕ_c) in a microemulsion. If two shells overlap, the corresponding spheres are considered to be connected and, hence, in the same cluster. Because of the shell, the ϕ_c can be reduced substantially. As shown in Fig. 16.4, the apparent ϕ_c ($\simeq$ 16%) is much lower than the typical static percolation threshold ($\phi_c \simeq$ 30%). The ϕ_c is also shown to be a strong function of temperature. The details are given elsewhere [25,26]. At high ϕ, the charge transport proceeds [12,27] via the anionic surfactants on the cluster of water droplets, which are mobile and rearrange in time. In this picture, the conductivity depends on a power law:

$$\sigma \sim |\phi - \phi_c|^{-\tilde{s}} \qquad \phi \leqslant \phi_c, \qquad (11)$$

where $\tilde{s}$ is the stirred percolation exponent, which depends on the characteristic time (τ) for charge carriers in a cluster and the rearrangement time (τ_R) for the cluster. The theory predicts the form

$$\tilde{s} = 2v - \beta, \tag{12}$$

where v is the exponent for the correlation length, and β is the probability of belonging to the infinite cluster. For $d = 3$, $\tilde{s}$ is 1.3 when τ_R is much less than τ. The exponent μ above ϕ is the same as that for static percolation.

At high ϕ, an independent determination of ϕ_c is necessary to determine the exponents. The quantity

$$\left| \left(\frac{1}{R} \right) \left(\frac{dR}{d\phi} \right) \right|,$$

evaluated from the data, shows a peak, which shifts with temperature. This peak position, ϕ_p, is then judged to be the percolation threshold (ϕ_c). Figure 16.5 shows the divergence of resistance as a function of $|\phi - \phi_p|$ for $\phi < \phi_p$ and $\phi > \phi_p$ at 25°C. The exponents that characterize the power laws are 1.6 and 1.2 for $\phi > \phi_p$ and $\phi < \phi_p$, respectively. This finding is very similar to other microemulsion findings and also confirms the theoretical approaches to understanding the conductivity near a percolation threshold of microemulsion systems.

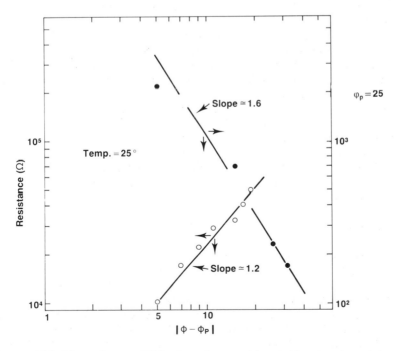

FIGURE 16.5. The resistance (Ω) is plotted versus $|\phi - \phi_p|$ for $\phi < \phi_p$ and $\phi > \phi_p$.

IV. Conclusions

We have measured various transport properties of dilute and concentrated water in oil microemulsions. In the dilute regime, we have found that the linear decrease in the mutual diffusion coefficient can be shown to be consistent with previous static (SANS) measurements within the model of hard-core and attractive square-well interactions.

We have also found that the electrical conductivity in the dilute regime gives a charge per droplet of about one, lending support to the idea of the exchange of surfactant molecules as the conduction mechanism in the concentrated regime. Since these surfactants can be exchanged over some distance, the conductivity percolation threshold is lowered from its expected value of the volume fraction of water. We also found that the exponent at which the conductivity diverges is more consistent with a dynamic rather than a static model.

This dynamic percolation model may also explain why we observed the self-diffusion coefficient to be constant with respect to the volume fraction above some critical value.

References

1. M. Kotlarchyk, S.-H. Chen, J.S. Huang, and M.W. Kim, *Phys. Rev.* **A29**, 2054 (1984).
2. P. Ekwall, L. Mandell, and K. Fontell, *J. Coll. and Int. Sci.* **33**, 215 (1970).
3. M. Zulauf and H.F. Eicke, *J. Phys. Chem.* **83**, 480 (1979).
4. M. Kotlarchyk, S.-H. Chen, and J.S. Huang, *J. Phys. Chem.* **86**, 3273 (1982).
5. J.S. Huang, S.A. Safran, M.W. Kim, G.S. Grest, M. Kotlarchyk, and N. Quirke, *Phys. Rev. Lett.* **53**, 592 (1984).
6. M.W. Kim, W.D. Dozier, and R. Klein, *J. Coll. and Int. Sci.* **87**, 1455 (1987).
7. B.V. Felderhof, *J. Phys.* **A11**, 929 (1978).
8. G.K. Batchelor, *J. Fluid Mech.* **131**, 155 (1983).
9. W. Hess and R. Klein, *Adv. Phys.* **32**, 173 (1983).
10. R. Finsey, A. Devriese, and H. Lekkerkerker, *JCS Faraday II* **76**, 767 (1980).
11. L.E. Scriven, in "*Micellization, Solubilization, and Microemulsions,*" Vol. 2, p. 877, K.L. Mittal, ed., Plenum, New York, 1977.
12. M. Laques, *J. Phys. Lett.* **40**, L331 (1979).
13. G.H. Brown, ed., "*Photochromism Techniques of Chemistry,*" Wiley, New York, 1971.
14. F. Rondelez, H. Hervet, and W. Urback, *Chem. Phy. Lett.* **53**, 138 (1978).
15. T. Assih, F. Larche, and P. Delord, *J. Coll. and Int. Sci.* **89**, 35 (1982).
16. B.J. Berne and R. Pecora, "*Dynamic Light Scattering,*" Wiley, New York, 1976.
17. J.A. Wesson, H. Takezoe, H. Yu, and S.P. Chen, *J. Appl. Phys.* **53**, 6513 (1982).
18. A.M. Cazabat, D. Chatenay, D. Langevin, J. Meunier, and L. Leger, in "*Surfactants in Solution,*" K.L. Mittal and B. Lindman, eds., Plenum, New York, 1984.
19. W.D. Dozier, H.M. Lindsay, and P.M. Chaikin, *J. de. Phys. Coll.* **46**, C3-165.
20. J.W. Essam, C.M. Place, and E.H. et Sondeheimer, *J. Phys.* **C7**, L-258 (1974).

21. S. Kirkpatrick, *Rev. Mod. Phys.* **45**, 574 (1973).
22. I. Webman, J. Jortner, and M.H. Cohen, *Phys. Rev.* **B11**, 2885 (1975).
23. P.D.I. Fletcher, B.H. Robinson, F. Bermejo-Barrera, D.G. Oakenfull, J.C. Dore, and D.C. Steyler, in *"Microemulsions,"* I.D. Robb, ed., Plenum, New York, 1982.
24. S.A. Safran, I. Webman, G.S. Grest, *Phys. Rev.* **A32**, 506 (1985).
25. S. Bhattacharaya, J. Stokes, M.W. Kim, and J.S. Huang, *Phys. Rev. Lett.* **55**, 1884 (1985).
26. M.W. Kim and J.S. Huang, *Phys. Rev.* **A34**, 719 (1986).
27. G.S. Grest, I. Webman, S.A. Safran, and A.L.R. Bug, *Phys. Rev.* **A33**, 2842 (1986).

Index